NATURAL PHARMACEUTICALS AND GREEN MICROBIAL TECHNOLOGY

Health Promotion and Disease Prevention

Edited by

Debarshi Kar Mahapatra, PhD

Ravindra S. Shinde, PhD

Aishwarya A. Andhare

A. K. Haghi, PhD

AAP | APPLE ACADEMIC PRESS

First edition published 2021

Apple Academic Press Inc.
1265 Goldenrod Circle, NE,
Palm Bay, FL 32905 USA

4164 Lakeshore Road, Burlington,
ON, L7L 1A4 Canada

CRC Press
6000 Broken Sound Parkway NW,
Suite 300, Boca Raton, FL 33487-2742 USA

2 Park Square, Milton Park,
Abingdon, Oxon, OX14 4RN UK

Library and Archives Canada Cataloguing in Publication

Title: Natural pharmaceuticals and green microbial technology : health promotion and disease prevention / edited by Debarshi Kar Mahapatra, PhD, Ravindra S. Shinde, PhD, Aishwarya A. Andhare, A.K. Haghi, PhD.

Names: Mahapatra, Debarshi Kar, editor. | Shinde, Ravindra S., 1979- editor. | Andhare, Aishwarya A., editor. | Haghi, A. K., editor.

Description: Includes bibliographical references and index.

Identifiers: Canadiana (print) 20200312103 | Canadiana (ebook) 20200312251 | ISBN 9781771888813 (hardcover) | ISBN 9781003003229 (ebook)

Subjects: LCSH: Drug development. | LCSH: Natural products. | LCSH: Pharmaceutical microbiology. | LCSH: Pharmaceutical chemistry. | LCSH: Health promotion. | LCSH: Medicine, Preventive.

Classification: LCC RM301.25 .N38 2020 | DDC 615.1/9—dc23

Library of Congress Cataloging-in-Publication Data

Names: Mahapatra, Debarshi Kar, editor. | Shinde, Ravindra S., 1979- editor. | Andhare, Aishwarya A., editor. | Haghi, A. K., editor.

Title: Natural pharmaceuticals and green microbial technology : health promotion and disease prevention / edited by Debarshi Kar Mahapatra, Ravindra S. Shinde, Aishwarya A. Andhare, A.K. Haghi.

Description: First edition. | Palm Bay, FL : Apple Academic Press, 2021. | Includes bibliographical references and index. | Summary: "This volume presents some of the latest research and applications in using natural substances and processes for pharmaceutical products. It presents an in-depth examination of the chemical and biological properties of selected natural products that are either currently used or have the potential for useful applications in the chemical and pharmaceutical industries. It covers emerging technologies and case studies and is a source of up-to-date information on natural products and microbial technology. It provides an applied overview of the field, from traditional medicinal targets to cutting-edge molecular techniques. Natural products have always been of key importance to drug discovery, but as modern techniques and technologies have allowed researchers to identify, isolate, extract, and synthesize their active compounds in new ways, they are once again coming to the forefront of drug discovery. Topics include the safe use of use of petrochemical for various chemical and organic processes, the antibiotic and bio-herbicidal potential of extracts of Abutilon indicum and Prosopis juliflora, the health benefits of eugenol, stem cells and their connection with cancer, and more. Scientists, industry professionals, and graduate students will gain a unique insight into green microbial technology and natural pharmaceuticals today along with their practical implementation for health promotion and disease prevention"-- Provided by publisher.

Identifiers: LCCN 2020037119 (print) | LCCN 2020037120 (ebook) | ISBN 9781771888813 (hardcover) | ISBN 9781003003229 (ebook)

Subjects: MESH: Biological Products--pharmacology | Biological Products--therapeutic use | Antineoplastic Agents--pharmacology | Microbiological Phenomena | Green Chemistry Technology

Classification: LCC RM301.12 (print) | LCC RM301.12 (ebook) | NLM QV 241 | DDC 615.1--dc23

LC record available at https://lccn.loc.gov/2020037119

LC ebook record available at https://lccn.loc.gov/2020037120

ISBN: 978-1-77188-881-3 (hbk)
ISBN: 978-1-00300-322-9 (ebk)

About the Editors

Debarshi Kar Mahapatra, PhD
Assistant Professor, Department of Pharmaceutical Chemistry, Dadasaheb Balpande College of Pharmacy, Rashtrasant Tukadoji Maharaj Nagpur University, Nagpur, Maharashtra, India

Debarshi Kar Mahapatra, PhD, is currently Assistant Professor in the Department of Pharmaceutical Chemistry at Dadasaheb Balpande College of Pharmacy, Rashtrasant Tukadoji Maharaj Nagpur University, Nagpur, Maharashtra, India. He was formerly Assistant Professor in the Department of Pharmaceutical Chemistry, Kamla Nehru College of Pharmacy, RTM Nagpur University, Nagpur, India. He has taught medicinal and computational chemistry at both the undergraduate and postgraduate levels and has mentored students in their various research projects. His area of interest includes computer-assisted rational designing and synthesis of low molecular weight ligands against druggable targets, drug delivery systems, and optimization of unconventional formulations. He has published research, book chapters, reviews, and case studies in various reputed journals and has presented his work at several international platforms, for which he has received several awards by a number of bodies. He has also authored the book titled *Drug Design*. Presently, he is serving as a reviewer and editorial board member for several journals of international repute. He is a member of a number of professional and scientific societies, such as the International Society for Infectious Diseases (ISID), the International Science Congress Association (ISCA), and ISEI.

Ravindra S. Shinde, PhD
Assistant Professor, Department of Chemistry and Industrial Chemistry, Dayanand Science College, Latur, Maharashtra, India

Ravindra S. Shinde, PhD, is presently serving as Assistant Professor, Department of Chemistry at Dayanand Science College, Latur in Maharashtra, India. He has around 14 years of teaching experience at BSc and MSc levels. Having more than 23 research publications to his credit in journals of national and international repute, he is also the author of many undergraduate- and postgraduate-level books. He has published two books with Apple Academic Press: *Modern Green Chemistry and Heterocyclic Compounds* and *Green Chemistry and Sustainable Technology: Biological, Pharmaceutical, and Macromolecular Systems;* and one with Lulu Press, Inc. Morrisville: *Practical Chemistry*. He has also written chapters in books published by several other international publishers. Dr. Shinde has delivered lectures and chaired sessions at national conferences and is a reviewer of number of international journals. In addition, he has completed minor research projects from different funding agencies. He is a university-approved recognized postgraduate teacher in chemistry. He has decade of administrative experience as a N.S.S Programme Officer, Member of University Exam Committee, Coordinator of UGC sponsored NET/SET coaching cell and One Teacher One Skill Committee of the college, and brand ambassador of online NPTEL examinations run by I.I.T and I.I.Sc. He was awarded a number of prestigious awards during his career, such as Vidyabushan Puraskar by the Indian NET/SET Association (2010) and National Teacher Award (2019). He also has working experience at S.R.T.M. University, Nanded in various departments, NSS District Area Co-coordinator, Internal and External Examiner of MSc, BSc Chemistry Examination, and Member of University Skill Development Committee.

Aishwarya A. Andhare
Department of Microbiology, Biotechnology and Chemistry,
Dayanand Science College, Latur-413512, Maharashtra, India

Aishwarya A. Andhare is studying for her bachelor's in microbiology at the Dayanand Science College, Latur, affiliated with Swami Ramanand Tirth Marathwada University, Nanded. She has a keen interest in research and has been chosen basic sciences as her career choice. She was actively doing research work from first year. Ms. Aishwarya has published eight papers in international journals (UGC, NAAS approved). She had actively participated in several research competitions and has won more than forty prizes in various competitions, including sports, drawing, debates, elocutions and many more. She has published an international book, *Astonishing Reviews of Nature towards Nature* (LAMBERT Academic Publishing of Germany), and she has also written many chapters in edited books. She had been selected for many summer research fellowship programs organized by the Jawaharlal

Nehru Center for Advanced Scientific Research, Jakkur P.O., Bengalore and under the CSIR-NCL SDP program held at the CSIR-National Chemical Laboratory, Pune, India. She has presented papers (both oral and poster) at more than 12 international conferences. The Dayanand Education Society has awarded her 50,000 Rs for research funding. Her research work is on "Seeking Hidden Potentials of Nature and Discovering Something New from Natural Ingredients." Along with her academic responsibilities, she is interested in social works and has been named "Best Volunteer of NSS" for two continuous and now trying to be an NCC cadet.

A. K. Haghi, PhD

Professor Emeritus of Engineering Sciences,
Former Editor-in-Chief, International Journal of Chemoinformatics and Chemical Engineering and Polymers Research Journal; Member, Canadian Research and Development Center of Sciences and Culture

A. K. Haghi, PhD, is the author and editor of 165 books, as well as 1000 published papers in various journals and conference proceedings. Dr. Haghi has received several grants, consulted for a number of major corporations, and is a frequent speaker to national and international audiences. Since 1983, he served as professor at several universities. He is former Editor-in-Chief of the *International Journal of Chemoinformatics and Chemical Engineering* and *Polymers Research Journal* and is on the editorial boards of many international journals. He is also a member of the Canadian Research and Development Center of Sciences and Cultures (CRDCSC), Montreal, Quebec, Canada.

Contents

Contributors .... *xi*

Abbreviations .... *xiii*

Preface .... *xv*

1. **Green Solvents for Pharmaceuticals 1**
   Gokul S. Talele and Hitesh V. Shahare

2. ***Abutilon indicum* and *Prosopis juliflora* as a Source of Antibiotic and Herbicidal Agents 11**
   Aishwarya A. Andhare and Ravindra S. Shinde

3. **Pharmacological Perspectives of Eugenol in Modern Context 31**
   Amena Ali and Abuzer Ali

4. **Predictive *In Silico* Global Metabolism Analysis: Disease Mechanism 49**
   Francisco Torrens and Gloria Castellano

5. **Intracellular Proteolysis and Neoplasia: Ubiquitin-Proteasome System, Cell Cycle, and Cancer 85**
   Francisco Torrens and Gloria Castellano

6. **Vitamin D, Analogues, Drugs, and their Relevance to Cancer Inhibition 117**
   Francisco Torrens and Gloria Castellano

7. **Stem Cells and Their Connection with Cancer: Prostate Disease 151**
   Francisco Torrens and Gloria Castellano

8. **MEK/ERK Pathway Overactivation in Liver Tumors: TGF-β Death 209**
   Francisco Torrens and Gloria Castellano

9. **Chalcone Scaffold Bearing Natural Antigout Agents 221**
   Debarshi Kar Mahapatra, Vivek Asati, and Sanjay Kumar Bharti

10. **Double-Pole Electrode in New Devices and Processes for Direct Electrochemical Oxidation of Blood Inside the Blood Vessel 231**
    V. A. Rudenok

11. Advanced Development in the Synthesis of Biologically Active Heterocyclic 1, 2, 3, 4 Tetrahydropyrimidine-2-Ones 239

Kalpana N. Handore, Vasant V. Chabukswar, and S. B. Sharma

Index 273

Contributors

Abuzer Ali
Department of Pharmacognosy, College of Pharmacy, Taif University, Taif, 21974, Saudi Arabia

Amena Ali
Department of Pharmaceutical Chemistry, College of Pharmacy, Taif University, Taif, 21974, Saudi Arabia

Aishwarya A. Andhare
Department of Microbiology, Biotechnology and Chemistry, Dayanand Science College, Latur-413512, Maharashtra, India

Vivek Asati
Department of Pharmaceutical Chemistry, NRI Institute of Pharmacy, Bhopal 462021, Madhya Pradesh, India

Sanjay Kumar Bharti
Institute of Pharmaceutical Sciences, Guru Ghasidas Vishwavidyalaya (A Central University), Bilaspur 495009, Chhattisgarh, India

Gloria Castellano
Departamento de Ciencias Experimentales y Matemáticas, Facultad de Veterinaria y Ciencias Experimentales, Universidad Católica de Valencia San Vicente Mártir, Guillem de Castro-94, E-46001 València, Spain

Vasant V. Chabukswar
Nowrosjee Wadia College, Chemistry Department, University of Pune, Pune, India

Kalpana N. Handore
Nowrosjee Wadia College, Chemistry Department, University of Pune, Pune, India

Debarshi Kar Mahapatra
Department of Pharmaceutical Chemistry, Dadasaheb Balpande College of Pharmacy, Nagpur 440037, Maharashtra, India

V. A. Rudenok
Izhevsk State Agricultural Academy

Hitesh V. Shahare
SNJB's Shriman Sureshdada Jain College of Pharmacy, Chandwad, Nashik, Maharashtra, India

S. B. Sharma
Modern Education Society's College of Engineering, Affiliated to Savitribai Phule Pune University, Bund Garden Road, Pune 411001, India

Ravindra S. Shinde
Department of Microbiology, Biotechnology and Chemistry, Dayanand Science College, Latur-413512, Maharashtra, India

Gokul S. Talele
NGSPM, College of Pharmacy, Anjaneri, Nashik, Maharashtra, India

Francisco Torrens
Institut Universitari de Ciència Molecular, Universitat de València, Edifici d'Instituts de Paterna, P. O. Box 22085, E-46071 València, Spain

Abbreviations

AC	algebraic chemistry
ALP	allopurinol
AML	acute myeloid leukemia
BAK	BCL-2-associated X-protein
BCL	B-cell lymphoma
BMF	BCL-2-modifying factor
BMPs	bone morphogenetic proteins
BrdU	bromodeoxyuridine
bRG	basal radial glia
CB	cord blood
CDK	cyclin-dependent kinase
CES™	cell electronic sensing (CES™)
CI	cell index
CIM	cellular invasion/migration
CNS	central nervous system
CSCs	cancer stem cells
CSF	cerebrospinal fluid
DFT	density functional theory
DHPMs	3,4-dihydropyrimidin-2(1H)-ones (DHPMs)
DNA	deoxyribonucleic acid
DRGs	dorsal root ganglia
EGF	epidermal growth factor
EGFR	epidermal growth factor receptor
ERK	extracellular signal-regulated kinase
ESCs	embryonic stem cells
EZH2	enhancer of zeste homolog 2
FA	Fanconi anaemia
FBS	foetal bovine serum
β-Gal	β-galactosidase
GABA	γ-aminobutyric acid
GFP	green fluorescent protein
HCC	human hepatocellular carcinoma
HNPCC	hereditary non-polyposis colorectal carcinoma
HLA	human leukocyte antigen

iPSCs	induced pluripotent stem cells
IRGC	intermediate radial glia cell
MAPK	mitogen-activated protein kinase
miRNA	microRNA
miRNome	microRNA genome
MK	megakaryocytic
ML1	myeloid leukaemia cell differentiation protein
mTOR	mammalian target and rapamycin
NADPH	nicotinamide adenine dinucleotide phosphate hydride
N-CoR	nuclear receptor corepressor
NDs	neurodegenerative disorders
NF	nuclear factor
NSCs	neural stem cells
OS	oxidative stress
OSVZ	outer subventricular zone
PCR	polymerase chain reaction
PEDF	pigment epithelium-derived factor
PIN	protein interaction network
PPM	precision personalized medicine
PRRs	pattern-recognition receptors
PSCs	pluripotent stem cells
RT	real-time
RMS	rostral migratory stream
ROS	reactive oxygen species
SA-β-Gal	senescence-associated β-galactosidase
SCs	stem cells
SEZ	subependymal zone
SGZ	subgranular zone
siRNA	short interfering ribonucleic acid
SVZ	subventricular zone
TET	ten–eleven translocation
TGF	transforming growth factor
TLRs	toll-like receptors
TNFRs	tumor necrosis factor α receptors
TNFα	tumor necrosis factor α
UPS	ubiquitin-proteasome system
XO	xanthine oxidase

Preface

Compounds derived from natural products have made a big impact on the pharmaceutical industry. Natural products have traditionally been an important source of pharmaceuticals. Although synthetic chemistry has also produced many new bioactive substances and combinatorial techniques have considerably expanded the number of compounds available for tests that, there is still a relatively high number of natural products and their derivatives among the best-selling drugs, and there has been a renewed interest in natural products as a source of pharmaceuticals.

This book presents an in-depth examination of the chemical and biological properties of selected natural products that are either currently used or have the potential for useful applications in the chemical and pharmaceutical industries.

The book covers emerging technologies and case studies and is a source of up-to-date information on the topical subject of natural products. It provides an applied overview of the field, from traditional medicinal targets, to cutting-edge molecular techniques. Natural products have always been of key importance to drug discovery, but as modern techniques and technologies have allowed researchers to identify, isolate, extract and synthesize their active compounds in new ways, they are once again coming to the forefront of drug discovery.

Combining the potential of traditional medicine with the refinement of modern chemical technology, the use of natural products as the basis for drugs can help in the development of more environmentally sound, economical, and effective drug discovery processes.

Microbes cover the surfaces of all other organisms (and occupy internal and even intracellular niches, in some) and influence diverse physiological activities of their hosts, including nutrition, health–disease status and hence well-being. Thus, the signature and almost unique characteristic of microbial technology is the exceptional diversity of applications it can address and the exceptional range of human activities and needs to which it is and can be applied. Precisely because sustainability goals have very diverse and complex components and requirements, microbial technology has the ability to contribute substantively on many levels in many arenas to global efforts to achieve sustainability. Indeed, microbial technology could be viewed as a unifying element in our progress toward sustainability. Microbial technology covers various aspects of microorganisms toward addressing some of the major challenges the world is facing, especially in the areas of human health and environment.

In the first chapter, it is shown that the use of petrochemical solvents is the key for various chemical and organic processes without severe implications on the environment. Green solvents were developed as an alternative to petrochemical solvents which reducing the use of petrochemical solvents in outlook of green chemistry. Green solvents or bio-solvents are environment friendly, which are derived from the processing of agricultural crops. The class of green solvents is involving water, supercritical fluids, ionic liquids, non-toxic liquid polymers and their varied combinations. These are characterized by low toxicity, convenient accessibility and possibility of reuse as well as great efficiency.

The purpose behind the study in Chapter 2 was to evaluate the Antibiotic and bio-herbicidal potential of extracts of *Abutilon indicum* and *Prosopis juliflora*. Effect of Extracts were assayed for control of growth and germination of invasive weed *Parthenium hysterophorus* and as Antibacterial agent against *Xanthomonas axonopodis pv. punicae.* Plant Extracts were tested separately for germination of Parthenium seeds and as Antibacterial agent against *Xanthomonas axonopodis pv. punicae.* Plant extracts has shown significant reduction in total germination percent (GP), germination index (GI),germination energy (GE), speed of emergence (SE) and seedling vigor index (SVI) of *P. hysterophorus* while, the antibacterial activity was tested by well diffusion assay, minimum inhibitory concentration (MIC) and minimum bactericidal concentration (MBC) and recorded the antibiotic activity by both plant extracts .

Result revealed that *P. juliflora* has strong phytotoxic and Antibiotic properties as compared with *Abutilon indicum.* The highest Zone of Inhibition (ZOI) was shown by *P. juliflora* while lowest ZOI was shown by *A. indicum.* The results infer that the extracts of *Prosopis juliflora* and Acacia Arabica are highly sensitive against the *Xanthomonas axonopodis pv. punicae*. Plant extracts exhibited antibacterial activity with a potential to be used in the management of many plant diseases as an alternative to chemical antibiotics. Further phytochemical analysis is required to identify the bioactive compounds responsible for antibacterial activity. Isolation and characterization of those phytotoxic substances from these plants may act as a tool for new natural, biodegradable herbicide development to manage invasive weeds like *P. hysterophorus*.

Chapter 3 provides a brief overview of the plethora of research regarding the health benefits of eugenol and also discusses the mechanisms involved.

Eugenol is the chief phytoconstituent present in the essential oil of clove and several other medicinal plants. Being non-carcinogenic and non-mutagenic, it is usually considered as safe by many organizations like Food and Agricultural Organization (FAO), United Nations (UN), etc. Eugenol exhibited various pharmacological properties like antimicrobial, antioxidant, anti-inflammatory and showed potential against life-threatening diseases including leishmaniasis, cancer, and sepsis.

Chapter 4 reviews the state of the art of metabolism platforms for discriminating disease (prostate neoplasia) mechanisms.

Intracellular Proteolysis and Neoplasia: Ubiquitin-Proteasome System, Cell Cycle and Cancer is discussed in Chapter 5 in detail.

Vitamin-D, analogues, drugs and their relevance to cancer inhibition are reported in Chapter 6.

In Chapter 7, it is shown that acute myeloid leukaemia enables research on different-oncogene biofunction, in leukaemic transformation/progression, and allows devising healing strategies, for example, gene/anti-sense therapies, etc., aimed at leukaemic stem cell. Chain net, in subventricular zone, carries neuronal precursors from caudal subventricular zone to rostral parts of adult brain. It should be determined what guides migration *via* subventricular zone net, which other vertebrate brains contain pathway array and biosignificance of such an extensive neuronal-precursor net, many destined for single brain region: olfactory bulb. Despite robust germinal capacity, no evidence of cells migrating in chains exists along subventricular zone or olfactory peduncle to bulb, which is explained by long distance that separates olfactory bulbs from cerebrum in man or micro-osmatic capabilities. Findings raise fate question of cells born in human subventricular zone. Identification of large, anatomically discrete population of proliferating, multipotent human astrocytes leads to understanding role that human neural stem cells act in tumorigenic demyelination/degeneration. Planar polarity of ciliated ependymal cells is essential for formation of chemorepulsive-factor gradients, which guide neuroblast migration in adult brain. Polarized ciliated cells contribute important vectorial information for body plan development. Polarized epithelia and motile cilia in brain serve as directional-information conveyors for neuronal migration.

MEK/ERK pathway overactivation in liver tumor is discussed in Chapter 8 in detail.

The chalcones obtained from natural and semi-synthetic sources have been seen to produce strikingly high anti-gout activity in comparison to the standard marketed formulation allopurinol by inhibiting the active enzyme XO that simultaneously reduces the serum uric acid level and hepatic XO levels. In addition to it, the molecules exhibit biological activity by suppressing the activation of NF-κβ and preventing the formation of pro-inflammatory factors. The number(s), position(s), and type(s) of substituents play a key role in the modulation of the therapeutic target by forming possible hydrogen bonding/Van der Waals interactions. Chapter 9 will provide unparallel information of chalcone scaffold bearing natural and semi-synthetic molecules having pharmacotherapeutic perspectives. However, at present these molecules are at nascent stages and only pre-clinical studies have been done so far and a large section of clinical studies are still pending. The structural, chemical, biological, pathophysiological, and miscellaneous aspects have an overall pharmaceutical interest which will motivate the active researchers in the exploration and development of anti-gout agents for treating the associated symptoms.

In Chapter 10, double-pole electrode in new devices and processes for direct electrochemical oxidation of blood inside the blood vessel is presented.

Chapter 11 discusses advanced developments in the synthesis of biologically active heterocyclic 1, 2, 3, 4 tetrahydropyrimidine-2-ones.

CHAPTER 1

Green Solvents for Pharmaceuticals

GOKUL S. TALELE[1] and HITESH V. SHAHARE[2*]

[1]*NGSPM, College of Pharmacy, Anjaneri, Nashik, Maharashtra, India*

[2]*SNJB's Shriman Sureshdada Jain College of Pharmacy, Chandwad, Nashik, Maharashtra, India*

[*]*Corresponding author. E-mail: hiteshshahare1@rediffmail.com*

ABSTRACT

The use of petrochemical solvents is the key for various chemical and organic processes without severe implications on the environment. Green solvents were developed as an alternative to petrochemical solvents which is reducing the use of petrochemical solvents in outlook of green chemistry. Green solvents or bio-solvents are environment friendly, which are derived from the processing of agricultural crops. The class of green solvents is involving water, supercritical fluids, ionic liquids, non-toxic liquid polymers and their varied combinations. These are characterized by low toxicity, convenient accessibility, and possibility of reuse as well as great efficiency.

1.1 INTRODUCTION

Nowadays, almost all manufacturing and processing industries depend on the extensive use of solvents, which is reflected in the nearly 15 billion kilograms of yearly produced organic and halogenated solvents worldwide. These horrendous numbers finally raised international concern in the 1960s and led to the U.S. Pollution Prevention Act in 1990. "Green Chemistry" or "Sustainable Technology," as it is known today, can be shortly defined as a chemical working process, utilizing raw materials, eliminating waste and avoiding the use of toxic and hazardous reagents and solvents. Furthermore, high product selectivity at an economical reaction rate should be achieved, thereby minimizing the amount of solvents, reagents and promoters. Although the most commonly used solvents cause serious environmental and health problems, solvents in general have been proven beneficial for temperature control of solutions by the boiling point, heat supply for exothermic and endothermic reactions, purification, extraction, recrystallization, azeotropic distillation, chromatography, alteration of reaction rates, and selectivity of the reaction.[1]

Generally speaking, there exist three different approaches to mitigate the release of solvents and contaminated water into the ecosystem.

1) *Solvent reduction or recycling.* Many industries made remarkable progress in implementing "closed-loop systems" that reduce solvent or water use and thus improve their ability to recycle.
2) *Switching to solvent free processes* could be another beneficial step towards a "greener environment". Elimination of organic solvents and other volatile organic compounds (VOC) from paints and coatings, resulting in dry powder coatings and solid ultraviolet curable coatings display a good alternative.
3) However, not many reactions can be carried out under solvent less condition, which leads to the third approach *to reduce the release of solvents into the ecosystem.* This especially applies to industries that are dependent on operations involving coatings, separations and reactions, as many processes have been shown to require "liquids" of some kind.

They include water, ionic liquids, fluorous solvents, organic carbonates, carbon dioxide as well as biosolvents, as shown in Figure 1.1, each with its own advantages and drawbacks.[2]

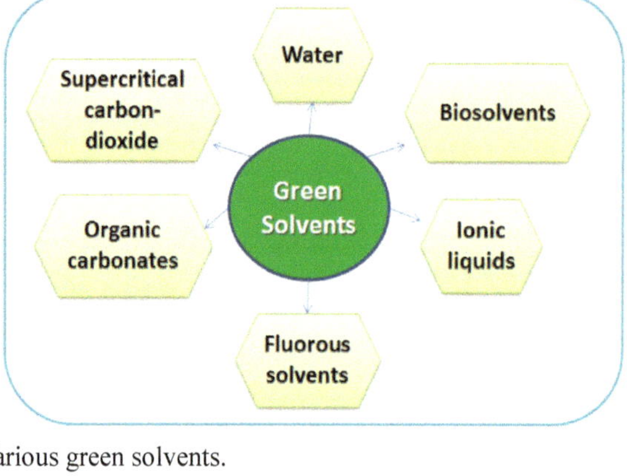

FIGURE 1.1 Various green solvents.

1.2 GREEN SOLVENTS

1.2.1 WATER

Water is most important solvent in green chemistry. It is non-flammable, nontoxic, and not expensive which is amply available. As a protic and polar solvent, it plays an important role in view of green chemistry. Supercritical water behaves like a nonpolar solvent and exists at pressures above 221 bar and temperatures above 374°C. The non-polarity is caused by the loss of hydrogen bonding; hence salts are no more soluble in supercritical water, whereas O_2 and non-polar solvents are. Therefore, oxidation reactions in supercritical water have been developed and studied since the 1970s, mainly for the disposal of organic waste. Unfortunately, there has not been a solution for the occurring corrosion problems due to halogen traces yet, leading to the fail of commercialization of this method. On the other hand, the high polarity of common water also has a lot of advantages in regard to reactivity and selectivity, like in organometallic catalysis. It is possible to recover and recycle the catalyst in an aqueous biphasic system via phase extraction whereas the product stays dissolved in the organic phase.[3]

1.2.2 IONIC LIQUIDS

Ionic liquids are molten salts with very low vapor pressure. Additionally, many of them have low combustibility, exceeding thermal stability as well as solvating qualities and they are electrically conductive. Hence, there is an increasing interest in replacing volatile organic solvents with ionic liquids.[4] They serve as solvents or reaction media for many separation or catalytic processes regarding the fact that there is a wide range of organic, inorganic as well as polymeric molecules which are well soluble in ionic liquids. Their solvating properties are depending on the smaller anions and larger organic cations.[5] The ionic liquids consist of:

1. Anions: Examples are formate, benzoate, acetate, halides, nitrate, hydrogensulfate, tosylate, hexafluorophosphate, phosphate, thiocyanate, tetrafluoroborate, and trifluoromethanesulfonate methanesulfonate anions.
2. Cations: Examples are phosphonium, pyridinium, ammonium, pyrrolidinium, and imidazolium cations.

They are stable over a wide range of temperature and can be used in biphasic systems due to the immiscibility with some organic solvents. After extraction with an organic solvent, the catalyst remains in the ionic liquid and can easily be reused.[6,7]

1.2.3 BIOSOLVENTS

Biosolvents have been developed as an alternative to VOC, which are usually harmful to the environment and to human health.[8] The most important chemical classes of biosolvents are shown in Figure 1.2:

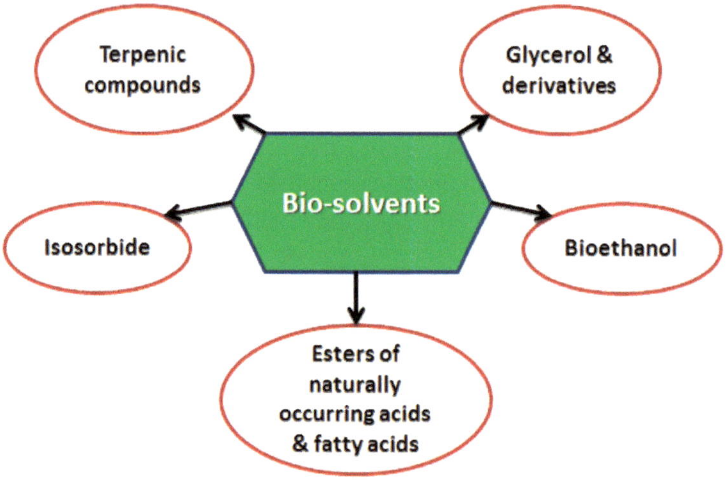

FIGURE 1.2 Various bio-solvents.

These compounds offer the advantage of being produced from renewable sources such as vegetable, animal or mineral raw materials by chemical and physical processes without the consumption of fossil resources. They are already widely used in cosmetics, cleaning agents, paint, inks, and agricultural chemicals.

A bio-based solvent has to meet certain criteria to be considered for application, such as

1) Optimal technical specifications (dissolution capability, volatility, flash point)
2) Environmental safety

3) Eco-compatible production
4) Availability and cost of the renewable raw materials.

Bio-based solvents have successfully been employed in multicomponent reactions with seemingly synergistic effects.[9]

1.2.4 ORGANIC CARBONATES

Organic carbonates represent esters of carbonic acids and are a class of compounds with a broad field of application due to their unusual properties. They are easily available in large amounts, inexpensive, possess low (eco) toxicity, and are complete biodegradable.[10]

Organic carbonates are widely used for extraction purposes, pharmaceutical and medical applications, and also in batteries.[11]

Two subclasses of carbonates are differentiated:

1. Open chain and
2. Cyclic organic carbonates

Cyclic organic carbonates have a wider temperature range in the liquid state and are therefore more suitable as solvents. Especially, such cyclic carbonates fulfill the requirements of green solvents which have low flammability, volatility and low toxicity.

Example, propylene carbonate (PC) is an aprotic, highly dipolar solvent with low viscosity and a very large liquid state range (bp. 243°C). Its stereogenic center could be exploited for influencing the enantioselectivity of a stereoselective reaction, if the pure enantiomeric carbonate was utilized, but to date organic carbonates are used in racemic form.

Pd-catalyzed allylic substitution reactions were successfully carried out in butylene carbonate, PC and diethyl carbonate using different phosphorous ligands.[12]

1.2.5 FLUOROUS SOLVENTS

The term "fluorous" solvents were first coined by Horváth and Rabai in analogy to "aqueous" to perfluorinated alkanes, dialkyl ethers, and trialkyl amines. Importantly, ethers and amines have no residual basicity and the lone pairs are very low in energy, which is explaining the lack of any intermolecular interactions.[13]

These perfluorinated liquids have useful and attractive properties for organic synthesis as shown in Figure 1.3.

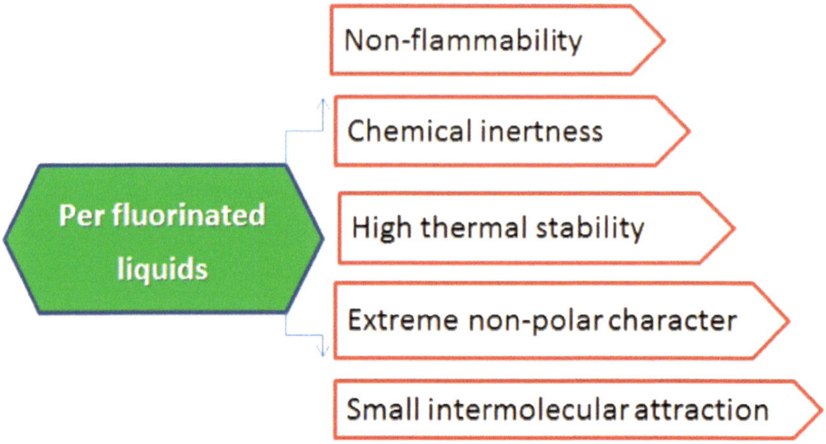

FIGURE 1.3 Properties of fluorinated liquids used as green solvents.

Their immiscibility with common organic solvents at room temperature allows the formation of biphasic systems which have been extensively studied in stoichiometric and catalytic transformations. For biphasic catalysis the catalyst must be solubilized in the fluorous phase which is achieved by attaching fluorocarbon moieties such as linear or branched perfluoroalkyl chains, so-called "fluorous ponytails". The reaction can take place either in the fluorous phase or at the interface between the phases, if the solubility of the reactants is low in the fluorous phase. Phase transfer agents may be added to facilitate the reaction. At higher temperature fluorous biphasic systems can become miscible forming a homogenous liquid phase reaction media. Thus, the advantages of both, a single-phase media for the reaction and a biphasic system for separation of the products can be exploited.

Fluorous solvents are not yet widely used commercially due to drawbacks like rather high cost, environmental persistence, and biological half-lives, especially of C7- and C8-perfluoroalkyl group containing compounds.[14]

1.2.6 SUPERCRITICAL CARBON DIOXIDE

Supercritical carbon dioxide ($scCO_2$) has been receiving increasing attention as an alternative reaction medium. Carbon dioxide, at room temperature, exists as a liquid with excellent wetting properties and a very low viscosity. Above its critical temperature and pressure (31°C and 73.8 bar) CO_2 is in the supercritical state, thereby featuring gas-like viscosities and liquid-like densities. Changing either the temperature or the pressure leads to dramatic changes in its viscosity, density and dielectric properties. This makes $scCO_2$ an unusually tunable, versatile and selective solvent.[15] Since carbon dioxide is renewable, nontoxic, nonflammable, readily evaporating and chemically inert toward many substances, it features outstanding characteristics for the utilization in Green Chemistry.[16] The potential of supercritical CO_2 as an environmentally preferable solvent is now being realized in several areas. The use of $scCO_2$ in the food and nutrition industry, it serves today as an extraction medium for decaffeination processes for coffee beans and tea, thus successfully replacing former used chlorinated organic solvents.[17] Further application fields, pioneering $scCO_2$ implementation, are for example dry cleaning, metal degreasing, polymer modification and pharmaceutical processing. But also industrial activities in the field of radical polymerization, heterogeneous hydrogenation and organometallic catalysis benefit from the outstanding characteristics of $scCO_2$. A feature of $scCO_2$ is its high miscibility with gases, which is especially useful in reactions such as hydrogenation with H_2, oxidation with O_2 and hydroformylation with syngas (CO/H_2), leading to high efficiency and often higher selectivity.[18,19]

1.2.7 GREEN SOLVENTS FROM PLANTS

Plants are considered renewable sources of energy but also a resource for various materials. Plant oils or vegetable oils derive from plant sources. Unlike petroleum which is the main source of chemicals in the petrochemical industry they are renewable sources.

There are three primary types of plant oil, differing both by the means of extraction and by the nature of the resulting oil: Vegetable oils can replace petroleum derived organic solvents, with better properties and more eco-friendly conditions as waste. Chemists have advanced recently techniques so that some vegetable oils to become solvents and replace hazardous organic solvents. As an example of plant-based oils are soybean oils and their esters. Vegetable oils can become a starting material for the production of eco-friendly solvents which are less toxic than the petrochemical industry's organic solvents. In the last decade, scientists were researching the use of "green" solvents in polymerization methods, since the polymer and plastics industries are using vast amounts of solvents. Polymers can be prepared under industrial scale production with the use of eco-friendly solvents. All these techniques aim at replacing toxic and hazardous solvents in many chemical processes in the synthetic laboratory and in the chemical industry.[20]

1.3 CONCLUSION

The obtained results from green solvents are similar and sometimes even better than these, resulting from conventional syntheses in organic solvents. Water, as a cheap, abundantly available, nontoxic and nonflammable solvent represents an ideal reaction medium for many processes, being mostly established in organometallic catalysis, hydroformylation processes and oxidations. Fluorous solvents and ionic liquids are attractive alternatives for performing reactions, which are not accomplishable in water or $scCO_2$. Organic carbonates mostly used for extraction purposes and pharmaceutical and medical applications, feature characteristics like low (eco) toxicity, complete biodegradability as well as inexpensiveness. $scCO_2$ also exhibits outstanding characteristics for the utilization in Green Chemistry, such as the possibility to separate it from the resulting product by simple pressure release. Reaction rates are very high in $scCO_2$, due to its intermediate properties, between gas and liquid state. Biosolvents, being produced from renewable sources are already widely used in cosmetics, cleaning agents, paint, inks, and agricultural chemicals and became to play an important role as an alternative to conventional solvents.

KEYWORDS

- **green chemistry**
- **bio-solvents**
- **green solvents**
- **sustainable chemistry**
- **ionic solvents**

REFERENCES

1. Jessop, P. G. Searching for Green Solvents. *Green Chem.* **2011,** *13*, 1391–1398.
2. Song, J.; Han, B. Green Chemistry: A Tool for the Sustainable Development of the Chemical Industry. *Natl. Sci. Rev.* **2014,** *2* (3), 255–256.
3. Kuntz, E. G. Homogeneous Catalysis…in Water. *Chemtech* **1987,** *17*, 570–575.
4. Suresh, S. J. S. Recent Advances in Ionic Liquids. Green Unconventional Solvents of this Century. Part 1. *Green Chem. Lett. Rev.* **2011,** *4*, 289–310.
5. Gorke, J.; Romas, S.; Kazlauskas, R. Toward Advanced Ionic Liquids. Polar, Enzyme-friendly Solvents for Biocatalysis. *Biotechnol. Bioprocess Eng.* **2010,** *1*, 40–53.
6. Patel, D. D.; Lee, J. M. Applications of Ionic Liquids. *Chem. Rec.* **2012,** *12* (3), 329–355.
7. Vekariya R. L. A Review of Ionic Liquids: Applications Toward Catalytic Organic Transformations. *J. Mol. Liquid.* **2017,** *227*, 44–60.
8. Berkeley, W. C.; Zhang, J. Green Process Chemistry in the Pharmaceutical Industry. *Green Chem. Lett. Rev.* **2009,** *2* (4), 193–211.
9. Sonali, R. S. Green Chemistry, Green Solvents and Alternative Techniques in Organic Synthesis. *Int. J. Chem. Phys. Sci.* **2015,** *1*, 516–520.
10. Vollmer, C.; Thomann, R.; Janiak, C. Organic Carbonates as Stabilizing Solvents for Transition-metal Nanoparticles. *Dalton Trans.* **2012,** *41*, 9722–9727.
11. Schäffner, B.; Verevkin, S. P.; Börner, A. Green Solvents for Synthesis and Catalysis. Organic Carbonates. *Chem. Unserer Zeit.* **2012,** *43*, 12–21.
12. Schäffner, B.; Andrushko, V.; Bayardon, J.; Holz, J.; Börner, A. Organic Carbonates as Alternative Solvents for Asymmetric Hydrogenation. *Chirality* **2009,** *21* (9), 857–861.
13. Yi, W.-B.; Cai, C.; Wang, X. A Novel Ytterbium/Perfluoroalkylated-pyridine Catalyst for Baylis-Hillman Reactions in a Fluorous Biphasic System. *J. Fluorine Chem.* **2007,** *128*, 919–924.
14. Dobbs, A. P.; Kimberley, M. R. Fluorous Phase Chemistry: A New Industrial Technology. *J. Fluorine Chem.* **2002,** *118*, 3–17.
15. Leitner, W. Supercritical Carbon Dioxide as a Green Reaction Medium for Catalysis. *Acc. Chem. Res.* **2002,** *35*, 746–756.
16. Nalawade, S. P.; Picchioni, F.; Janssen, L. P. B. M. Supercritical Carbon Dioxide as a Green Solvent for Processing Polymer Melts: Processing Aspects and Applications. *Progress Polym. Sci.* **2006,** *31* (1), 19–43.
17. Jutz, F.; Adanson, J. M.; Balker, A. Ionic Liquids and Dense Carbon Dioxide: A Beneficial Biphasic System for Catalysis. *Chem. Rev.* **2011,** *111* (2), 322–353.
18. Hyatt, J. A. Liquid and Supercritical Carbon Dioxide as Organic Solvents. *J. Org. Chem.* **1984,** *49*, 5097–5101.
19. Rayner, C. M. The Potential of Carbon Dioxide in Synthetic Organic Chemistry. *Org. Process Res. Dev.* **2007,** *11*, 121–132.
20. Li, Ch. J.; Trost, B. M. Green Chemistry for Chemical Synthesis. *Proc. Natl. Acad. Sci. USA* **2008,** *105* (36), 13197–13202.

CHAPTER 2

Abutilon indicum and *Prosopis juliflora* as a Source of Antibiotic and Herbicidal Agents

AISHWARYA A. ANDHARE and RAVINDRA S. SHINDE*

Department of Microbiology, Biotechnology and Chemistry, Dayanand Science College, Latur-413512, Maharashtra, India

**Corresponding author. E-mail: rss.333@rediffmail.com*

ABSTRACT

The purpose behind this study was to evaluate the antibiotic and bio-herbicidal potential of extracts of *Abutilon indicum* and *Prosopis juliflora*. The effects of extracts were assayed for control of growth and germination of invasive weed *Parthenium hysterophorus* and as antibacterial agent against *Xanthomonas axonopodis pv. punicae*. Plant extracts were tested separately for germination of Parthenium seeds and as an antibacterial agent against *Xanthomonas axonopodis pv. punicae*. Plant extracts have shown a significant reduction in total germination percentage (GP), germination index (GI), germination energy (GE), speed of emergence (SE) and seedling vigor index (SVI) of *P. hysterophorus,* while the antibacterial activity was tested by well diffusion assay, minimum inhibitory concentration (MIC) and minimum bactericidal concentration (MBC), and the antibiotic activity by both plant extracts was recorded.

Result revealed that *P. juliflora* has strong phytotoxic and antibiotic properties as compared with *A. indicum*. The highest Zone of Inhibition (ZOI) was shown by *P. juliflora* while the lowest ZOI was shown by *A. indicum*. The results infer that the extracts of *P. juliflora* and Acacia arabica are highly sensitive against the *Xanthomonas axonopodis pv. punicae*. Plant extracts exhibited antibacterial activity with a potential to be used in the management of many plant diseases as an alternative to chemical antibiotics. Further phytochemical analysis is required to identify the bioactive compounds responsible for antibacterial activity. Isolation and characterization of those phytotoxic substances from these plants may act as a tool for new natural, biodegradable herbicide development to manage invasive weeds like *P. hysterophorus*.

2.1 INTRODUCTION

Pomegranate (*Punica granatum* L.) is an ancient fruit, belonging to the smallest botanical family punicaceae. Pomegranate is a native of Iran, where it was first cultivated in about 2000 BC, but spread to the Mediterranean countries at an early date. It is extensively cultivated in Spain, Morocco and other countries around the Mediterranean, Egypt, Iran, Afghanistan, Arabia and Baluchistan. Pomegranate is a good source of carbohydrates and minerals such as calcium, iron and sulfur. It is rich in vitamin C and citric acid is the most predominant organic acid in pomegranate.[1] Apart from the fleshy portion of the fruit, the crop residues are also finding place in industries. The rind of the fruit is a good source of dye, which gives yellowish brown to khaki shades and is being used for dying wool and silk. The flower and buds yield light red dye, which is used for dying of cloths in India. The bark of the stem and root contains a number of alkaloids belonging to pyridine group. The bark is used as a tanning material especially in Mediterranean countries in the East.[2]

The causal organism of blight is *Xanthomonas axonopodis pv. Punicae*[3] Vauterin, Hoste, Kersters, and Swings.[4,5] The plant is susceptible to blight during all stages of growth and results in huge economic loss. Bacterial blight primarily affects the above ground plant parts, especially leaves, twigs, and fruits. While the leaves how early water soaked lesions to late necrotic blighting, the fruits show isolated or coalesced water soaked lesions followed by necrosis with small cracks and splitting of the entire fruit.[6] Stems show lesions around nodes or injuries, forming cankers in later stages. Suspected symptoms on floral parts have also been reported.[7,8] It is also presumed that the stem canker could be the outcomes of systemic spread of bacterium from leaf.[7] Although reported by[7,8] attempts to reproduce the field symptoms of blight on detached leaves, twigs and fruits were unsuccessful in the artificial inoculations. The management of bacterial blight of pomegranate is a major concern. This disease could not be effectively managed with conventional antibiotics like streptocycline in field conditions.

Parthenium hysterophorus L. an alien invasive weed is becoming a major weed of cropped and non-cropped areas in Pakistan.[12] Since the existing local weed flora is already a threat to the crop productivity, thus introduction of another alien species, like parthenium weed will further reduce the crop yield drastically and consequently increase the cost of production. Parthenium weed not only competes with desirable crops and pasture species but also causes farmers and stock animals to suffer an allergic skin condition while in contact with it.[13] Parts of parthenium weed plant are allelopathic, exhibiting strong competitive ability for soil moisture and nutrients while inhibiting the germination and growth of neighboring plant species.[11]

In India, parthenium weed reduced yield up to 40% in several crops[14] and it was reported to reduce forage production by up to 90%.[16] In India, parthenium weed is widely spread and infests about two million hectares of land.[18] Parthenium weed is generally unpalatable, but cattle and sheep eat it when feed is scarce. Consumption of large amounts will produce taints in mutton.[19] In non-cropped situations, various methods are being used to manage parthenium weed but manual removal is most prevalent in Pakistan. However, manual and mechanical methods for controlling parthenium weed are not effective.[20] Manual cutting results in rapid regeneration, which is quickly followed by flowering with abundant seed production.[21]

Overuse of synthetic herbicides to control weeds lead to an increased risk of herbicide resistant weed biotypes and harsh environmental pollutions. Alternative weed management strategies that are eco-friendly and cost-effective are therefore a time demanding issue throughout the world. In this backdrop, phytotoxic plants might help in resolving the problems created by synthetic herbicides as they possess growth retarding substances.

Thus this investigation was carried out on management of bacterial blight disease and to check weedicide potential by extracts of *Abutilon indicum* and *Prosopis juliflora* plants. Current research was undertaken to investigate and identify the properties of the extracts of *A. indicum* and *P. juliflora* under control laboratory conditions.

2.2 MATERIAL AND METHOD

2.2.1 COLLECTION AND EXTRACT PREPARATION OF PLANT SPECIES

A. indicum and *P. juliflora* species were collected from grassy area of farm near Krishi Vigyan Kendra (KVK), Latur. Leaves of Disease Free plants were selected for test. Database of these plants are shown in (Table 2.1).

TABLE 2.1 Database on Tropical Medicinal Plants.

Scientific names	Family	Habit	Part used	Properties	References
A. indicum	Malvaceae	Herb	Seed, root, bark, leaves	Hypothermic, CNS active, analgesic, aphrodisiac	[17]
P. juliflora	Leguminosae	Shurb			[17]

Weighed quantity of 50 g of each species was taken. Collected leaves of *A. indicum* and *P. juliflora* were oven dried till complete dehydration; then they were pulverized into a granular form by grinding in a mixer. Leaves of each species were pulverized into powder form separately. 50 g powder of each species were treated with 200 mL of water for 24 h in stirrer and stirred. After stirring, all extracts were filtered by using Whatman no.1 filter paper. The filtrate was concentrated at 35°C and stored at 4°C until further use.

2.2.2 COLLECTION, ISOLATION, AND IDENTIFICATION OF XANTHOMONAS AXONOPODIS PV. PUNICAE

2.2.2.1 FIELD VISIT AND COLLECTION OF BACTERIAL SAMPLES

The field visit was undertaken in major pomegranate-growing regions of Latur, Maharashtra, India viz., Murud and Harangul during the month of June to August. 2018. During the field survey, the randomly selected plant parts were inspected at the fields of for the incidence of bacterial blight. Distribution of bacterial blight of pomegranate was recorded in these areas. Plants were diagnosed as infected based on typical symptoms of bacterial blight, viz., yellow water soaked lesions at early stages and corky, dark oily spots at later stages of infection. The suspected plant leaves and fruits (Fig. 2.1) were collected and transferred into sterilized plastic bags and brought to Research laboratory of Microbiology Department of Dayanand Science College, Latur for the further studies.

2.2.2.2 STORAGE OF SAMPLE

Samples were surface sterilized and stored in laboratory condition.

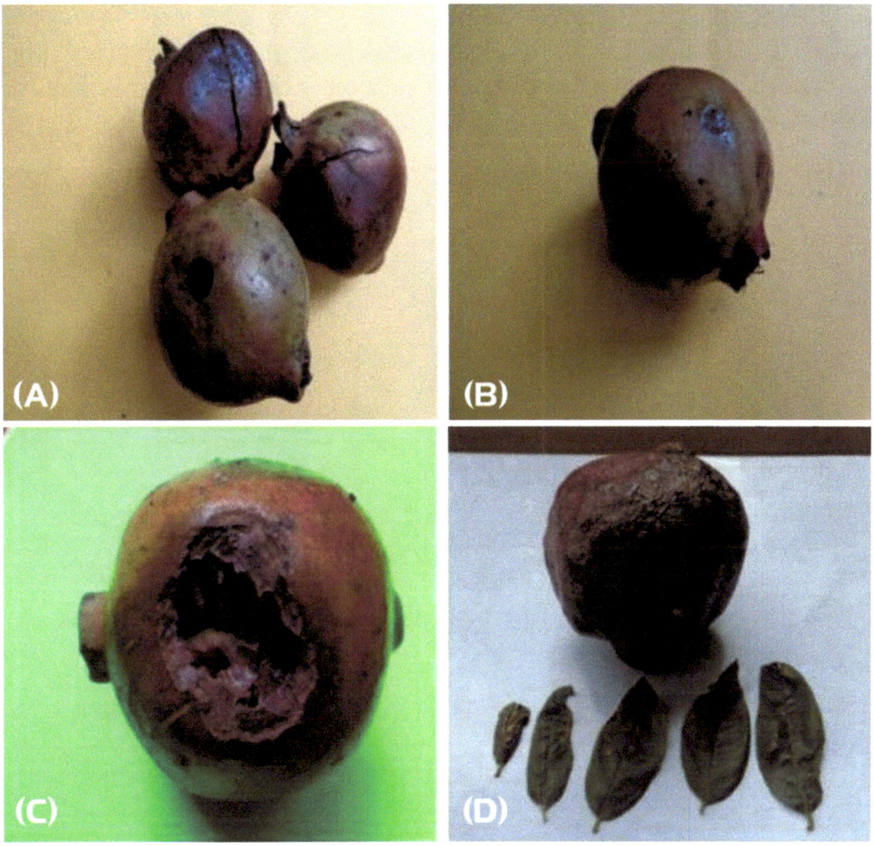

FIGURE 2.1 Bacterial blight symptoms on pomegranate fruits and leaves. (A) bacterial blight symptoms on pomegranate fruit; (B) bacterial blight symptoms on fruit; (C) bacterial blight symptoms on pomegranate fruit; (D) bacterial blight symptoms on fruit and leaves.

2.2.2.3 ISOLATION OF THE BACTERIA

The bacteria were isolated from the infected leaves, and fruits of pomegranate collected from regions of Latur, Maharashtra, India, that is, Murud and Harangul during the month of June to August. 2018. These samples were washed, air dried, and then disinfected with 0.1% $HgCl_2$ for about 30–60 s and washed thrice with sterile water to remove traces of $HgCl_2$. They were macerated with sterilized blade in a sterile Petri dish containing few drops of sterile distilled water in order to allow the bacteria to diffuse out. A loop full of suspension was then transferred with the help of sterilized bacteriological needle to sterilized Petri plates filled with nutrient agar medium with (NA) and incubated at 28°C for 24–72 h. After 2–3 days, incubated plates were observed for the presence of typical pale yellow, glistening colonies (Fig. 2.2) which were transferred to the NA slants and maintained in laboratory condition for further studies.

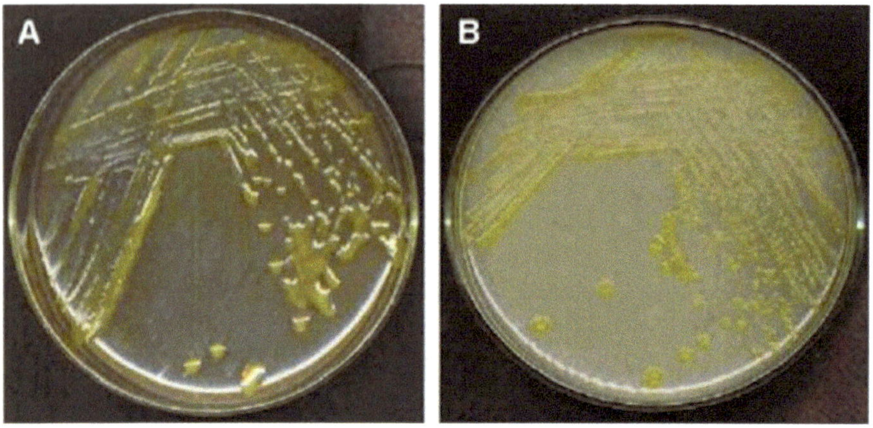

FIGURE 2.2 Pale yellow, glistening colonies of Xanthomonas on NA medium.

2.2.2.4 BIOCHEMICAL VARIABILITY

Methodology followed for the following experiments on biochemical variability is according to Schaad (1992)

i. *Lactose utilization:* Carbon source (lactose) was filter sterilized and mixed with autoclaved, cooled Dye's medium along with 1.2% purified agar. The pH was adjusted to 7.2. Bacterial isolate were spot inoculated with replica plating method and incubated at 30°C for 3, 7, and 14 days. Growth was compared with control, where carbon source was not supplemented.[22]
ii. *Starch hydrolysis:* Medium used for Starch hydrolysis was sterilized by autoclaving and poured into sterilized Petri plates. These plates were inoculated and incubated at 30°C for 7 days. The plates were flooded with Lugol's iodine and allowed to act for few minutes. The presence of starch hydrolysis was indicated by the presence of clear zones and vice-versa. The zone hydrolysed was measured for each isolate.
iii. *Acid production from Sucrose, Maltose, Dextrose:* The acid production by the isolate of *Xanthomonas axonopodis pv. punicae* wastested

by using Peptone water medium of Dye. Ten ml of medium was dispensed in each test tube. This medium was sterilized in an autoclave for 15 min. To these tubes, filter sterilized carbohydrates viz., Sucrose, Maltose and Dextrose were added at 0.14%concentration. The tubes were inoculated with 0.1 mL of 24 h old bacterial culture and incubated at room temperature for 3 days. Change in the color of the medium confirmed the acid production (Fig. 2.3).

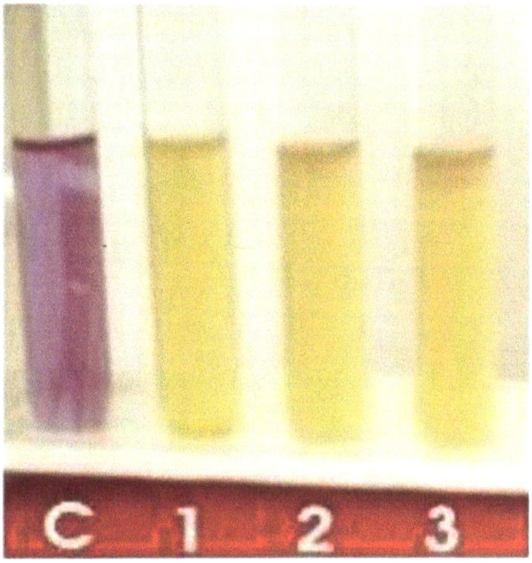

FIGURE 2.3 Acid production from sucrose, maltose, dextrose.
C: control, 1: sucrose, 2: maltose, 3: dextrose.

2.2.2.5 TESTING ANTIBACTERIAL ACTIVITY

After conformation of *Xanthomonas axonopodis pv. punicae*, the antibacterial activity of plant extract was preliminarily screened by well diffusion assay. LB agar plates were spread plated with 20 μL of bacterial strain (1 × 108 cfu/mL). Wells of 6 mm diameter were made in the agar plates. Each plant extract was tested for antibacterial activity by adding 40 μL of extracts in different concentrations viz., 50, 100, and 200 mg/mL. The experiment was repeated thrice. The plates were incubated at 37°C for 24 h. Subsequently, the plates were examined for zone of inhibition (ZOI) and diameter was measured in mm after subtracting well diameter (Ahmad et al. 1998). To determine minimum inhibitory concentration (MIC), required quantity of extracts were added in to the LB broth of 4 mL to bring initial concentration of 20 mg/mL. In each test tube 0.1 mL of standardized inoculum (1 × 108 cfu/mL) was added. Two control tubes were maintained for each test batch namely extract control (tube containing plant extract and LB medium without inoculum) and organism control (tube containing LB medium and inoculum). The test tubes were incubated at 37°C for 24 h. The lowest concentration (highest dilution) of plant extract that produced no visible growth (no turbidity) recorded as MIC. Minimum bactericidal concentration (MBC) was assayed by sub-culturing test dilutions on a drug free solid medium. The plates were incubated for 24 h at 37°C. The lowest concentration of the antimicrobial at which no single colony observed after sub-culturing is regarded as MBC.[23]

2.2.3 COLLECTION, GERMINATION AND PRESERVATION OF PARTHENIUM HYSTEROPHORUS

2.2.3.1 PREPARATION OF GERMINATION AGAR

We prepared 40% agar solution by dissolving 1 g of agar agar in 100 mL of warm distilled water, pH was balanced to 7.1 by adding NaOH in media. Media was autoclaved at 15 psi pressure at 121°C temperature for 15 min. After autoclaving, media was poured in two sterile Petri plates and kept for cooling until it forms a stuff jelly. Plates were named as T1, T2, Tc.

2.2.3.2 SEEDING PARTHENIUM HYSTEROPHORUS SEEDS ON AGAR

After sterilizing with 1% HgCl, seeds were washed properly with distilled water and were inoculated on agar plates in proportion of nine seeds/plate. Agar plates were incubated for 96 h at 27°C. Germination medium have maintained moist condition due to its Jelly like Nature. As moisture and temperature conditions were maintained, after 4 days, seedling occurred. After 4 days, plates were removed out from incubator and kept in laminar air flow. Seedlings were kept away from external environment to maintain seedling purity and avoid contamination of media. To maintain the moist condition of agar media and to provide water to newly born seedlings, irrigation was done in very low quantity.

2.2.3.3 TREATMENT OF PARTHENIUM HYSTEROPHORUS SEED WITH EXTRACTS

Extracts of *A. indicum* and *P. juliflora* were prepared by using distilled water. After 24 h stirring process, all extracts were filtered by using Whatman no.1 filter paper. Test was started on fifth day of seedling germination. Extracts were treated with Seedlings of *P. hysterophorus* as follows:

- The test was carried out for 6 days/144 h.
- Plates were labeled as T1, T2, and Tc.
- Tc plate was set as control.
- 2 mL Extracts of *A. indicum* and *P. juliflora* were poured in plates T1, T2 respectively and further labeled as Ai-T1, Pj-T2
- Extracts were poured on seedlings three times at the interval of 2 days.
- After sixth day of test, germination indices were assayed.

2.3 OBSERVATION

2.3.1 MORPHOLOGICAL CHARACTERS OF XANTHOMONAS AXONOPODIS PV. PUNICAE

The morphological characters such as shape, gram reaction, and pigmentation characters were studied as described by Society of American Bacteriologists, Bradbury[9] and Schaad and Stall (Table 2.2).[10]

TABLE 2.2 Cultural and Morphological Characteristics of *Xanthomonas axonopodis pv. Punicae* Isolate on Nutrient Agar Media.

Sr. no	Colony characters	*Xanthomonas axonopodis pv. punicae*
1	Color	Yellowish
2	Size of colony	Medium to large
3	Shape of colony	Small circular colonies
4	Cell shape	Single rod
5	Appcarance	Slightly raised, glistening
6	Elevation	Convex
7	Margin	Entire margin
8	Texture	Highly mucoid

2.3.2 GROWTH RATE AT 28°C AND 37°C OF XANTHOMONAS AXONOPODIS PV. PUNICAE

The effect of varied temperature levels on the growth of *Xanthomonas axonopodis pv. punicae* was studied and data so obtained is presented in (Table 2.3). Isolate of *Xanthomonas axonopodis pv. punicae* were tested at temperature 28°C and 37°C on NA agar medium. The data clearly indicated that the temperature of 28°C was found optimum for the growth of the pathogen as significantly maximum number of colonies was observed at this temperature. Isolate grew well at 28°C but no growth was observed at temperature of 37°C. Growth of *Xanthomonas axonopodis pv. punicae* started 72 h after incubation. Maximum growth was observed after 120 h of incubation at 28°C but no growth was observed after 48 h of incubation. At 37°C no growth was observed up to 120 h. Similar work on temperature requirement was carried out by Hingorani and Mehta.[24] They found that the pomegranate bacterium grows well at a cordial temperature of 30°C and can tolerate a minimum and maximum temperature of 5°C and 40°C, respectively. Gour et al.[25] also got the similar results while working with *Xanthomonas axonopodis pv. vignicola*, the causal agent of leaf blight of cowpea. They have recorded the maximum growth of the pathogen at a temperature level of 30°C, whereas, Manjula[26] recorded the highest number of colonies of *Xap.* at a temperature of 27°C.

TABLE 2.3 Growth Rate of *Xanthomonas axonopodis pv. Punicae* Isolate.

Sr. no.	Isolate	Growth rate							
		28°C				37°C			
		48 h	72 h	96 h	120 h	48 h	72 h	96 h	120 h
1	*Xanthomonas axonopodis pv. punicae*	+	++	+++	++++	–	–	–	–

2.3.3 EFFECTS OF PLANT EXTRACTS ON SEEDLING OF PARTHENIUM HYSTEROPHORUS

It has taken 11 days which was started with seeding and ends with final test assay. After eleventh day of test, germination indices viz., germination percent (GP), germination index (GI), germination energy (GE), speed of emergence (SE) and seedling vigor index (SVI) of *P. hysterophorus* were assayed. Significant reduction in germination indices were observed when compared with Control. Clearly visible reduction was observed in seedlings (Fig. 2.4). Germination indices were assayed by using equations given in Table 2.4.

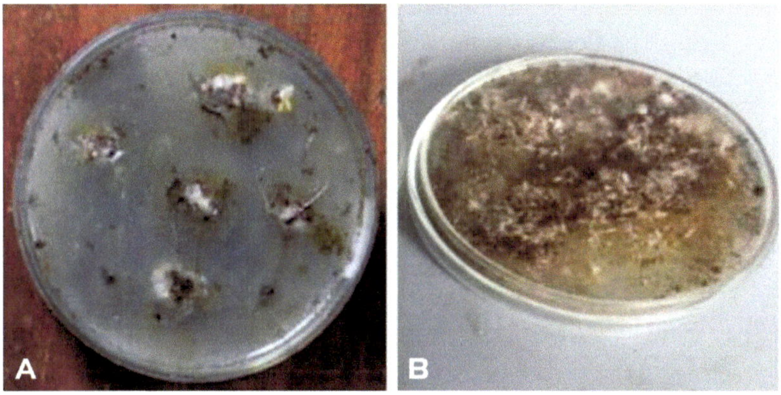

FIGURE 2.4 Phytotoxic effects on *Parthenium* seeds. (A) *A. indicum*; (B) *P. juliflora*.

TABLE 2.4 The Equations Used to Calculate Different Germination Indices.

Germination parameters	Equations	References
Germination percent	$\frac{\text{Number of germinated seeds at final count}}{\text{Total number of seeds sets for bioassay}} \times 100$	[15]
Germination index	$\frac{G_T}{T_t}$ Or $\frac{\text{Number of germinated seeds}}{\text{Days of first count}} + .. + \frac{\text{Number of germinated seeds}}{\text{Days of last or final count}}$	[15]
Speed of emergence	$\frac{\text{Number of germinated seeds at the starting day of germinationa}}{\text{Number of germinated seeds at the final days of measurement}} \times 100$	[27]
Germination energy	$\frac{\text{Percentage of germinated seeds at the starting day of germination}}{\text{Total number of seeds sets for bioassay}} \times 100$	[28]
Seedling vigor index	$\frac{\text{Seedling length (mm)} \times \text{Germination percent}}{100}$	[27]

2.4 RESULTS

2.4.1 WELL DIFFUSION ASSAY

The preliminary screening of selected three plant extracts against the *Xanthomonas axonopodis pv. punicae* was done using well diffusion method. The ZOI greater than 5 mm diameter is found to be having significant activity against particular bacteria.[29] The aqueous extracts were sensitive against *Xanthomonas axonopodis pv. punicae* tested at different concentrations (Table 2.5)

P. juliflora extract arrested the growth of *Xanthomonas axonopodis pv. punicae* (24 mm) at 200 mg/mL. The *P. juliflora* extract shows a ZOI about 21 and 19 mm at 100 and 50 mg/mL respectively. Based on ZOI, Satish et al. (1999) reported significant antibacterial activity *of P. juliflora* and *A. arabica* against *X. campestris pathovars*. The extracts of *A. indicum was* exhibited the antibacterial activity of (11 mm) ZOI at 200 mg/mL

TABLE 2.5 Antibacterial Activity of Plant Extracts Showing Zone of Inhibition.

Sr. no	Plant species	Conc. mg/mL	Zone of inhibition (mm)
1	*P. juliflora*	50	19
		100	21
		200	24
2	*A. indicum*	50	6
		100	9
		200	11

2.4.2 MINIMUM INHIBITORY AND BACTERICIDAL CONCENTRATION

The antibacterial activity of *A. Arabica* and *P. juliflora* extracts are found to be high against all the bacteria. The maximum activity recorded in *P. juliflora* (MIC = 1.03 and MBC = 0.15 mg/mL). Raghavendra et al.[30] reported significant antibacterial activity of *A. nilotica* extracts against *Xanthomonas pathovars* and human pathogenic bacteria tested, while the lowest activity was recorded by *A. indicum* (MIC = 0.619 and MBC = 0.923 mg/mL). These findings indicate that the extracts of *P. juliflora* and *A. indicum* are potential to use in the management of plant diseases. Further phytochemical analysis is required to identify the active components of plant extracts showing antimicrobial activity (Table 2.6) (Fig. 2.5).

TABLE 2.6 MIC and MBC of Plant Extracts Against *Xanthomonas axonopodis pv. Punicae*.

Sr. no	Plant species	MIC (mg/mL)	MBC (mg/mL)
1	*P. juliflora*	1.03	0.15
2	*A. indicum*	0.619	0.923

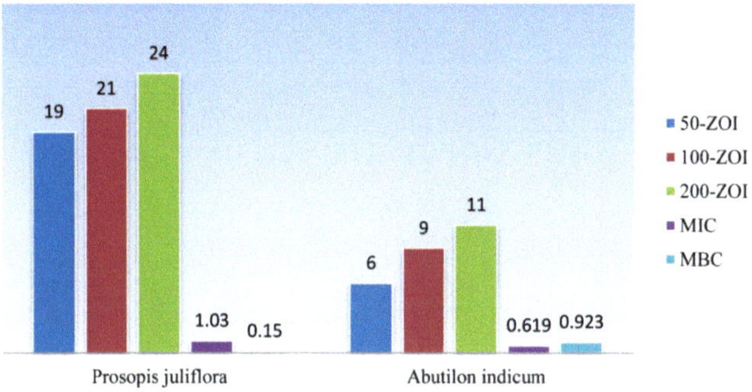

FIGURE 2.5 Graphical representation of susceptibility of plant extracts.

2.4.3 GERMINATION INDICES OF PARTHENIUM HYSTEROPHORUS

Results revealed that the extracts of *A. indicum* and *P. juliflora* has phytotoxic activity which significantly reduces Germination of *P. hysterophorus*. The data generated in this study show that the extracts of *A. indicum* and *P. juliflora* have significant effects on all calculated germination indices. Result revealed that *P. juliflora* has strong phytotoxic properties as compared with *A. indicum* (Fig. 2.6) (Table 2.7).

TABLE 2.7 Significant Reduction in Germination Indices of *Parthenium hysterophorus*.

Treatments	Germination indices (11 days after sowing)				
	Germination percentage %	Germination energy %	Germination index	Speed of emergence %	Seedling vigor index
Ai-T1	11.23	12.45	14.78	16.67	17.21
Pj-T3	9.87	8.43	12.56	13.96	12.97
Tc	17.42	19.23	21	20.22	23
SEm±	1.32	1.52	0.64	1.47	1.19
CD @ 5%	3.98	4.50	4.08	3.23	0.49
CV	4.62	3.41	3.50	2.56	3.74

Treatment details:

Ai-T1: Seedling treated with *A. indicum* extract for 6 days at intervals of 2 days.

Pj-T3: Seedling treated with *P. juliflora* extract for 6 days at intervals of 2 days.

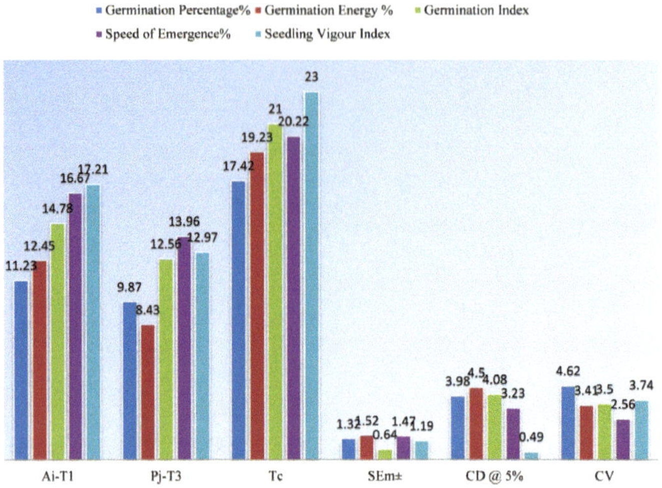

FIGURE 2.6 Graphical representation of germination indices of *Parthenium hysterophorus.*

Highest reduction in GP% was recorded by *P. juliflora* (9.87%) while lowest was recorded by *A. indicum* (11.23%). Strongest Phytotoxic activity was shown by *P. juliflora*. Highest reduction in GE% was recorded by *P. juliflora* (8.43%) while lowest was recorded by *A. indicum* (12.45%). Highest reduction in SVI was recorded by *P. juliflora* (12.97) while lowest was recorded by *A. indicum* (17.21).

These results indicate the inhibitory potential of *A. indicum* and *P. juliflora* plant extracts. The delay or inhibition of germination caused by phytotoxic plant extracts or substances was also reported by Anjum and Bajwa[32] and Hussain et al.[33] Although germination bioassay is the most widely used method to inspect the phytotoxic activity,[34,35] early seedling growth is reported to be most sensitive parameter to test the phytotoxicity.[36] Hence, we have conducted the growth bioassay using the same test species to confirm the phytotoxic properties of *A. indicum* and *P. juliflora* plants extracts. The bioassay results showed a significant reduction of shoot and root growth of all test species at 25% concentration. However, the sensitivity to the *P. hysterophorus* was varied among the plant extracts.

2.5 DISCUSSION

Many synthetic antibiotics are used to control several phytopathogens. The increased awareness of environmental problems with these chemical antibiotics has led to the search for non-conventional chemicals of biological origin for the management of these diseases. Bactericides of plant origin can be one approach to disease management because of their eco-friendly nature.[31] Total GP is a commonly used index to measure the effects of phytotoxic substances.[37,38] It is the maximum percentage of germination that mainly depends on final measurements. However, this index cannot interpret the possible delayed germination caused by phytotoxic plant extracts or substances. Therefore, GP is considered to be suitable for ecological studies rather than physiological process like germination.[39,40]

A number of indices over GP have been proposed by many researchers to study the inhibitory activity of phytotoxic substances on germination process.[41,32] To investigate the actual inhibition (either direct inhibition or delayed effect) by *A. indicum* and *P. juliflora* plant extracts on germination of parthenium weed, we analyzed few important germination indices: GI, SE, GE, SVI together with GP. We observed a significant reduction of GI, SE, GE, SVI, and GP of extract. Phytotoxic substances, therefore inhibitory effects are more visible on roots rather than on shoots. In summary, the extracts of *A. indicum* and *P. juliflora* inhibited the seed germination of *P. husterophorus*. These results indicated that *A. indicum* and *P. juliflora* plants extracts have phytotoxic properties and thus contain phytotoxic substances. As the strong phytotoxic activity is shown by *P. juliflora,* thus, *P. juliflora* could be a potential starting point of green weedicide and bioherbicides. Along with Weedicide activity, both the plant species has shown a significant reeducation in growth of *Xanthomonas axonopodis pv. punicae.*

Therefore, the plant could be served as an important candidate for isolation and identification of allelopathic substances and can be used as good natural antibiotic substances, which may promote the development of new natural herbicides, pesticides and antibiotic agents. Besides this, the plant extracts or their residues could be directly used as bioherbicides and strong antibiotics.

2.6 CONCLUSION

The products of plant origin are of greater advantage to user, the public, and the radical environmentalist. Laboratory screening of plant extracts has given encouraging results, indicating their potential use in the management of disease caused by *Xanthomonas* species. Plant extracts resulted in antibacterial activity is potential to use in the management of plant diseases as an alternative to chemical antibiotics. Further phytochemical analysis is required to identify the bioactive compounds responsible for antibacterial activity. Yet the results gained by these methods provide simple data that make it possible to classify extracts in respect to their antioxidant potential. As can be observed also from the present data, antibiotic activity does not necessarily correlate with high amounts of phytochemicals, and that is why both phytochemicals content and antibiotic activity information must be discussed when evaluating the phytochemical potential of extracts. Weed management is one of the most challenging tasks in crop production. Over use of synthetic herbicides causes severe environmental pollution besides being developed herbicide resistant weed biotypes.

On the basis of this study, the most potent Finnish plant sources for natural weedicide and antibiotic agents are medicinal plants and vegetable peels, and different tree materials. Further work is under way to confirm the antibiotic and herbicidal effect of these promising plant extracts by using other types of bacterial strains and to characterize the active phytochemicals, their mechanism of action, and their possible interactive effects together with harmful effects to environment. Natural Plant Products could serve as an alternative to synthetic herbicides, Pesticide, Antibiotics that are biodegradable and environment friendly. In this regard, *A. indicum* and *P. juliflora* have a promising role. Isolation and characterization of phytotoxic substances from *A. indicum* and *P. juliflora* may promote the development of plant product based natural herbicides.

ACKNOWLEDGMENT

The authors thankfully acknowledge Dr. J. S. Dargad, Principle, Dayanand Science College, Latur for his constant support, encouragement and for providing microbiology lab to work. Also wants to extend thanks to Dr. S. S. Digrase, Senior Scientist, Krishi Vigyan Kendra, Latur-413512, MH, India for providing laboratories and chemicals to work. The authors also wish to acknowledge Mr. Rajendra Biradar for kindly providing the dried plant material and all respected teachers Dr. R. S. Shinde, Mrs. S. D. Patil, Dr. R. S. More sir and all other teachers for their constant support

KEYWORDS

- **bio-herbicide**
- ***Abutilon indicum***
- **minimum bactericidal concentration**
- ***Prosopis juliflora***
- **synthetic antibiotics**

REFERENCES

1. Malhotra, N. K.; Khajuria, H. N.; Jawanda. Studies on Physio-chemical Characters of Pomegranate Cultivars II. Chemical Characters. *Punjab Horti. J.* **1983,** *23*, 158.
2. Bose, T. K. Pomegranate. In *Fruit of India—Tropical and Subtropical*; Mitra Naya, B., Ed.; Prakash Publications: Calcutta, 1985; p. 637.
3. Hingorani, M. K.; Singh, N. J. *Xanthomonaspunicae* sp. Nov. on Pomegranate (*Punicagranatum* L.). *Indian J. Agri. Sci.* **1959,** *29*, 45–48.
4. Hingorani, M. K.; Mehta, P. P. Bacterial Leaf Spot of Pomegranate. *Indian Phytopath.* **1952,** *5*, 55–56.
5. Vauterin, L.; Haste, B.; Kersters, K.; Swings, J. Reclassification of *Xanthomonas*. *Int. J. Syst. Bacterial* **1995,** *45*, 475–489.
6. Petersen, Y.; Mansvelt, E. L.; Venter, E.; Langenhoven, W. E. Detection of *Xanthomonas axonopodis pv. punicae* Causing Bacterial Blight on Pomegranate in South Africa. *Australas. Plant Pathol.* **2010,** *39*, 544–546.
7. Chand, R.; Kishun, R. Studies on Bacterial Blight (*Xanthomonas campestris* pv. *punicae*) on Pomegranate. *Indian Phytopathol.* **1991,** *44*, 370–372.
8. Rani, U.; Verma, K. S.; Sharma, K. K. Pathogenic Potential of *Xanthomonas axonopodis pv. punicae* and Field Response of Different Pomegranate Cultivars. *Plant Dis. Res.* **2001,** *16*, 198–202.
9. Bradbury, J. F. Isolation and Preliminary Study of Bacteria from Plants. *Rev. Plant Pathol.* **1970,** *49*, 213–218.
10. Schaad, N. W.; Stall, R. E. *Xanthomonas*. In *Laboratory Guide for Identification of Plant Pathogenic Bacteria*, 2nd ed., APS Press: St Paul, MN, 1988; pp. 81–94.
11. Adkins, S. W.; Sowerby, M. S. Allelopathic Potential of the Weed *Parthenium hysterophorus* L. in Australia. *Plant Prot. Quart.* **1996,** *11*, 20–23.
12. Adkins, S. W.; Navie, S. C. Parthenium Weed: A Potential Major Weed for Agro-ecosystems in Pakistan. *Pak. J. Weed Sci. Res.* **2006,** *12* (1–2), 19–36.
13. Chippendale, J. F.; Panetta, F. D. The Cost of *Parthenium* Weed to the Queensland Cattle Industry. *Plant Prot. Quart.* **1994,** *9*, 73 76.
14. Khosla, S. N.; Sobti, S. W. Effective Control of *Parthenium hysterophorus* L. *Pesticides* **1981,** *15*, 18–19.
15. Association of Official Seed Analysis (AOSA). *Seed Vigor Testing Handbook.* Handbook on Seed Testing, Contribution no. 32, 1983.
16. Nath, R. *Parthenium hysterophorus* L. A General Account. *Agric. Rev.* **1988,** *9*, 171–179.

17. Joy, P. P.; Thomas, J.; Samuel, Mathew, Baby, P. Skaria. Medicinal Plants, Database on Tropical Plants. *Tropical Horticulture* **1998,** 186–210.
18. Dwivedi, P.; Vivekan, V.; Ganguly, R.; Singh, R. P. *Parthenium* sp. As a Plant Biomass for the Production of Alkalitolerant Xylanase from Mutant *Penicillium oxalicum* SAUE-3.510 in Submerged Fermentation. *Biomass Energ.* **2009,** *33*, 581–588.
19. Tudor, G. D.; Ford, A. L.; Armstrong, T. R.; Bromage, E. K. Taints in Meat from Sheep Grazing *Parthenium hysterophorus*. *Aust. J. Exp. Agric. Anim. Husb.* **1982,** *22*, 43–46.
20. Muniappa, T. V.; Prasad, T. V. R.; Krishnamurthy, K. Comparative Efficacy and Economics of Mechanical and Chemical Methods of Control of *Parthenium hysterophorus* L. *Ind. J. Weed Sci.* **1980,** *12*, 137–144.
21. Dhawan, S. R.; Dhawan, P. Regeneration in *Parthenium hysterophorus* L. *World Weeds* **1996,** *3*, 181–182.
22. Schaad, N. W. Laboratory Guide for the Identification of Plant Pathogenic Bacteria; 2nd ed. *Am. Phytopath. Soc.* **1992,** *138*.
23. Akinyemi, K. O.; Oladapo, O.; Okwara, C. E.; Ibe, C. C.; Fasure, K. A. Screening of Crude Extracts of Six Medicinal Plants Used in South West Nigerian Unorthodox Medicine Foranti- methicillin Resistant *Staphylococcus aureus* Activity. *BMC Complem. Altern. M.* **2005,** *6*.
24. Hingorani, M. K.; Mehta, P. P. Bacterial Leaf Spot of Pomegranate. *Indian Phytopath.* **1952,** *5*, 55–56.
25. Gour, H. N.; Ashiya, J.; Mali, B. L.; Nath, Ranjan. Influence of Temperature and pH on the Growth and Toxin Production by *Xanthomonas axonopodis pv. Vignicola* Inciting Leaf Blight of Cowpea. *J. Mycol. Plant Pathol.* **2000,** *30*, 389–392.
26. Manjula, C. P. Studies on Bacterial Blight of Pomegranate (*Punicagranatum* L.) Caused by *Xanthomonas axonopodis pv. punicae*. M. Sc. (Agri.) Thesis; submitted to University of Agricultural Sciences, Bangalore, 2002; p. 98.
27. Steel, R. G. D.; Torrie J. H. *Principle and Procedures of Statistics: A Biometrical Approach*, 2nd ed.; McGraw Hill: NewYork, NY, USA, 1980.
28. Haugland, E.; Brandsaeter, L. O. Experiments on Bioassay Sensitivity in the Study of Allelopathy. *J. Chem. Ecol.* **1996,** *22* (10), 1845–1859.
29. Palombo, E. A.; Semple, S. J. Antibacterial Activity of Traditional Australian Medicinal Plants. *J. Ethanopharmacol.* **2001,** *77*, 151–157.
30. Raghavendra, M. P.; Satish, S.; Raveesha, K. A. In vitro Evaluation of Antibacterial Spectrum and Phytochemical Analysis of *Acacia nilotica*. *J. Agri. Tech.* **2006,** *2*, 77–88.
31. Bolkan, H. A.; Reinert, W. R. Development and Implementing IPM Strategies to Assist Farmers; An Industry Approach. *Plant Dis.* **1994,** *78*, 545–550.
32. Anjum, T.; Bajwa, R. Importance of Germination Indices in Interpretation of Allelo Chemical Effects. *Int. J. Agric. Biol.* **2005,** *7*, 417–419.
33. Hussain, M. I.; Gonzalez-Rodriguez, L.; Reigosa, M. J. Germination and Growth Response of Four Plant Species to Different Allelo Chemicals and Herbicides. *Allelopathy J.* **2008,** *22* (1), 101–110.
34. Rice, E. L. *Allelopathy*, 2nd ed.; Academic Press: Orlando, FL, USA, 1984.
35. Putnam, A. R.; Tang, C. S. Allelopathy: State of Science. In *The Science of Allelopathy*; Putnam, A. R.; Tang, C. S., Eds.; 1986, 1–19.
36. Williams, R. D.; Hoagland, R. E. The Effects of Naturally Occurring Phenolic Compounds on Seed Germination. *Weed Sci.* **1982,** *30*, 206–212.

37. Haugland, E.; Brandsaeter, L. O. Experiments on Bioassay Sensitivity in the Study of Allelopathy. *J. Chem. Ecol.* **1996,** *22* (10), 1845–1859.
38. Hoffman, M. L.; Weston, L. A.; Snyder, J. C.; Regnier, E. E. Allelopathic Influence of Germinating Seeds and Seedlings of Cover Crops on Weed Species. *Weed Sci.* **1996,** *44* (3), 579–584.
39. Chiapusio, G.; S′anchez, A. M.; Reigosa, M. J.; Gonz′alez, L.; Pellissier, F. Do Germination Indices Adequately Reflect Allelo Chemical Effects on the Germination Process? *J. Chem. Ecol.* **1997,** *23*(11), 2445–2453.
40. Haugland, E.; Brandsaeter, L. O. Experiments on Bioassay Sensitivity in the Study of Allelopathy. *J. Chem. Ecol.* **1996,** *22* (10), 1845–1859.
41. Bewley, J. D.; Black, M. *Seeds: Physiology of Development and Germination*; Plenum Press: New York, NY, USA, 1985.

CHAPTER 3

Pharmacological Perspectives of Eugenol in Modern Context

AMENA ALI[1*] and ABUZER ALI[2]

[1]*Department of Pharmaceutical Chemistry, College of Pharmacy, Taif University, Taif, 21974, Saudi Arabia*

[2]*Department of Pharmacognosy, College of Pharmacy, Taif University, Taif, 21974, Saudi Arabia*

[*]*Corresponding author. E-mail: ankitarathore23@gmail.com*

ABSTRACT

Eugenol is the chief phytoconstituent present in the essential oil of clove and several other medicinal plants. Being non-carcinogenic and non-mutagenic, it is usually considered as safe by many organizations like Food and Agricultural Organization (FAO), United Nations (UN), etc. Eugenol exhibited various pharmacological properties like antimicrobial, antioxidant, anti-inflammatory and showed potential against life-threatening diseases including leishmaniasis, cancer, and sepsis. The current chapter provides a brief overview of the plethora of research regarding the health benefits of eugenol and also discuss the mechanisms involved.

3.1 INTRODUCTION

Eugenol is one of the important phytochemicals extracted from certain essential oils, especially from clove oil. It is also present in cinnamon, soybean, coffee, basil, banana, and other foods.[1–6] Eugenol is a phenolic compound which is chemically 4-allyl-2-methoxyphenol, and belongs to the class of phenylpropanoids. Synthetic EGL can be prepared by allylation of guaiacol with allyl chloride in laboratory or in industry, possessing similar chemical properties.[7] Being a chief constituent in the non-polar extracts of several medicinal plant, it received too much attention of scientists in extensive area of research to develop it as medicine against numerous diseases. Various studies testified that the EGL possesses a number of pharmacological properties; that is, antioxidant, anaesthetic, antimicrobial, anti-inflammatory, anti-cancer, anti-fumigant, anti-helmintic, and anti-repellent properties.[8–15] EGL had been used as part of traditional remedy for toothache and for culinary purposes. It is a versatile molecule present as an ingredient in perfumes, cosmetics, flavoring agents for foods and cosmetics, preservative, irritant, and sensitizer.[16,17]

For EGL, World Health Organization (WHO) and FAO have given limit for daily intake which is 2.5 mg/kg body weight.[18] Moreover, it has been proclaimed safe by United States Food and Drug Administration (FDA) and has no carcinogenic and mutagenic actions. During recent years, it has received the researcher's attention owing to its chemopreventive, anti-inflammatory, and excellent anti-oxidant activity.[19–22] Studies on the effects of EGL still considered as research priority due to its multiple biological and pharmacological activities. Thus, it is worthy to rationally unite significant research findings associated with EGL including its importance in human health and to explicate its mode of action where possible.

3.2 SOURCE OF EUGENOL

Eugenol can be extracted from various aromatic plants. Apart from the *Eugenia caryophyllata* (clove), it can be isolated from *Ocimum grattisimum* (clove basil/African basil), *Myristica fragrans* (nutmeg), *Cinnamomum tamala* (cinnamon), *Ocimum basilicum* (basil), *Ocimum tenuiflorum* (holi basil/ tulsi), *Pimenta racemose* (bay rum tree), etc. Nevertheless, the principal source of eugenol is clove oil that contains 45%–90% eugenol of its constituent.[7,23,24]

3.2.1 PHYSICAL AND CHEMICAL PROPERTIES OF EUGENOL

EGL ($C_{10}H_{12}O_2$) has a phenolic group and comes under the class of phenylpropanoids. 4-allyl-2-methoxyphenol is the IUPAC name of this compound (Fig. 3.1), with molecular mass (164.2 g/mol) and pKa (10.19) at 25°C. It is present in two isoforms; eugenoland isoeugenol. It is also known by other names as allylguaiacol, caryophyllic acid, 2-methoxy-4-(2-propenyl) phenol or eugenic acid. Phenolic function is responsible for the antioxidant potential of EGL. This compound is freely solubility in organic solvents and has partially solubility in aqueous solutions. EGL occurs as a colorless to pale-yellow aromatic liquid.[7,25] After oral administration, EGL absorbs through small intestine while on intraperitoneal administration it quickly distributed to almost all organs. Glucuronic acid, sulfate, and glutathione conjugates are formed on its metabolism.[26]

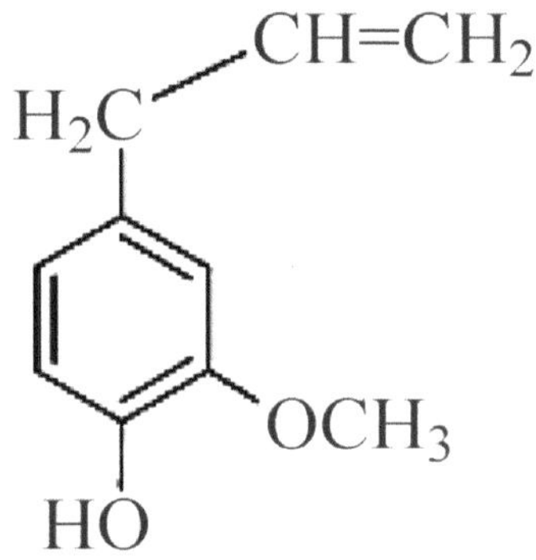

FIGURE 3.1 Chemical structure of eugenol.

3.2.2 *ISOLATION AND IDENTIFICATION OF EUGENOL FROM PLANTS*

In 1929, it was first isolated and commercial production began in the U.S. in 1940s.[7] EGL is mainly isolated from natural sources by treating and shaking the essential oil with potassium or sodium hydroxide (3%) solution to form alkaline phenolic salt. Further, non-phenolic part is extracted through steam distillation or using a suitable solvent. EGL is liberated by the acidification of alkaline portion at low temperature and purified either by fractional distillation, thin layer chromatography (TLC) or high pressure liquid chromatography (HPLC). The purity of isolated EGL can be checked by FTIR, NMR, and mass spectroscopy.[22,23,27]

3.2.3 *PHARMACOLOGICAL ACTIVITIES OF EUGENOL*

EGL is a versatile phytoconstituent which exhibits numerous therapeutic potentials (Fig. 3.2).

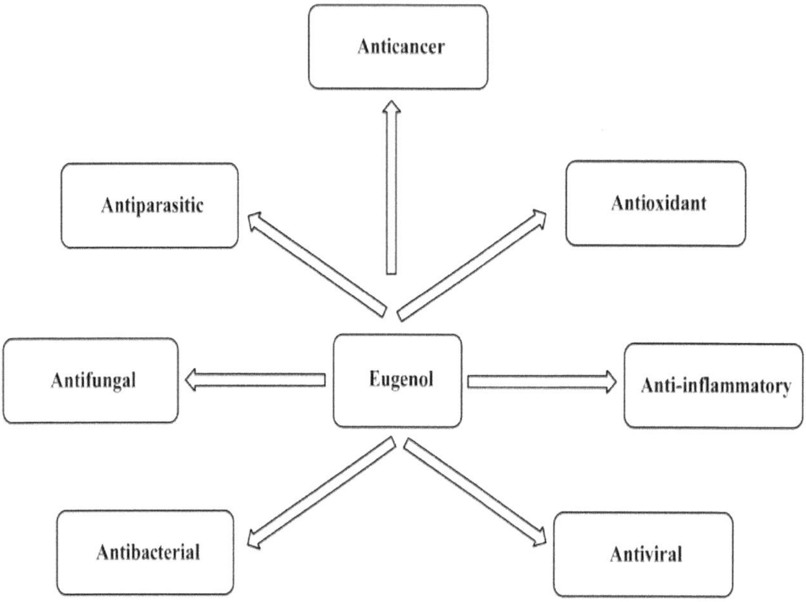

FIGURE 3.2 Pharmacological activities of eugenol.

3.2.4 ANTIBACTERIAL ACTIVITY OF EUGENOL

Studies have shown that eugenol displayed potent antibacterial activity against several strains of Gram-positive (*Bacillus subtilis*, *Bacillus cereus*, *Staphylococcus epidermidis*, *Staphylococcus aureus*, *Streptococcus pyogenes*, *Streptococcus pneumonia*, *Listeria monocytogenes* and *Enterococcus faecalis*) and Gram-negative (*Salmonella typhi*, *Escherichia coli*, *Pseudomonas aeruginosa*, *Salmonella choleraesuis*, *Yersinia enterocolitica*, *Helicobacter pylori*, and *Proteus vulgaris*) bacteria.[28–34] Eugenol causes cell lysis of Gram-positive and Gram-negative bacteria through disrupting the cell wall and membrane which lead to leakage of protein and lipid contents (Fig. 3.3).[35] Various in vitro and in vivo studies have been done on biofilms which revealed that potent inhibitory and eradication properties have been displayed by eugenol. It has shown significant inhibition against the formation of biofilms by methicillin-resistant and methicillin sensitive *S. aureus* (MRSA and MSSA) strains. Eugenol exhibited 50% inhibition and decreased the biomass of MRSA and MSSA strains at a concentration of 0.5 × MIC. Eugenol has potentially decreased 88% *S. aureus* colonization in rat middle ear at sub-MIC. Eugenol and carvacrol have shown the synergistic effects on eradication of biofilms. The combinations of eugenol and carvacrol have been shown to decrease the established biofilms by 99% called minimum biofilm eliminating concentration (MBEC).[36]

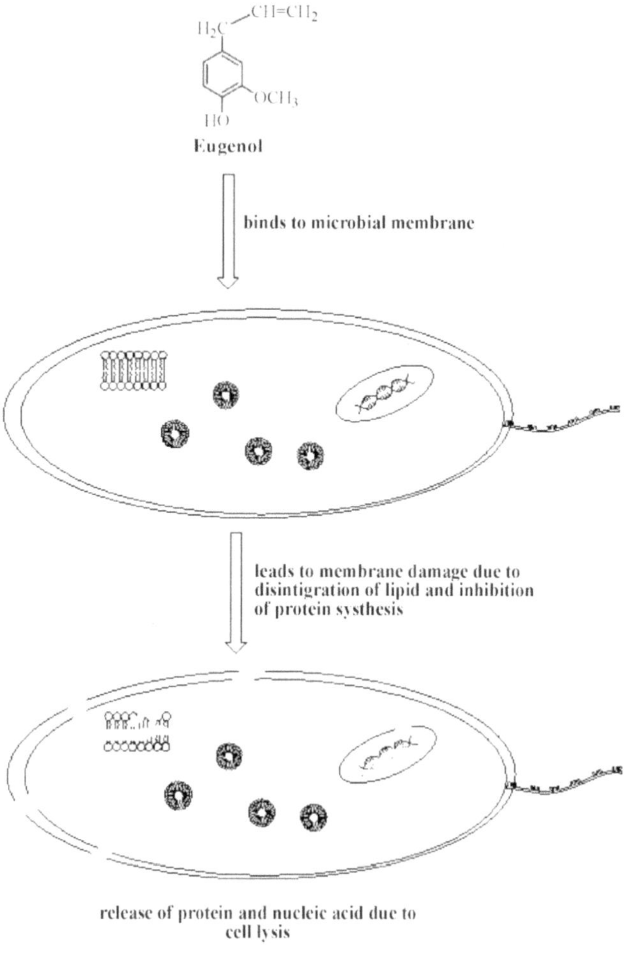

FIGURE 3.3 Schematic presentation of antimicrobial activity of eugenol.

Eugenol has potential to induce ≥90% inhibition of *P. aeruginosa* biofilms at 0.5 MIC.[37] Eugenol combination therapy with other antibiotics aids to lessen the risk of resistant microbes. Various study results revealed that eugenol showed synergistic interactions with various antibiotics such as vancomycin, gentamicin, and β-lactams resulting in superior antimicrobial action.[32,38] Similar synergistic interactions have been displayed by eugenol with cinnamaldehyde, cinnamate, carvacrol, and thymol which lead to greater anti-bacterial activity.[39,40] Eugenol has shown to reduce the necrosis factor-inducing and hemolytic activities of culture supernatants in a dose-dependent manner and decreased the production of staphylococcal enterotoxin A significantly at sub-inhibitory concentrations (16–128 μg/mL).[41] Low solubility, strong odor, and liability to sublimation are few drawbacks of eugenol which might be overcome by glycosylation of eugenol to eugenol α-D-glucopyranoside (α-EG), which is tested with *S. aureus* and *E. coli* and found to be more effective than that of pure eugenol.[42]

3.2.5 ANTIOXIDANT ACTIVITY OF EUGENOL

Eugenol has promising potential to scavenge the free radicals.[21,43–45] Various studies reported that EGL displayed 2,2-diphenyl-1-picrylhydrazyl (DPPH) and hydroxyl radical inhibition at concentrations dependent manner.[46] The antioxidant property of eugenol and its isomerisoeugenol has been tested by the autooxidation of Fe^{2+} and iron-mediated lipid peroxidation.[47] Besides, the FRS activity, EGL also exhibited nitric oxidescavenging activity, while reducing power was tested by Griess reagent and FTC method.[21] EGL also has the ability to reverse the neuronal excitotoxic or oxidative injury and also shown protective effect against *N*-methyl-D-aspartate-induced neurotoxicity.[48] It also protected reactive oxygen species (ROS) generation and ROS-induced lipid-protein and DNA damage in in vitro and in vivo models (Fig. 3.4) along with increased cellular anti-oxidant activity, specifically, glutathione system[20,49–51].

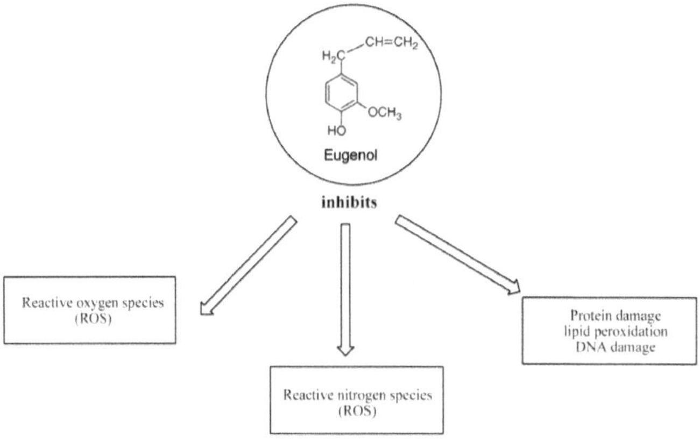

FIGURE 3.4 Schematic presentation of antioxidant activity of eugenol.

3.2.6 ANTI-FUNGAL ACTIVITY OF EUGENOL

Disc diffusion method was used to screen EGL against 53 strains of human pathogenic yeasts. The results revealed anti-fungal potential of EGL against each tested fungal strains.[52] In another study, eugenol-Mannich base derivatives were synthesized and screened for their anti-fungal activity. Compounds 4-allyl-2-methoxy-6-(morpholin-4-ylmethyl) phenylbenzoate and 4-{5-allyl-2-[(4-chlorobenzoyl)oxy]-3-methoxybenzyl}morpholin-4-ium chloride were appeared to be the most potent antifungal agents as compared with standard drug fluconazole. The most significant IC_{50} values against *C. albicans*, *C. Glabrata*, and *C. Krusei* were ranging between 0.063–1.23 μM.[53] Moreover, fractional inhibitory concentration indices against the formation of *C. albicans* biofilm for eugenol-FCZ and carvacrol-FCZ combinations were found to be 0.25 and 0.311, respectively.[54] On the other hand, EGL significantly reduced the adherence and metabolic performance of *C. albicans* biofilms isolated from the HIV patients.[55] Similarly, there is reduction of ergosterol biosynthesis followed by apoptosis resulting from exposure of Candida cells to eugenol. EGL also showed the potential to alter the morphogenesis pattern of *C. albicans*. Synergistic effects were reported by different dose combinations of EGL-thymol against *C. albicans* colonization and infectivity.[57]

3.2.7 ANTI-PARASITIC ACTIVITY

EGL possessed anti-parasitic activities, such as anti-leishmanial, anti-giardial, anti-malarial, and trypanocidal potential at higher concentrations showed by various in vitro studies worldwide.[58–61] It didn't induce cell death but inhibits the adherence of *G. lamblia* trophozoites since the third hour. Modifications on the cell shape, autophagic vesicles, presence of precipitates in the cytoplasm, membrane blebs, intracellular/nuclear clearing, and internalization of flagella and ventral disc were the key morphological alterations.59 Eugenol concentration ranging 100–1000 μg/mL restricted the growth of the *L. amazonensis*. Treatment of the promastigotes with eugenol (IC_{50} value: 80 μg/mL) resulting in various ultra structural changes such as inner membrane collapse, swelling, and increased number of cristae. When eugenol has been treated with promastigotes and amastigotes; approximately 30% promastigotes and amastigotes were found to contain two or moreflagella or nuclei signifying the cell division arrest.[62] Moreover, anti-leishmanial activity against L. major promastigote has been displayed by eugenol (IC_{50} value: 47.2 μg/mL).[63] Similarly, eugenol along with the methanolic extract of *Piper betle* displayed anti-leishmanial potential against *L. donovani*.[64] Another study revealed that the acetylated and benzylated derivatives of EGL exhibited activity against *L. infantum* chagasi.[65] Essential oil of clove containing eugenol has shown promising trypanocidal activity with IC_{50} value 99.5 μg/mL for epimastigotes and 57.5 μg/mL for trypomastigotes.[67] Eugenol also exhibited remarkable anti-malarial potential against the chloroquine-resistant strain of *P. falciparum* (FCR-3) with IC_{50} value of 753 μM.[34]

3.2.8 ANTI-VIRAL ACTIVITY

Eugenol has shown potential against viral replication and reduced viral infection. Various studies revealed the antiviral activity of eugenol especially against herpes simplex-1 (HSV-1) and herpes simplex-2 (HSV-2) with IC_{50} values ranging 16.2–25.6 μg/mL determined by plaque reduction assay.[68,69] Eugenol exhibited significant potential against clinical isolates of HSV-1.[69] Conversely, another study results further revealed that cytotoxicity of eugenol alone is insignificant against HSV-1. Nevertheless, it exhibited promising anti-viral action in combination with acyclovir.[70] EGL inhibits the activation of several signal pathways resulting in inhibition of autophagy and influenza-A virus (IAV) replication. EGL also inhibits the autophagic gene expressions via amelioration of the oxidative stress. It is revealed as a promising inhibitor of autophagy and IAV replication.[71] Moreover, eugenol also has the ability against human cytomegalovirus (CMV), murine cytomegalovirus (MCMV), and hepatitis C virus (≥90% inhibition at 100 μg/mL).[72,73]

3.2.9 ANTI-INFLAMMATORY POTENTIAL OF EUGENOL

Eugenol exerted promising anti-inflammatory activity by targeting various pathways. In vitro studies revealed that it inhibited nuclear factor-κB (NF-κB) activation induced by tumor necrosis factor (TNFα) and blocked cyclooxygenase activity (COX-2) in LPS stimulated macrophages. The expression of COX-2 is triggered by cytokines, growth factors and LPS[74]. Eugenol inhibits the prostaglandin synthesis and neutrophil/macrophage chemotaxis and exerts the anti-inflammatory action. It displayed diminished inflammation by reducing TNF-α and infiltration of neutrophils during pulmonary infection in animals. Eugenol at a dose of 160 mg/kg body weight has shown reduction in alveolar collapse and PMN infiltration in lungs.[75] Both the studies (in vitro and in vivo) exhibited that EGL declined the number of leukocytes migrated towards perivascular tissue.[76] It also reduces the interleukin 6 (IL-6), TNF-α expression, NF-κB signalling, and enhanced the levels of associated enzymes.[77] In allergic asthma model, EGL inhibited the ovalbumin-induced eosinophil count, levels of cellular proteins (P-IκBα, NF-κBP65, p-NF-κBP65), improved IL-4 and IL-5 levels, and increased Vitamin D3 upregulated protein-1 and IκBα protein levels[78]. Eugenol has the ability to interfere with cell growth by reducing the COX-2 and IL-1β expression.[79] Eugenol has also shown to reduce the incidence of pain, alveolar osteitis, inflammation, infection, and better wound healing.[80] Further study suggested that eugenol causes the reduction of NO, TNF-α, IL-1β, NF-κB, and iNOS expression, also decreased the ERK1/2 and p38 MAPK signaling pathways and modulate the inflammatory response.[81] Eugenol has also exhibited protective effect against chemical-induced cellular dysfunction of macrophages and balanced the pro/anti-inflammatory mediators in mouse peritoneal macrophages (Fig. 3.5).[19]

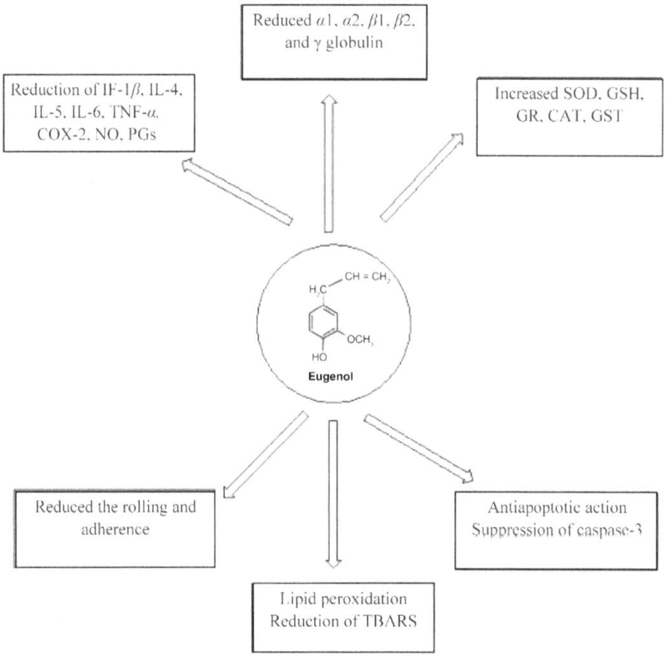

FIGURE 3.5 Schematic presentation of eugenol effects to control inflammation.

3.2.10 ANTI-CANCER ACTIVITY

The anticancer treatment lies in prohibition of the cell proliferation and destruction of the malignant cells. Eugenol together with its natural synthetic derivatives was screened to inhibit cell growth on primary melanoma cell lines established from patients' tissue samples. In vitro anticancer studies revealed that eugenol, isoeugenol monomers and their respective O-methylated forms haven't shown the inhibition of the cell proliferation. However, the biphenyl forms of eugenol found to have some effect. Biphenyl(S)-6,6'-dibromo-dehydrodieugenol elicited antiproliferative activity against neuroectodermal tumor cells via triggering apoptosis partially.[82] ROS played a crucial role in eugenol and eugenol loaded nano-emulsion for antiproliferative activity by inducing apoptosis in HB8065 and HTB37 cell lines.[83]

The epoxide form of eugenol emerged as a promising candidate against human breast cancer cells by inducing apoptosis.[84] Essential oil extracted by hydrodistillation of roots and bark of *U. angustifolium* contained 85.3% and 68.3% of methyleugenol, respectively, which exhibited cytotoxic properties againstMCF-7 human breast cancer cell lines.[85] In another study, EGL displayed specific cytotoxicity against BCCL and breast cancer linked oncogenes (NF-κB and cyclin D1).[86] The compound, ericifolin [eugenol 5-O-β-(6'-galloylglucopyranoside)] displayed pro-apoptosis, antiproliferative and anti-androgen receptor transcription activities indicating thatit can be potentially used against prostate cancer.[87] EGL exerted non-apoptotic cell death in oral squamous cell carcinoma and normal oral cell lines.[88]

EGL showed inactivation of extracellular signal-regulated kinase (ERK) and inhibited the matrixmetalloproteinase-9 activities in HT1080 cells (PMA-stimulated) indicating that it can be utilized against metastasis linked to oxidative stress.[89] To overcome toxicity and drug-induced resistance, combination therapy can be applied against cancer. EGL and 5-fluorouracilin combination showed greater cytotoxicity against cervical cancer cells (HeLa).[90] Jaganathan et al. extensively reviewed and reported that the eugenol and its synthetic derivatives exhibited activity against skin tumors, melanoma, gastric cancer, leukemia, and prostate cancer via caspase dependent pathway and oncogene regulation.[91]

3.3 CONCLUSION

Eugenol is a bioactive phytoconstituent exhibited promising potential as a therapeutic agent which can be incorporated in the treatment of inflammation, cancer, leishmaniasis, and several other disorders. Eugenol also appeared as a potential candidate having broad spectrum activity against various bacterial, fungal, viral, and parasitic infections. The combination therapy of eugenol with other standard drugs has shown remarkable potential against the various resistant strains. It has much fewer drawbacks as compared to other chemically synthesized compounds due to a natural bioactive component of essential oil. Although eugenol showed concentration dependant activity in most of the cases. Eugenol and its derivatives have kindled the research interests and lightened up a new era in the field of pharmacology.

KEYWORDS

- **eugenol**
- **pharmacology**
- **anti-oxidant**
- **anti-viral**
- **anti-inflammatory**
- **anti-cancer**
- **anti-microbial**
- **anti-parasitic**

REFERENCES

1. Lee, K. G.; Shibamoto, T. Antioxidant Properties of Aroma Compounds Isolated from Soybeans and Mung Beans. *J. Agri. Food Chem.* **2000,** *48* (9), 4290–4293.
2. Charalambous, G., Ed. *The Quality of Foods and Beverages V1: Chemistry and Technology*; Elsevier, 2012.
3. Marotti, M.; Piccaglia, R.; Giovanelli, E. Differences in Essential Oil Composition of Basil (*Ocimum basilicum* L.) Italian Cultivars Related to Morphological Characteristics. *J. Agri. Food Chem.* **2005,** *44* (12), 3926–3929.
4. Bezerra, D. P.; Militão, G. C. G.; de Morais, M. C.; deSousa, D. P. The Dual Antioxidant/Prooxidant Effect of Eugenol and Its Action in Cancer Development and Treatment. *Nutrients* **2017,** *9* (12), 1367.
5. Jordán, M. J.; Tandon, K.; Shaw, P. E.; Goodner, K. L. Aromatic Profile of Aqueous Banana Essence and Banana Fruit by Gas Chromatography-mass Spectrometry (GC-MS) and Gas Chromatography-olfactometry (GC-O). *J. Agri. Food Chem.* **2001,** *49* (10), 4813–4817.
6. Diógenes, M. J. N.; Matos, F. J. A. Dermatite De Contatopor Plantas (DCP). *Anais Brasileiros De Dermatologia* **1999,** *74* (6), 629–634.

7. Barceloux, D. G. *Medical Toxicology of Natural Substances. Foods, Fungi, Medicinal Herbs, Plants and Venomous Animals*. Wiley: Hoboken, NJ, USA, 2008.
8. Bezerra, D. P.; Militao, G. C. G.; de Morais, M. C.; de Sousa, D. P. The Dual Antioxidant/Prooxidant Effect of Eugenol and Its Action in Cancer Development and Treatment. *Nutrients* **2017,** *9*, 1367.
9. Chung, G.; Oh, S. B. Eugenol as Local Anesthetic. In *Natural Products*; Ramawat, K., Mérillon, J. M., Eds.; Springer: Berlin, Heidelberg, 2013.
10. Huang, X.; Liu, Y.; Lu, Y.; Ma, C. Anti-inflammatory Effects of Eugenol on Lipopolysaccharide-induced Inflammatory Reaction in Acute Lung Injury via Regulating Inflammation and Redox Status. *Int. Immunopharmacol.* **2015,** *26* (1), 265–271.
11. Fadilah, F.; Yanuar, A.; Arsianti, A.; Andrajati, R. Phenylpropanoids, Eugenol Scaffold, and Its Derivatives as Anticancer. *Asian J. Pharm. Clin. Res.* **2017,** *10* (3), 41–46.
12. Barboza, J. N.; da Silva Maia Bezerra Filho, C.; Silva, R. O.; Medeiros, J. V. R.; de Sousa, D. P. An Overview on the Anti-inflammatory Potential and Antioxidant Profile of Eugenol. *Oxid. Med. Cell. Longev.* **2018,** Article ID 3957262, 9.
13. Pavesi, C.; Banks, L. A.; Hudaib, T. Antifungal and Antibacterial Activities of Eugenol and Non-Polar Extract of *Syzygium aromaticum* L. *J. Pharm. Sci. Res.* **2018,** 10 (2), 337–339.
14. Raja, M. R. C.; Srinivasan, V.; Selvaraj, S.; Mahapatra, S. K. Versatile and Synergistic Potential of Eugenol: A Review. *Pharm. Anal. Acta* **2015,** *6*, 5.
15. Trongtokit, Y.; Rongsriyam, Y.; Komalamisra, N.; Apiwathnasorn, C. Comparative Repellency of 38 Essential Oils Against Mosquito Bites. *Phytother. Res.* **2005** Apr, *19* (4), 303–309.
16. Chatterjee, D.; Bhattacharjee, P. Use of Eugenol-lean Clove Extract as a Flavoring Agent and Natural Antioxidant in Mayonnaise: Product Characterization and Storage Study. *J. Food Sci. Technol.* **2015,** 52 (8), 945–4954.
17. Zhang, H.; Chen, X.; He, J. J. Pharmacological Action of Clove Oil and Its Application in Oral Care Products. *Oral Care Ind.* **2009,** *19*, 23–24.
18. World Health Organization. Evaluation of Certain Food Additives and Contaminants; Twenty-sixth report of the Joint FAO/WHO Expert Committee on Food Additives, WHO Technical Report Series, No. 683; Geneva, Switzerland, 1982.
19. Mahapatra, S. K.; Bhattacharjee, S.; Chakraborty, S. P.; Majumdar S.; Roy, S. Alteration of Immune Functions and Th1/Th2 Cytokine Balance in Nicotine-induced Murine Macrophages: Immunomodulatory Role of Eugenol and *N*-acetylcysteine. *Int. Immunopharmacol.* **2011,** *11*, 485–495.
20. Mahapatra, S. K.; Chakraborty, S. P.; Majumdar, S.; Bag, B. G.; Roy, S. Eugenol Protects Nicotine-induced Superoxide Mediated Oxidative Damage in Murine Peritoneal Macrophages In Vitro. *Eur. J. Pharmacol.* **2009,** *623*, 132–140.
21. Mahapatra, S. K.; Roy, S. Phytopharmacological Approach of Free Radical Scavenging and Anti-oxidative Potential of Eugenol and *Ocimum gratissimum* Linn. *Asian Pac. J. Trop. Med.* **2014,** *7S1*, S391–S397.
22. Yogalakshmi, B.; Viswanathan, P.; Anuradha, C. V. Investigation of Antioxidant, Anti-inflammatory and DNA-protective Properties of Eugenol in Thioacetamide-induced Liver Injury in Rats. *Toxicology* **2010,** *268*, 204–212.
23. Bedoukian, P. Z. *Perfumery and Flavouring Synthetics*, 3rd ed.; Allured Publishing Corporation: Carol Stream, IL, USA, 1986.

24. Zheng, G. Q.; Kenney, P. M.; Lam, L. K. Sesquiterpenes from Clove (Eugenia Caryophyllata) as Potential Anticarcinogenic Agents. *J. Nat. Prod.* **1992,** *55*, 999–1003.
25. Harborne, J. B.; Baxter, H. *Phytochemical Dictionary*; Taylor and Francis: London, UK, 1993.
26. Thompson, D. C.; Constantin-Teodosiu, D.; Moldéus, P. Metabolism and Cytotoxicity of Eugenol in Isolated Rat Hepatocytes. *Chem. Biol. Interact.* **1991,** 137–147.
27. Sudarma, I. M.; Ulfa, M.; Sarkono, S. Chemical Transformation of Eugenol Isolated from Clove Oil to 4-allyl-2-methoxy-6-sulfonicphenol and 4-allyl-2-methoxy-6-aminophenol. *Indo. J. Chem.* **2009,** *9*, 267–270.
28. Ali, S. M.; Khan, A. A.; Ahmed, I.; Musaddiq, M.; Ahmed, K. S.; Polasa, H.; Rao, L. V.; Habibullah, C. M.; Sechi, L. A.; Ahmed, N. Antimicrobial Activities of Eugenol and Cinnamaldehyde Against the Human Gastric Pathogen Helicobacter Pylori. *Ann. Clin. Microbiol. Antimicrob.* **2005,** *4*, 20–24.
29. Laekeman, G. M.; van Hoof, L.; Haemers, A.; Berghe, D. A. V.; Herman, A. G. et al. Eugenol a Valuable Compound for In Vitro Experimental Research and Worthwhile for Further In Vivo Investigation. *Phytother. Res.* **1990,** *4*, 90–96.
30. López, P.; Sánchez, C.; Batlle, R.; Nerín, C. Solid- and Vapor-phase Antimicrobial Activities of Six Essential Oils: Susceptibility of Selected Foodborne Bacterial and Fungal Strains. *J. Agric. Food Chem.* **2005,** *53*, 6939–6946.
31. Leite, A. M.; Lima, E. D. O.; de Souza, E. L.; Diniz, M. D. F. F. M.; Trajano, V. N. et al. Inhibitory Effect of ß -Pinene, a -Pinene and Eugenol on the Growth of Potential Infectious Endocarditis Causing Gram-positive Bacteria. *Braz. J. Pharm. Sci.* **2007,** *43*, 121–126.
32. Hemaiswarya, S.; Doble, M. Synergistic Interaction of Eugenol with Antibiotics Against Gram Negative Bacteria. *Phytomedicine* **2009,** *16*, 997–1005.
33. Singh, G.; Maurya, S.; DeLampasona, M. P.; Catalan, C. A. A Comparison of Chemical, Antioxidant and Antimicrobial Studies of Cinnamon Leaf and Bark Volatile Oils, Oleoresins and Their Constituents. *Food Chem. Toxicol.* **2007,** *45*:1650–1661.
34. van Zyl, R. L.; Seatlholo, S. T.; van Vuuren, S. F.; Viljoen, A. M. The Biological Activities of 20 Nature Identical Essential Oil Constituents. *J. Essent. Oil. Res.* **2006,** *18*:129–133.
35. Oyedemi, S. O.; Okoh, A. I.; Mabinya, L. V.; Pirochenva, G.; Afolayan, A. J. The Proposed Mechanism of Bactericidal Action of Eugenol, a a-terpineol and ?-terpinene Against *Listeria monocytogenes*, *Streptococcus pyogenes*, *Proteus vulgaris* and *Escherichia Coli*. *Afr. J. Biotechnol.* **2009,** *8*, 1280–1286.
36. Yadav, M. K.; Chae, S. W.; Im, G. J.; Chung, J. W.; Song, J. J. Eugenol: A Phyto-compound Effective against Methicillin-resistant and Methicillin-sensitive *Staphylococcus aureus* Clinical Strain Biofilms. *PLoS One* **2015,** *10*, e0119564.
37. El abed, S.; Houari, A.; Latrache, H.; Remmal, A.; Koraichi, S. I. In Vitro Activity of Four Common Essential Oil Components Against Biofilm-producing *Pseudomonas aeruginosa*. *Res. J. Microbiol.* **2011,** *6*, 394–401.
38. Moon, S. E.; Kim, H. Y.; Cha, J. D. Synergistic Effect Between Clove Oil and Its Major Compounds and Antibiotics Against Oral Bacteria. *Arch Oral Biol* **2011,** *56*, 907–916.
39. Pei, R. S.; Zhou, F.; Ji, B. P.; Xu, J. Evaluation of Combined Antibacterial Effects of Eugenol, Cinnamaldehyde, Thymol, and Carvacrol Against *E. coli* with an Improved Method. *J. Food Sci.* **2009,** *74*, M379–383.

40. Rico-Molina, D.; Aparicio-Ozores, G.; Dorantes-Alvarez, L.; Hernández-Sanchez, H. Antimicrobial Activity of Cinnamate-eugenol: Synergistic Potential, Evidence of Efflux Pumps and Amino Acid Effects. *Am. J. Food Technol.* **2012,** 7: 289–300.
41. Qiu, J.; Feng, H.; Lu, J.; Xiang, H.; Wang, D. et al. Eugenol Reduces the Expression of Virulence-related Exoproteins in *Staphylococcus aureus*. Appl. Environ. Microbiol. **2010,** *76*, 5846–5851.
42. Zhang, P.; Zhang, E.; Xiao, M.; Chen, C.; Xu, W. Enhanced Chemical and Biological Activities of a Newly Biosynthesized Eugenol Glycoconjugate, Eugenol Î±-d-glucopyranoside. *Appl. Microbiol. Biotechnol.* **2013,** *97*, 1043–1050.
43. Slamenová, D.; Horváthová, E.; Wsólová, L.; Sramková, M.; Navarová, J. Investigation of Anti-oxidative, Cytotoxic, DNA-damaging and DNA-protective Effects of Plant Volatiles Eugenol and Borneol in Human-derived HepG2, Caco-2 and VH10 Cell Lines. *Mutat. Res.* **2009,** *677*, 46–52.
44. Mahboub, R.; Memmou, F. Antioxidant Activity and Kinetics Studies of Eugenol and 6-bromoeugenol. *Nat. Prod. Res.* **2015,** *29*, 966–971.
45. Tominaga, H.; Kobayashi, Y.; Goto, T.; Kasemura, K.; Nomura, M. DPPH Radical Scavenging Effect of Several Phenylpropanoid Compounds and their Glycoside Derivatives. *Yakugaku Zasshi- Pharm. Soc. Jpn.* **2005,** *125*, 371–375.
46. Chogo, J. B.; Crank, G. Chemical Composition and Biological Activity of the Tanzanian Plant *Ocimum suave*. *J. Nat. Prod.* **1981,** *44*, 308–311.
47. Ito, M.; Murakami, K.; Yoshino, M. Antioxidant Action of Eugenol Compounds: Role of Metal Ion in the Inhibition of Lipid Peroxidation. *Food Chem. Toxicol.* **2005,** *43*, 461–466.
48. Wie, M. B. 1.; Won, M. H.; Lee, K. H.; Shin, J. H.; Lee, J. C.; Suh, H. W.; Song, D. K.; Kim, Y. H. Eugenol Protects Neuronal Cells from Excitotoxic and Oxidative Injury in Primary Cortical Cultures. *Neurosci. Lett.* **1997,** Apr 4, *225* (2), 93–96.
49. Ali, S.; Prasad, R.; Mahmood, A.; Routray, I.; Shinkafi, T. S. et al. Eugenolrich Fraction of *Syzygium aromaticum* (Clove) Reverses Biochemical and Histopathological Changes in Liver Cirrhosis and Inhibits Hepatic Cell Proliferation. *J. Cancer Prev.* **2014,** *19*, 288–300.
50. Rao, M.; Kumar, M. M.; Rao, M. A. In Vitro and In Vivo Effects of Phenolic Antioxidants Against Cisplatin-induced Nephrotoxicity. *J. Biochem.* **1999,** *125*, 383–390.
51. Singh, V.; Panwar, R. In Vivo Antioxidative and Neuroprotective Effect of 4-Allyl-2-methoxyphenol Against Chlorpyrifos-induced Neurotoxicity in Rat Brain. *Mol. Cell. Biochem.* **2014,** *388*, 61–74.
52. Chaieb, K.; Zmantar, T.; Ksouri, R.; Hajlaoui, H.; Mahdouani, K. et al. Antioxidant Properties of the Essential Oil of *E. caryophyllata* and Its Antifungal Activity Against a Large Number of Clinical *Candida* species. *Mycoses* **2007,** *50*, 403–406.
53. Abrão, P. H.; Pizi, R. B.; de Souza, T. B.; Silva, N. C.; Fregnan, A. M. et al. Synthesis and Biological Evaluation of New Eugenol Mannich Bases as Promising Antifungal Agents. *Chem. Biol. Drug Des.* **2014,** *86*(4), 459–465.
54. Doke, S. K.; Raut, J. S.; Dhawale, S.; Karuppayil, S. M. Sensitization of *Candida albicans* Biofilms to Fluconazole by Terpenoids of Plant Origin. *J. Gen. Appl. Microbiol.* **2014,** *60*, 163–168.
55. de Paula, S. B.; Bartelli, T. F.; Di Raimo, V.; Santos, J. P.; Morey, A. T. et al. Effect of Eugenol on Cell Surface Hydrophobicity, Adhesion, and Biofilm of *Candida tropicalis* and *Candida dubliniensis* Isolated from Oral Cavity of HIV-Infected Patients. *Evid. Based Complem. Alternat. Med.* **2014,** *2014*, 505204.

56. Khan, M. S.; Ahmad, I.; Cameotra, S. S. Phenyl Aldehyde and Propanoids Exert Multiple Sites of Action Towards Cell Membrane and Cell Wall Targeting Ergosterol in *Candida albicans*. *AMB Express* **2013,** *3*, 54.
57. Braga, P. C.; Sasso, M. D.; Culici, M.; Alfieri, M. Eugenol and Thymol, Alone or in Combination, Induce Morphological Alterations in the Envelope of *Candida albicans*. *Fitoterapia* **2007,** *78*, 396–400.
58. Audrey, R.; Silva, A. R. S. T.; Scher, R.; Santos, F. V.; Ferreira, S. R.; Cavalcanti, Sócrates C. H.;. Correa, Cristiane B.; Bueno, L. L.; Alves, R. J.; Souza, D. P.; Fujiwara, R. T.; Silvio, S. D. Leishmanicidal Activity and Structure-Activity Relationships of Essential Oil Constituents. *Molecules* **2017,** *22*, 815–824.
59. Machado, M.; Dinis, A. M.; Salgueiro, L.; Custódio, J. B.; Cavaleiro, C. et al. Anti-Giardia Activity of *Syzygium aromaticum* Essential Oil and Eugenol: Effects on Growth, Viability, Adherence and Ultrastructure. *Exp. Parasitol.* **2011,** *127*, 732–739.
60. Kayembe, J. S.; Taba, K. M.; Ntumba, K.; Kazadi, T. K. In Vitro Antimalarial Activity of 11 Terpenes Isolated from *Ocimum* gratissimum and *Cassia alata* Leaves. Screening of their Binding Affinity with Haemin. *J. Plant Stud.* **2012,** *1* (2), 168–172.
61. Azeredo, Camila M. O.; Soares, Maurilio J. Combination of the Essential Oil Constituents Citral, Eugenol and Thymol Enhance their Inhibitory Effect on *Crithidia fasciculata* and *Trypanosoma cruzi* Growth. *Rev. Bras. Farmacogn.* **2013,** *23*, 762–768.
62. Ueda-Nakamura, T.; Mendonça-Filho, R. R.; Morgado-Díaz, J. A.; Korehisa Maza, P.; Prado Dias Filho, B. et al. Antileishmanial Activity of Eugenol-rich Essential Oil from *Ocimum gratissimum*. *Parasitol. Int.* **2006,** *55*, 99–105.
63. Fabri, R. L.; Coimbra, E. S.; Almeida, A. C.; Siqueira, E. P.; Alves, T. M. et al. Essential Oil of *Mitracarpus frigidus* as a Potent Source of Bioactive Compounds. *An. Acad. Bras. Cienc.* **2012,** *84*, 1073–1080.
64. Misra, P.; Kumar, A.; Khare, P.; Gupta, S.; Kumar, N. et al. Pro-apoptotic Effect of the Landrace Bangla Mahoba of Piper Betle on *Leishmania donovani* May Be Due to the High Content of Eugenol. *J. Med. Microbiol.* **2009,** *58*, 1058–1066.
65. de Morais, S. M.; Vila-Nova, N. S.; Bevilaqua, C. M.; Rondon, F. C.; Lobo, C. H.; et al. Thymol and Eugenol Derivatives as Potential Antileishmanial Agents. *Bioorg. Med. Chem.* **2014,** *22*, 6250–6255.
66. Santoro, G. F.; Cardoso, M. G.; Guimarães, L. G.; Mendonça, L. Z.; Soares, M. J. Trypanosoma Cruzi: Activity of Essential Oils from *Achillea millefolium* L., *Syzygium aromaticum* L. and *Ocimum basilicum* L. on Epimastigotes and Trypomastigotes. *Exp. Parasitol.* **2007,** *116*, 283–290.
67. Benencia, F.; Courrèges, M. C. In Vitro and In Vivo Activity of Eugenol on Human Herpesvirus. *Phytother. Res.* **2000,** *14*, 495–500.
68. Astani, A.; Reichling, J.; Schnitzler, P. Screening for Antiviral Activities of Isolated Compounds from Essential Oils. *Evid. Based Complem. Alternat. Med.* **2011,** *2011*, 253643.
69. Tragoolpua, Y.; Jatisatienr, A. Anti-herpes Simplex Virus Activities of *Eugenia caryophyllus* (Spreng.) Bullock & S. G. Harrison and Essential Oil, Eugenol. *Phytother. Res.* **2007,** *21*, 1153–1158.
70. Kurokawa, M.; Nagasaka, K.; Hirabayashi, T.; Uyama, S.; Sato, H. et al. Efficacy of Traditional Herbal Medicines in Combination with Acyclovir Against Herpes Simplex Virus Type 1 Infection In Vitro and In Vivo. *Antiviral Res.* **1995,** *27*, 19–37.

71. Dai, J. P.; Zhao, X. F.; Zeng, J.; Wan, Q. Y.; Yang, J. C. et al. Drug Screening for Autophagy Inhibitors Based on the Dissociation of Beclin1-Bcl2 Complex Using BiFC Technique and Mechanism of Eugenol on Anti-influenza A Virus Activity. *PLoS One* **2013,** *8*, e61026.
72. Yukawa, T. A.; Kurokawa, M.; Sato, H.; Yoshida, Y.; Kageyama, S. et al. Prophylactic Treatment of Cytomegalovirus Infection with Traditional Herbs. *Antiviral Res.* **1996,** *32*, 63–70.
73. Hussein, G.; Miyashiro, H.; Nakamura, N.; Hattori, M.; Kakiuchi, N. et al. Inhibitory Effects of Sudanese Medicinal Plant Extracts on Hepatitis C Virus (HCV) Protease. *Phytother. Res.* **2000,** *14*, 510–516.
74. Kim, S. S.; Oh, O. J.; Min, H. Y.; Park, E. J.; Kim, Y. et al. Eugenol Suppresses Cyclooxygenase-2 Expression in Lipopolysaccharide-stimulated Mouse Macrophage RAW264.7 Cells. *Life Sci.* **2003,** *73*, 337–348.
75. Raghavenra, H.; Diwakr, B. T.; Lokesh, B. R.; Naidu, K. A. Eugenol—the Active Principle from Cloves Inhibits 5-lipoxygenase Activity and Leukotriene-C4 in Human PMNL Cells. *Prostaglandins Leukot Essent. Fatty Acid.* **2006,** *74*, 23–27.
76. Estevão-Silva, C. F.; Kummer, R.; Fachini-Queiroz, F. C. et al. Anethole and Eugenol Reduce In Vitro and In Vivo Leukocyte Migration Induced by fMLP, LTB4, and Carrageenan. J. Nat. Med. **2014,** *68* (3), 567–575, 2014.
77. Huang, X.; Liu, Y.; Lu, Y.; Ma, C. Anti-inflammatory Effects of Eugenol on Lipopolysaccharide-induced Inflammatory Reaction in Acute Lung Injury via Regulating Inflammation and Redox Status. *Int. Immunopharmacol.***2015,** *26* (1), 265–271.
78. Pan, C.; Dong, Z. Antiasthmatic Effects of Eugenol in a Mouse Model of Allergic Asthma by Regulation of Vitamin D3 Upregulated Protein 1/NF-κB Pathway. *Inflammation* **2015,** *38* (4), 1385–1393.
79. Hussain, K.; Brahmbhatt, A.; Priyani, M.; Ahmed, T. A. Rizvi; Sharma, C. Eugenol Enhances the Chemotherapeutic Potential of Gemcitabine and Induces Anticarcinogenic and Anti-inflammatory Activity in Human Cervical Cancer Cells. *Cancer Biother. Radiopharm.* **2011,** *26* (5), 519–527.
80. Jesudasan, J. S.; Wahab, P. U. A.; Sekhar, M. R. M. Effectiveness of 0.2% Chlorhexidine Gel and a Eugenol-based Paste on Postoperative Alveolar Osteitis in Patients Having Third Molars Extracted: A Randomised Controlled Clinical Trial. *Br. J. Oral Maxillofac. Surg.* **2015,** 53 (9), 826–830.
81. Yeh, J. L.; Hsu, J. H. ; Hong, Y. S. et al. Eugenolol and Glycerylisoeugenol Suppress LPS-induced iNOS Expression by Downregulating NF-κB AND AP-1 through Inhibition of MAPKS and AKT/IκBα Signaling Pathways in Macrophages. *Int. J. Immunopathol. Pharmacol.* **2011,** *24* (2), 345–356.
82. Pisano, M.; Pagnan, G.; Loi, M.; Mura, M. E.; Tilocca, M. G. et al. Antiproliferative and Pro-apoptotic Activity of Eugenol-related Biphenyls on Malignant Melanoma Cells. *Mol. Cancer* **2007,** *6*, 8.
83. Majeed, H.; Antoniou, J.; Fang, Z. Apoptotic Effects of Eugenol-loaded Nanoemulsions in Human Colon and Liver Cancer Cell Lines. *Asian Pac J Cancer Prev* **2014,** *15*, 9159–9164.
84. Behbahani, M. Evaluation of In Vitro Anticancer Activity of *Ocimum basilicum*, *Alhagi maurorum*, *Calendula officinalis* and their Parasite *Cuscuta campestris*. *PLoS One* **2014,** *9*, e116049.

85. Noudogbessi, J. P.; Gary-Bobo, M.; Adomou, A.; Adjalian, E.; Alitonou, G. A. et al. Comparative Chemical Study and Cytotoxic Activity of *Uvariodendron angustifolium* Essential Oils from Benin. *Nat. Prod. Commun.* **2014,** *9*, 261–264.
86. Al-Sharif, I.; Remmal, A.; Aboussekhra, A. Eugenol Triggers Apoptosis in Breast Cancer Cells through E2F1/Survivin Down-Regulation. *BMC Cancer* **2013,** *13*, 600.
87. Shamaladevi, N.; Lyn, D. A.; Shaaban, K. A.; Zhang, L; Villate, S. et al. Ericifolin: A Novel Antitumor Compound from Allspice that Silences Androgen Receptor in Prostate Cancer. *Carcinogenesis* **2013,** *34*, 1822–1832.
88. Koh, T.; Murakami, Y.; Tanaka, S.; Machino, M.; Onuma, H. et al. Changes of Metabolic Profiles in an Oral Squamous Cell Carcinoma Cell Line Induced by Eugenol. *In Vivo* **2013,** *27*, 233–243.
89. Nam, H.; Kim, M. M. Eugenol with Antioxidant Activity Inhibits MMP-9 Related to Metastasis in Human Fibrosarcoma Cells. *Food Chem. Toxicol.* **2013,** *55*, 106–112.
90. Hemaiswarya, S.; Doble, M. Combination of Phenylpropanoids with 5-fluorouracil as Anti-cancer Agents Against Human Cervical Cancer (HeLa) Cell Line. *Phytomedicine* **2013,** *20*, 151–158.
91. Jaganathan, S. K.; Supriyanto, E. Antiproliferative and Molecular Mechanism of Eugenol-induced Apoptosis in Cancer Cells. *Molecules* **2012,** *17*, 6290–6304.

CHAPTER 4

Predictive *In Silico* Global Metabolism Analysis: Disease Mechanism

FRANCISCO TORRENS[1*] and GLORIA CASTELLANO[2]

[1]*Institut Universitari de Ciència Molecular, Universitat de València, Edifici d'Instituts de Paterna, P. O. Box 22085, E-46071 València, Spain*

[2]*Departamento de Ciencias Experimentales y Matemáticas, Facultad de Veterinaria y Ciencias Experimentales, Universidad Católica de Valencia San Vicente Mártir, Guillem de Castro-94, E-46001 València, Spain*

[*]*Corresponding author. E-mail: torrens@uv.es*

ABSTRACT

We witness an explosion in bionetwork data along with computational approaches development for analyses. Interdisciplinary research area is integral part of systems biology, promising to provide new insights into life, evolution, and disease organizational principles. Prostate neoplasia aggressiveness is classified into benign, localized, and metastatic via a study of hundreds of thousands ion features. Metabolism software shows that sarcosine modulates invasiveness. Mitochondrial dysfunction plays a central role in a wide range of age-related disorders and various cancer forms. Urine and blood are ideal fluids in search for new markers able to discriminate between prostate cancer and health. Presence of prostate-specific antigen in blood is a probe for cancer detection and, although sometimes when it rises it can indicate cancer, when it is not present it can overlook tumors that endanger life. This report reviews the state of the art of metabolism platforms for discriminating disease (prostate neoplasia) mechanisms.

4.1 INTRODUCTION

Metabolomics is *nonbiased quantification/identification of all metabolites present in biosystem*, but broad term covers[1] global identification of as many as possible to subsets targeted analysis.[2–4] Our goal is understanding the processes occurring as perturbation.[5,6] Since metabolites are critical, the best way to accomplish them is to detect, quantify, and identify as many small molecules as possible. Challenging aspects exist associated with creating metabolomics platform (cf. Fig. 4.1): to generate quantitative analytical method, which would be capable of producing data with low process variability as it is able to measure small biochanges. It must be able to detect molecules, ranging in chemical diversity and dynamic range common in biosystems from small organic acids, for example, in Krebs tricarboxylic (citric) acid (TCA) cycle, to large lipids. Metabolomics provides information to identify quick/accurately detected molecule; unfortunately ability is nontrivial.[7]

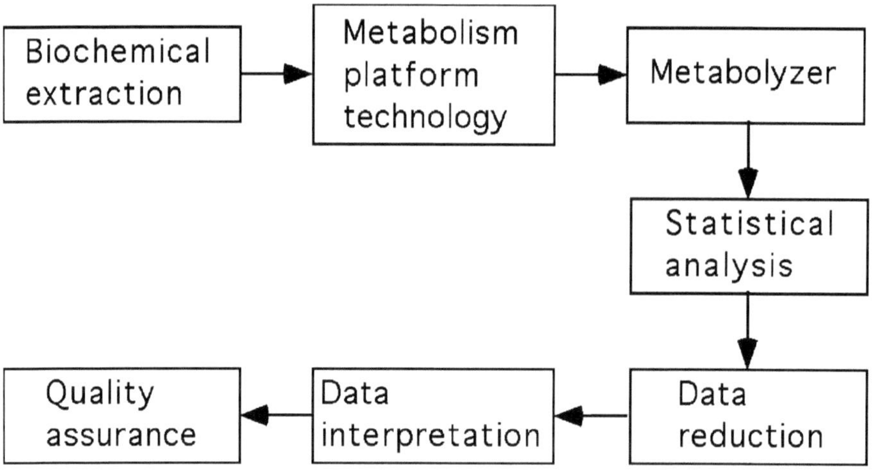

FIGURE 4.1 Chemical analysis-integrated platform: identification and relative quantification, etc.

Liquid chromatography, coupled with mass spectrometry (LC/MS), became standard for metabolomics because of its ability to separate, ionize and detect chemical range.[8–12] It benefits from detecting small-molecules range without derivatization, which is needed in gas chromatography (GC)/MS.[13–15] Another metabolomics use nuclear magnetic resonance (NMR) because of fundamental quantitative nature; however, they are restricted by poor sensitivity and dynamic range.[16] Fingerprint NMR successfully predicted treatment/disease versus control groups, based on pattern recognition; however, analyses are based on most abundant molecules.[17–20] Most successful approach to increase small-molecules breadth was to diversify LC/MS varying ionization sources, monitoring cat/anions,[21] varying chromatographic methods[22] and via ultrahigh-performance LC (UHPLC).[23,24] Targeted methods exist, which are based on multiple reaction monitoring (MRM),[25] nontargeted ones and combinations.[26] The former presents benefits of high-sensitivity/large-dynamic range measurements and eliminating need for compound identification selecting metabolites to monitor; however, only molecules that were targeted are detected. Drawbacks are as follows: (1) Treating all ion features independently without regard to biochemical origin, it is possible to skew statistical outcome by given-biochemical over-representation.[27] (2) Ion-centric approach suffers from increase in false discoveries number, because of increased number of observations processed in statistical analysis.[28] (3) It is time consuming. Approach involves generating list of potential molecular formulae, from accurate mass data, and searching methods versus databases to make possible-structure catalogue.[29,30] Misconception results from assumption that molecule is identified, which is fallacy because a chemical formula can present many structures[31] (e.g., $C_9H_{11}NO_2$, $>6 \times 10^6$ as calculated via Molgen Online[32]). Although process is only route available when true unknown identification is needed, it is inefficient and alternative identification is needed for high-throughput metabolomics platform.[33] Liu et al. reviewed oxidative stress (OS) metabolomics in endo/exogenously administered intermediate metabolites.[34] Laronga and Drake revised proteomic approach to breast cancer.[35] Choo and Nielsen examined markers for basal-like disease.[36] Drake et al. analyzed challenges to developing proteomic-based illness diagnostics.[37]

In earlier publications it was informed the modeling of complex multicellular systems and tumor-immune cells competition,[38] information theoretic entropy for molecular classification of oxadiazolamines as potential therapeutic agents,[39] molecular classification of 5-amino-2-aroylquinolines and 4-aroyl-6,7,8-trimethoxyquinolines as highly potent tubulin polymerization inhibitors,[40] polyphenolic phytochemicals in cancer prevention, therapy, bioavailability versus bioefficacy,[41] molecular classification of antitubulin agents with indole ring binding at colchicine-binding site,[42] molecular classification of 2-phenylindole-3-carbaldehydes as potential antimitotic agents in human breast cancer cells,[43] cancer, its hypotheses,[44] precision personalized medicine (PPM) from theory to practice and cancer.[45] It was reported how human immunodeficiency virus/acquired immunodeficiency syndrome (HIV/AIDS) destroy immune defences, hypothesis,[46] 2014 emergence, spread, uncontrolled Ebola outbreak,[47,48] Ebola virus disease, questions, ideas, hypotheses, models,[49] clinical translational research, cancer, diabetes, cardiovascular disease (CVD),[50] primary health, smoking habits, immunotherapy in cancer,[51] epigenomics, epithelial plasticity, clinical genetics and rare diseases.[52] The present report reviews the state of the art of metabolism platform technology for elucidating disease mechanisms.

4.2 COMPUTATIONAL METHOD

Network topology–biofunction relation received attention.[53] Methods were reviewed for system-based prediction of protein function[54] and protein nets role in disease.[55] Approach was to find correlations between connectivity in protein–protein interaction network (PIN) and essentiality[56] but the technique failed on newer data[57,58]. Connectivity–protein function correlations were used:[59] proteins that are closer in PIN undergo similar function[60,61] or use of graph theoretic methods,[62] for example, cut/network flow-based tactics.[63] Functional homogeneity of protein groups that shows PIN *coherence* was used for function prediction.[64–67] Network analyses of drug action belong to *systems pharmacology*, which aims to develop action understanding across hierarchy from cell to tissue to organism.[68] Network types were constructed that link biochemical systems, for example, PINs, with drug similarity, interaction or therapeutic indication nets.[69] Graphlets were used to isolate structural characteristics of individual nodes and relate them to protein involvement in disease.[70,71] *Degree* of a node is number of edges it touches. Edge is the only graphlet on two nodes, for example, graphlet G_0. One can define *graphlet degree* of node v with respect to each graphlet G_i: G_i-degree of v counts *how many type-G_i graphlets touch node v*.[72] Traditional degree is G_0-degree. There are 30 graphlets in 2–5-node ones, which provides vector of 30 *graphlet degrees*; however, specificity increases: not all nodes in graphlet are topologically equivalent; for example, middle node is topologically distinct from end ones in G_1. There are 73 topologically distinct nodes across 2–5-node graphlets. Each is called (*automorphism*) *orbit* and one labels them 0,… 72. *Graphlet degree vector* (GDV) or *GD-signature* of node v contains 73 elements: element i represents number of times node v *touches* graphlet at orbit i across all graphlets in its neighborhood. Protein node degree in PIN was used as biofunction prediction but link between such a simple measure of topological similarity, between proteins in PIN and their biofunctions, was questioned. The GD-signature is superior to such a simple measure since it is based on all ≤5-node graphlets: it corresponds to similarity in disease involvement and function predictions were phenotypically validated.[73] Topographic indices were interchanged between molecular graphs and biological/computer networks, for example, proximity index from computer ones.[74–76]

Smart/dumb thieves exist: smart ones try to fit swiftly right keys into appropriate locks, knowing which doors they wish to open; dumb ones attempt to match any key into any lock no matter what door, and they get caught by police.[77] In drug discovery, right keys (compounds) must fit appropriate locks (targets) to open suitable doors (cure diseases) and avoid bad ones (adverse events, AEs). Approaches are used in drug design (cf. Fig. 4.2).[78]

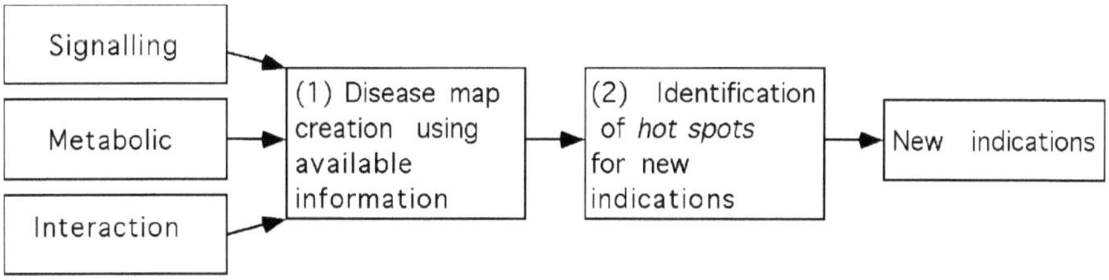

FIGURE 4.2 Successful modeling/simulation via use of a system biology predictive *in silico* approach.

Major attrition causes in clinic are lack of efficacy (30% failures)/ safety (30%).[79] Some AE effectors accomplish; for example, CVD drugs terazosin/prazosin cause dry mouth because of effects on salivary gland α_1-adrenoreceptors; however, AEs generally occur by mechanisms indirectly linked to primary target action or target profile is uncharacterized.[80] Similar AEs of unrelated drugs are caused by common off-target actions, which was exploited to infer medicine molecular activities, which are not implicit in chemical/sequence similarity.[81] Extensive chemical repositories, for example, DrugBank, PubChem, KEGG, ChemSpider, etc., which include relations between molecules, targets, AEs and pathways, are paradigmatic;[82,83] for example, *systems chemical biology* combines AE to chemical structures and pathway knowledge-based annotation.[84] Another approach is based on creating predictive models founded on relation between proteins for condition to be treated, drug targets and AE *motives*.[85] Chemical toxicities are predicted considering compound functional proximity, in network, and proteins that cause undesired AEs if their function is affected.[86]

4.3 RESULTS AND DISCUSSION

Machine learning (expert systems, ES) presented bioinformatics applications, for example, decision tree; two uses in solving bioproblems follow.[87] (1) Learning decision tree and ensemble classifier, biologists get ideas of how ES work. (2) Being exposed to decision tree and ensemble algorithm applications in bioinformatics, computer scientists obtain insight of which topics they work on. Platform is provided to bridge biologist/computer scientist gap. Low specificity of prostate-specific antigen (PSA) test makes it poor marker for prostate cancer detection; because single markers are not expressed by all genetic forms of disease, proteomic approach was developed for simultaneous analysis of multiple proteins, for non/ill patients differentiation.[88] It was analyzed by surface-enhanced laser desorption/ionization (SELDI)-MS, serum samples from 386 men [197 with prostate cancer, 92 with benign prostatic hyperplasia (BPH) and 97 healthy ones], randomly divided into training (TS, $n = 326$) and prediction (PS, $n = 60$) sets. 124 peaks detected by computer analyses were examined in TS, boosting tree algorithm to develop classifier for separating prostate non/cancer groups. Classifier was challenged with PS (30 prostate cancer, 15 BPH, 15 samples from healthy men): two classifiers were developed. (1) AdaBoost categorizer separated prostate non/cancer samples achieving 100% sensitivity/specificity. (2) Boosted Decision Stump Feature Selection organizer resulted easier, and used 21/74 peaks and combination of 21/500 base cataloguers to achieve 97% sensitivity/specificity for PS. Sensitivity/specificity provided SELDI support, coupled with bioinformatics-learning algorithm, to improve diagnosis.

4.3.1 OXIDATIVE STRESS, O_2 TOXICITY, RADICALS, METALS, DISEASES AND MITOCHONDRIA

Epigenetic mechanisms involved in ageing play physiopathological roles in OS presence (cf. Fig. 4.3).

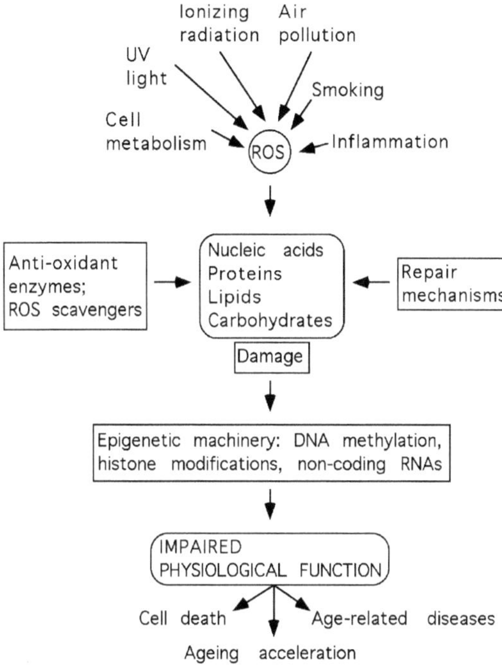

FIGURE 4.3 Oxidative stress, epigenetics and ageing. ROS, reactive oxygen species.

Figure 4.4 shows most frequent radicals and their effects.

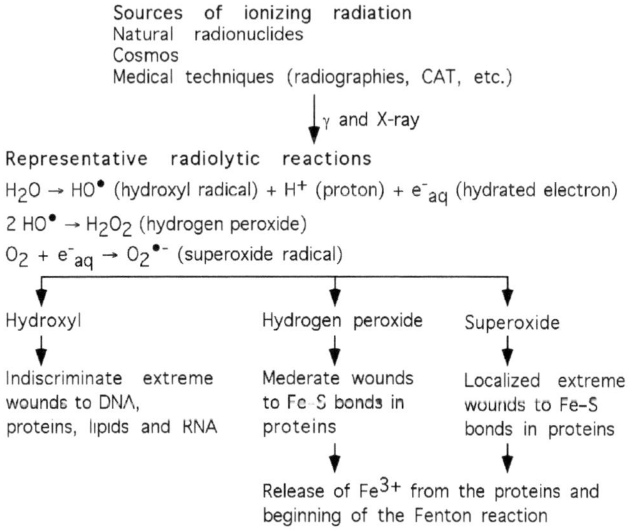

FIGURE 4.4 Most frequent radicals and their effects.

Intracellular reactive oxygen species (ROS)–antioxidant balance (cf. Table 4.1) determines cell survival;[89] for example, thiol reductants are involved in defence.[90,91]

TABLE 4.1 Some Reactive Oxygen and Nitrogen Species in Oxidative Stress.

Oxidant	Description
$O_2^{\bullet-}$, superoxide anion	One-electron reduction state of O_2, formed in many autoxidation reactions and by ETC. Rather unreactive but can release Fe^{2+} from Fe–S proteins and ferritin. Undergoes dismutation to form H_2O_2 spontaneously or by enzymatic catalysis, and is a precursor for metal-catalyzed $OH^{\bullet}$ formation.
H_2O_2, hydrogen peroxide	Two-electron reduction state, formed by $O_2^{\bullet-}$ dismutation or direct O_2 reduction. Lipid soluble and able to diffuse across membranes.
$OH^{\bullet}$, hydroxyl radical	Three-electron reduction state, formed by Fenton reaction and $ONOO^-$ decomposition. Extremely reactive, will attack most cellular components.
ROOH, organic hydroperoxide	Formed by radical reactions with cellular components, for example, lipids/nucleobases.
$RO^{\bullet}$, alkoxy and $ROO^{\bullet}$, peroxy radicals	O-centered organic radicals. Lipid forms participate in lipid peroxidation reactions. Produced in O_2 presence by radical addition to double bonds or H-abstraction.
HOCl, hypochlorous acid	Formed from H_2O_2 by myeloperoxidase. Lipid soluble and highly reactive. Will readily oxidize protein constituents, for example, thiol/amino groups and methionine.
$ONOO^-$, peroxynitrite	Formed in a rapid reaction between $O_2^{\bullet-}/NO^{\bullet}$. Lipid soluble and similar in reactivity to HOCl. Protonation forms peroxynitrous acid, which can undergo homolytic cleavage to form $OH^{\bullet}/NO_2$.

Cell exposure to oxidants results in apoptosis, which is inhibited by free-radical scavengers and thiols.[92] Apoptotic agents evoke ROS production.[93] Apoptotic thymocytes after dexamethasone (DEX) stimulation decrease thiols and increase $O_2^{\bullet-}$.[94] Glutathione (GSH)/*N*-acetylcysteine (*N*-Ac-Cys) prevent cell death. The OS evokes apoptosis and was proposed as mediator. Amatore and co-workers showed that in murine thymocytes, OS/apoptosis (cf. Fig. 4.5)[95] occur in the same cell; they identified apoptotic subpopulations that appeared sequentially: catalase (CAT) protected from DEX-induced death in initial stages while Fe chelators and vitamin (Vit)-E did not. Earlier production of intracellular ROS intermediate, for example, H_2O_2, is obligatory for apoptosis. They reviewed secretion electrochemical monitoring.[96] They investigated OS responses of MG63 osteosarcoma mechanically stressed, by amperometry at platinized (Pt)C-fiber electrodes, for monitoring

ROS/reactive nitrogen species (RNS): MG63 released prominent RNS and small H_2O_2 amounts.[97] Species resulted from 13:28 fmol production for $O_2^{\bullet-}$:$NO^{\bullet}$ per cell: high $NO^{\bullet}/H_2O_2$ and $NO^{\bullet}/O_2^{\bullet-}$ ratios are consistent with relation of malignant bone formation ability of MG63 to high $NO^{\bullet}$ and small $O_2^{\bullet-}$ production. Damages of OS cause pathologies [deoxyribonucleic acid (DNA) alteration, ageing, cancers, HIV and Parkinson (PD)/Alzheimer's (AD) diseases, etc.]. The OS is used positively by aerobic complex organisms to manage organs size/shape via apoptotic control of cell population; it is central to macrophage functions in preventing bacterial infection. Via PtC-fiber ultramicroelectrodes positioned in *artificial synapse* configuration, they established that *OS cellular bursts* consist in H_2O_2, $NO^{\bullet}$, $ONOO^-$ and NO_2^- femptomoles release; they showed that burst stems from coupled initial production of $O_2^{\bullet-}/NO^{\bullet}$ by two pre-existing enzymatic pools: nicotinamide (Nam) adenine dinucleotide (NAD^+) phosphate hydride (NADPH)-oxidases and NO-synthases. Method allows kinetic characterization of functional hyperaemia: it, investigated on stellate neurones within perfused rat cerebellum slices, relies on OS via $NO^{\bullet}$ local release by neurone while it is active. They reported hyperaemia kinetics; local $NO^{\bullet}$ waves in brain are deconvoluted, which evidences that neurone emits $NO^{\bullet}$ waves containing up to 500 nM and distributed within spherical volume of 30–50 μm radius. Process activates blood capillaries in active-neurone vicinity: since neurones cooperate along chain, larger blood supplies irrigate brain area while being active, which is reported by NMR/positron emission tomography (PET)-scan imaging.

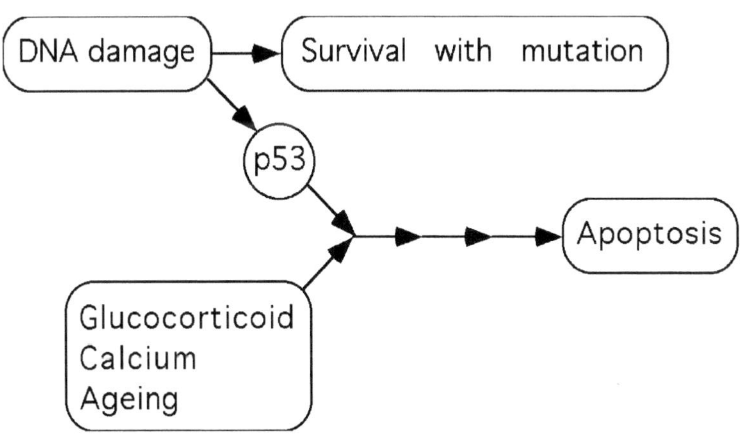

FIGURE 4.5 Two independent pathways lead to apoptosis in thymocytes: not/p53 mediated.

Frequency with which $O_{2(g)}$ is used in clinical medicine must not blind toxicity: O_2, at concentrations greater than air (21%), damages plants, animals and aerobic bacteria, for example, *Escherichia coli.*[98–100] Double-logarithmic plots of survival time versus O_2 pressure showed inverse, linear relations for protozoa, mice, fishes, rats, rabbits and insects; air damages vary with organism, age, physiological state and diet, for example, Vits-A/E/C, transition metals, antioxidants (added to foods) and polyunsaturated lipids. Gershman and Gilbert (1954) attributed O_2 damage to ROS.[101] Fridovich developed hypothesis into $O_2^{\bullet-}$ theory of O_2 toxicity in which $O_2^{\bullet-}$ formation in vivo plays major role.[102–104] Cu^{2+}/Fe^{3+} lead to harmful ROS generation, which damage cells at lipids, proteins and nucleic acids:

$$Fe^{3+} + O_2^{-} \rightarrow Fe^{2+} + O_2$$

$$Fe^{2+} + H_2O_2 \rightarrow Fe^{3+} + OH^{\bullet} + OH^{-}$$

$$O_2^{-} + H_2O_2 \rightarrow O_2 + OH^{\bullet} + OH^{-}$$

Cu/Zn-superoxide dismutase (SOD) protects cells versus OS, for example, Cu^{2+}. The ATX1-type Cu-chaperone antioxidant-1 (Atox1) functions as Cu^{2+}-sensitive transcriptional regulator, involved in Cu^{2+}-mediated effects on cell proliferation.[105] Keunen et al. reviewed metal-induced OS and plant mitochondria.[106] Cu^{2+}–circadian rhythm connection exists in organisms, for example, *Arabidopsis thaliana.*[107–109] In mammals, expression of pineal adenosine triphosphate (ATP)ase, Cu^{2+} transporter variant, displays rhythm.[110] Metal-binding prion proteins are involved in Cu^{2+} metabolism, and absence was associated with alterations in hormone rhythms.[111] Should mutual Cu^{2+} homeostasis/rhythm influence be extended feature among higher eukaryotes, experiments are envisaged to address questions, for example, Cu^{2+} role in neurodegenerative disorders, introducing temporal dimension in Cu^{2+}-dependent cell processes.

Park et al. reviewed protein sirtuin-3 (Sirt3), mitochondrial ROS, ageing and carcinogenesis.[112] Mitochondrial dysfunction is associated with insulin resistance in humans/animals, which is associated with decay in mitochondrial number, oxidative capacity, size and density; mechanisms that decrease mitochondrial activity/biogenesis, in insulin-sensitive tissues, are important to developing therapies to reverse insulin resistance. Lipid exposure is associated with increased accumulation of intracellular fatty acids, and decays in mitochondrial function in liver and skeletal muscle tissue; fatty-acid contact leads to accumulation of toxic proinflammatory lipids, which activate stress kinase signalling pathways resulting in antagonism of insulin receptor signalling; alternatively, increased rates of fatty-acid metabolism exceed capacity of TCA cycle and electron transport chain (ETC), resulting in incomplete oxidation and toxic metabolites. The OS is implicated in lipotoxicity pathogenesis; chronic OS is linked to insulin resistance in multiple tissues. The ROS produced by multiple cellular processes and controlled by cellular antioxidant mechanisms, for example, enzymatic scavengers or antioxidant modulators, mediate OS; when their production exceeds cellular antioxidant capacity, OS damage occurs. Multiple pathways can contribute to cellular ROS, for example, mitochondrial ETC, inflammatory signalling and endoplasmic reticulum (ER) stress; the former is major source of endogenous $O_2^{\bullet-}$ in cell; as electrons pass via it, fraction escape and prematurely react with O_2 resulting in $O_2^{\bullet-}$. Nutrients excess increases electron flux via ETC augmenting $O_2^{\bullet-}$; excessive ROS induce insulin resistance, activating stress-triggered signalling cascades or damaging mitochondrial proteins/DNA.

There are >100 neurodegenerative diseases grouped into:[113] (1) dementia syndrome, for example, AD, (2) movement/motor co-ordination disorders, for example, PD and (3) disorders of peripheral nervous system (PNS), for example, Charcot–Marie–Tooth type 4A (CMT4A).[114] Connections exist among mitochondria, ageing and diseases, for example, degenerations.[115] Mitochondrial functions in neurones are: (1) to provide energy to cell, (2) Ca^{2+} regulation, (3) to keep membrane potential $\Delta\psi_m$, (4) axonal/dendritic transports and (5) to provide energy to re-synthesize more neurotransmitter from breakdown products in synapses. Mitochondrial dysfunction is common in degenerative PNS neuropathies, for example, Friedreich's ataxia (FRDA)[116]/CMT4A. González-Cabo et al. analyzed disease molecular bases in neurobiology: they performed identification of routes involved in degenerative PNS neuropathies, comparing FRDA/CMT4A.[117] (1) The FRDA is autosomal recessive congenital ataxia caused by mutation in gene FXN on chromosome-9, which codes for protein frataxin, which is essential for mitochondria functioning (Fe removal from surrounding cytoplasm). Motor neurones (axon) are affected but sensory ones (dendrites), not. (2) The CMT4A is inherited neuropathy that takes different forms, characterized by muscle tissue and touch sensation loss in feet/legs and hands/arms. Commons of FRDA/CMT4A are: (1) peripheral axonal neuropathies, (2) degenerative diseases and (3) deficiency in mitochondrial proteins; differences show in: (1) target cells, (2) CMT4A does not present affectation of central nervous system (CNS) and (3) mitochondrial localization.

The OS was linked to deleterious effects responsible for neuronal cell death, which lead to, for example, AD/PD. At molecular level, role is played by unfolding/aggregation processes undergone by neuronal proteins upon oxidation, nitration and other covalent modifications by reactive species. In PD, selective damage occurs at neurones producing neurotransmitter dopamine (DA) in small brain compartment, *substantia nigra pars compacta*, where neuronal damage is produced by ROS from DA, for example, DA quinone (DAQ).[118] Casella studied modification of globin family proteins, for example, human myoglobin, hemoglobin and neuroglobin, to establish modification sites caused by ROS/RNS and DA reactions.[119] Most efficient ROS/RNS source is endogenous haeme group, and their modification results from intramolecular reactions. He studied structural effects accompanying protein modification/unfolding experiments in denaturing agent presence: limited alterations number at protein surface residues causes drastic reduction in unfolding energy barrier, facilitating protein denaturation/aggregation.

Tumor metabolism is different because it accommodates cell proliferation. Mitochondrial DNA (mtDNA) mutations in cancer were described; however, mitochondrial DNA/function alterations could be consequence. Fatty-acid mitochondrial/peroxisomal oxidations produce free radicals in liver and OS, which result in mtDNA alterations in oxidative phosphorylation (OXPHOS), causing structural abnormalities and ATP depletion; however, mitochondrial defects could be pre-existing, which allow for ROS overproduction.[120] Cancer is heterogeneous disease and dissecting cell/cell variations is important in tumor initiation, progression, metastasis and therapeutics.[121] In cases, mitochondrial dysfunction cause in cancer was demonstrated.[122] Mutations in mtDNA fulfilled criteria for pathogenic ones producing prostate cancer; focusing on gene cytochrome oxidase subunit-I (COI), 11.9% disease patients harbored COI mutations, which altered conserved amino acids (AAs, conservation index CI = 83%), whereas 1.9% of non-illness controls and 7.8% of general population presented COI ones, the latter altering less-conserved AAs (CI = 71%).[123] Four conserved prostate cancer COI mutations were found in multiple independent patients on different mtDNA backgrounds; three others contained heteroplasmic COI mutations, one of which created stop codon; the latter included germ ATP6 one. Germ/somatic mtDNA mutations contribute to prostate cancer; tumors produce increased ROS and mtDNA mutations which, in turn, inhibit OXPHOS, augment ROS and contribute to tumorigenicity. In order to determine if mutant tumors increased ROS and tumor growth rates, pathogenic mtDNA mutation ATP6 T8993G was introduced into prostate cancer PC3 cells, via cytoplasmic hybrid (cybrid) transfer, and tested for tumor growth in athymic (nude) mice: mutant cybrids generated tumors seven times larger than wild-type (WT) ones, which barely grew in mice; the former generated more ROS.

The ROS-mediated OS causes cellular damage and contributes to inflammation, ageing, cancer, arteriosclerosis, hypertension and diabetes; ROS elevation comes from imbalance between ROS production and scavenging by endogenous antioxidants, which disturb physiological functions.[124–126] The ROS induce cell death by apoptosis/necrosis.[127] Mitochondrial dysfunction results in increased ROS, which enhances OS if cellular defences are overwhelmed.[128] The ROS alter intracellular redox states, change inner mitochondrial $\Delta\psi_m$ and release soluble intermembrane proteins, for example, cytochrome complex (cyt *c*), from mitochondria into cytosol;[129,130] they trigger mitochondria-mediated apoptotic pathway, which is associated with activation of Cys-aspartic (Asp) protease (caspase) cascade and B-cell lymphoma type-2 (Bcl-2) protein family.[131–133] 7,3',4'-Trihydroxydihydroflavone (butin) protects cells versus H_2O_2-induced apoptosis by: (1) scavenging ROS activating antioxidant enzymes, for example, SOD/CAT, (2) decreasing OS-induced 8-hydroxy-2'-deoxyguanosine via oxoguanine glycosylase-1 (OGG1) activation and (3) reducing OS-induced mitochondrial dysfunction.[134] Butin cytoprotection was determined on OS-induced mitochondria-dependent apoptosis.[135] Butin reduced H_2O_2-induced $\Delta\psi_m$ loss as determined by confocal image analysis and flow cytometry, alterations in Bcl-2 family, for example, decay in Bcl-2 and rise in Bcl-2-associated X protein (Bax) and phospho-Bcl-2 expressions, cyt *c* release from mitochondria into cytosol and caspases 9/3 activation (cf. Fig. 4.6). It exerted anti-apoptotically via inhibition of mitogen-activated protein kinase (MAPK) kinase-4 (MKK4), c-jun NH_2-terminal kinase (JNK) and activator protein-1 (AP-1) cascades induced by H_2O_2; it protected versus H_2O_2-induced apoptosis shown by decreased apoptotic bodies, sub-G_1 hypodiploid cells and DNA fragmentation, which were exerted via $\Delta\psi_m$ depolarization blockade, JNK pathway inhibition and mitochondria-involved caspase-dependent apoptosis.

4.3.2 DIAGNOSTIC DILEMMA IN PROSTATE CANCER: CYP2B6, RECEPTORS AND GENES

Molecular profiles of endometrial cancer were reported.[136] Prostate cancer diagnostic is based on triad: PSA levels analysis in serum, digital rectal examination (DRE) and prostate biopsy.[137] In order to detect prostate cancer, men ages >50 carry out regularly PSA analyses in serum and, when marker levels >4 μM or patient presents disease history, DRE will be performed to determine biopsy necessity; problem is: while women are regularly subjected to exploratory gynaecological examinations, in men there exists refusal. Prostate cancer would benefit from more specific screening markers and lesser invasive diagnostic, which avoid biopsies/cost. Given prostate position in organism under bladder and encircling urethra, secretions and normal/malign gland cells are found in urine, which is source to determine gland situation: studies evidenced potential markers in urine, which improve diagnostic. Reynolds and co-workers reviewed molecular markers for disease,[138] its alterations[139] and prostate cancer gene-3 (PCA3).[140] One deals with proteins, which are present in urine secreted by prostate cells, or those that flake off to urine; if gland massage enrich urine sample in protein types, hypothesis supports that this results fluid in search for markers to discriminate between patients with/out gland cancer. The objective is to determine proteomic profile in urine, able to differentiate between prostate cancer/health patients; synthetic protein use as disease *reporter* was proposed.[141]

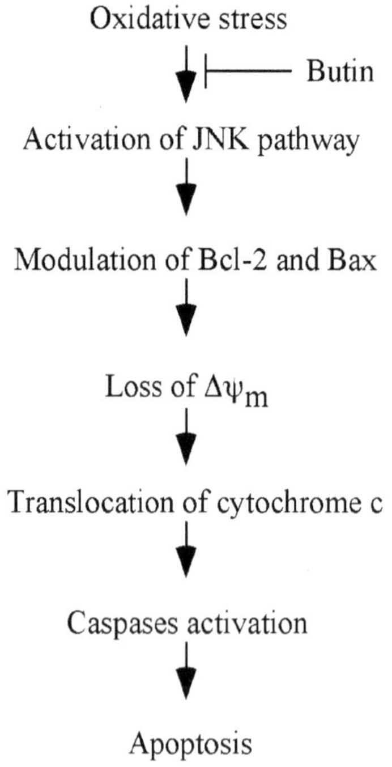

FIGURE 4.6 Butin cyto-protective pathway versus OS-induced mitochondrial-involved apoptosis.

Cytochrome P450s (CYPs) influence carcinogen, drug and hormone (testosterone) bioeffects; for example, CYP2B6 plays role in testosterone deactivation (cf. Table 4.2).

TABLE 4.2 Summary of Some CYP Families in Humans: Some of the Genes and Proteins They Encode.

Family	Function	Members	Names
CYP2	drug and steroid metabolism	13 subfamilies, 16 genes, 16 pseudogenes	CYP2A6, CYP2A7, CYP2A13, CYP2B6, CYP2C8, CYP2C9, CYP2C18, CYP2C19, CYP2D6, CYP2E1, CYP2F1, CYP2J2, CYP2R1, CYP2S1, CYP2U1, CYP2W1
CYP24	Vit-D degradation	1 subfamily, 1 gene	CYP24A1
CYP27	*varied*	3 subfamilies, 3 genes	CYP27B1 (Vit-D_3 1α-hydroxylase, activates Vit-D_3), CYP27C1 (unknown function)

Kumagai et al. examined CYP2B6 expression in human prostate tissues/cancer; they performed immunohistochemical analysis in 98 benign and 106 malignant tissues, and reviewed patients' charts for clinical, pathologic and survival data.[142] They investigated whether stable CYP2B6 expression in human prostate cancer cell line LNCaP influences proliferation; CYP2B6 was abundantly expressed in normal epithelial cells compared to ill ones. They found CYP2B6 immunostaining in 75/106 samples (71%) in cytoplasm of cancerous tissue samples; they inversely correlated CYP2B6 immunoreactivity with high score in Gleason scale (2–10). Decreased CYP2B6 immunoreactivity correlated with poor prognosis; uni/multivariate hazard analyses revealed correlation of decreased CYP2B6 expression with poor cancer-specific survival: its overexpression in LNCaP decreased testosterone-induced proliferation; underexpression plays a role in disease development and is useful as prognosis.

Reduction by dutasteride (Avodart®) of prostate cancer events (REDUCE) 4-year trial was designed to evaluate drug use for chemoprevention.[143] Dutasteride prescribed to shrink benign, enlarged prostates reduced cancer risk by 23% in increased-danger men.[144] Trial found that dutasteride decreased chances that men be diagnosed with excessively treated tumors: those that fall in aggressiveness mid-range, which account for most prostate cancers and grow unpredictably, leading men to opt for surgery/radiation therapy, which conducts to incontinence/impotence. It is first to evaluate prostate cancer chemoprevention in men at increased disease risk; it involved 8231 men ages 50–75 who were randomly assigned to receive either a placebo or daily 0.5 mg drug dose. Men presented elevated PSA 2.5–10 μM but no cancer evidence on biopsies performed within six months. Scheduled biopsies were performed two years after registration and following four years. 19.9% on dutasteride were diagnosed with prostate cancer (25.1% on placebo); none died of disease; among men with family history of prostate cancer, dutasteride reduced risk by 31.4%. Dutasteride was most effective reducing medium-grade tumors risk: Gleason 5/6; 70% men diagnosed with prostate cancer presented score 5/6 (same as in clinical practice): 18.1% on placebo versus 13.2% on dutasteride, which was approved for BPH treatment, which causes frequent urination, which is painful because swollen gland blocks urine flow. No increase was found in aggressive, high-grade tumors (Gleason 7–10) among men who took dutasteride over four years: there were 6.7% on drug with aggressive, high-grade tumors (6.8% on placebo); however, disparity was noted in most aggressive tumors (score 8–10) among men taking medicine in years 3/4 of study: 12 on prescription (one on placebo).

Men are commonly screened for prostate cancer risk via PSA measurements; however, test suffers from low specificity/sensitivity as diagnostic: according to US National Cancer Institute (NCI), 30% of men who have biopsy performed as follow-up to elevated PSA level show disease. Urine-based test was developed to detect TMPRSS2:ERG gene fusion, which is present in 50% prostate cancers, and described as most specific marker; ERG scores correlate with clinical disease on biopsy/prostatectomy: scores are higher in patients with cancerous/benign prostate lesions. The ERG improves PCA3 assay predictability, which detects PCA3 messenger ribonucleic acid (RNA) (mRNA) in male urine after DRE; PCA3 assays guide performing biopsy following positive PSA test; marker, in combination with PCA3, enhances serum PSA utility for predicting prostate cancer and clinical disease on biopsy.

Andreoiu and Cheng reviewed multifocal prostate cancer, which clinical management was challenged by intratumoral heterogeneity, on genomic/pathological levels, and limited understanding of genetics governing disease progression.[145] Gene-based test determined probability that prostate cancer either keep latent in gland or spread deadly over other organs; in 85% cases, gland presented more than one tumor focus, each containing different malignant cells with distinct set of genetic abnormalities, which makes difficult to identify genes/characteristics that indicate tumor potential to extend.

Murine experiments were exploited, in order to test hypothesis that pathways constraining progression is activated in indolent phosphatase/tensin homologue (*Pten*)-null prostate tumors, and progression barrier inactivation would engender metastasis-prone condition.[146] Transcriptomic–canonical pathway comparative analyses, plus biochemical confirmation, of normal prostate epithelium versus poorly progressive *Pten*-null gland cancers revealed robust activation of transforming growth factor-β bone morphogenetic protein-homologue-4 of *Drosophila* mothers-against-decapentaplegic protein (TGF-β/BMP-SMAD4) signalling axis, which functional relevance was supported by invasive, metastatic and lethal diseases emergence with 100% penetrance upon *Smad4* genetic deletion in *Pten*-null murine prostate; pathological/molecular analyses, and transcriptomic knowledge-based pathway profiling of emerging tumors, identified cell invasion as cardinal biofeatures in metastatic *Smad4*/*Pten*-null disease model. Follow-on pathological/functional assessment confirmed cell cycle regulator protein G1/S-specific cyclin-D1, and secreted phosphoprotein-1 (SPP1), as key bioprocess mediators, which together with PTEN/SMAD4 form 4-gene signature, which is prognostic of PSA biochemical recurrence and lethal metastasis in human prostate cancer; model-informed progression analysis and genetic, functional and translational studies established SMAD4 as regulator of murine/human disease progression.

Test was developed that detected specific protein in men urine, which is associated with prostate cancer; small sample is used to find out if protein is excreted in urine and is detected via test, which is simple, rapid and presents potential to be used in consulting rooms of family doctors. 194 urine samples were investigated to detect homeodomain-containing transcrition factor protein Engrailed-2 (EN2): there were men with lower-urinary tract symptoms (LUTS), persons worried about asymptomatic prostate cancer, for example, showing positive family history, and men with abnormal PSA test; two control groups were examined. Samples were analyzed via enzyme-linked immunosorbent assay (ELISA), which uses two murine monoclonal antibodies; another studies were performed to detect EN2, for example, Western blot and semi-quantitative reverse transcription (RT) polymerase chain reaction (PCR). Key limitations with PSA currently standard detection test justified new-marker evaluation. It was assessed EN2 diagnostic potential, expressed in prostate cancer cell lines and secreted into urine by disease in men; EN2 expression in disease cell/tissue was determined by PCR/immunohistochemistry.[147] Urine without prior DRE was collected from men presenting LUTS, to confirm prostate cancer presence, and controls. The EN2 was measured by ELISA in urine from men with prostate cancer (n = 82) and controls (n = 102); it was expressed/secreted by ill cell/tissue but not by normal gland tissue/stroma. The EN2 in urine was prostate cancer predictive with 66% sensitivity and 88.2% specificity without DRE; there was no correlation with PSA. Second academic center confirmed results: urinary EN2 is specific/sensitive candidate marker of prostate cancer.

4.3.3 SPATIAL AND TEMPORAL ORGANIZATION OF PROTEINS AND mRNAs IN CELL

Ability to sort proteins to specific regions within cell is conserved in life domains, for example, bacteria. Two parameters dictate how a protein arrives at destination: (1) location harbors unique feature that distinguishes it from other subcellular sites and (2) specific signal typically embedded within protein directs this to target and is recognized at end; for example, in *E. coli*, localization signal for proteins is not contained solely within protein,[148–150] but lies within mRNA that encodes this (cf. Fig. 4.7).[151–154] Focus was on mRNAs encoding proteins, which either remain in cytosol or are inserted into plasma membrane of *E. coli*. Localization of mRNA in cells was monitored via fusing it with specific nucleotide sequence, which is recognized by RNA-binding protein, which is fused to green fluorescent protein (GFP); engineered RNA molecules produced functional proteins, which were tracked in cells by fluorescence microscopy. The mRNAs encoding membrane proteins were observed at cells periphery, and those encoding soluble proteins, in helical pattern in cytosol; biochemical analysis of fractionated cell lysates confirmed locations. According to signal peptide hypothesis, shortly after ribosome begins synthesizing integral membrane protein, corresponding mRNA translation temporarily arrests; ternary complex (ribosome, mRNA and nascent polypeptide) is recruited to cell's secretion machinery which, in bacteria, resides at plasma membrane; translation arrest is relieved and polypeptide is inserted into membrane as synthesis resumes. Translation was inhibited to see what mRNAs fate would be; in its absence, mRNAs still localized to ultimate destination of encoded proteins. Van der Lelij et al. published human individual with biallelic mutations in *DDX11*.[155] Farràs et al. reported jun-B degradation in mitosis and aberrant consequences of deregulated control mechanisms.[156]

a mRNA-Ribosome-Polypeptide → Protein has localization signal → Cellular destination (cytoplasm, membrane, organelle)

b mRNA has localization signal → Cellular destination (cytoplasm, membrane, organelle)

FIGURE 4.7 (A) Old paradigm. (B) Shooting mechanism.

4.3.4 MICROFLUIDIC-BASED MULTIPLEX QRT-PCR, [–2]PRO-PSA- AND AUTOANTIBODY-PSA

Screening trials based on PSA indicated need for non-invasive marker identification strategies to improve prostate cancer prediction; non-coding microRNAs (miRNA) in serum/plasma present potential as non-invasive markers for physio/pathologic conditions, for example, cancer. In order to identify serum miRNAs that diagnose/correlate with prostate cancer prognosis, Moltzahn et al. developed multiplex quantitative RT (qRT)-PCR: purification of multiplex PCR products, followed by uniplex analysis on microfluidics chip to evaluate 384 human miRNAs.[157] Via *Dgcr8*/*Dicer* knockout (KO) murine ES cells as benchmark, they validated and uncovered accuracy lack in established methods. Profiling 48 sera from healthy and untreated ill men, with differing cancer of prostate risk assessment (CAPRA) scores, they identified miRNA signatures that allowed diagnosing unwell patients and correlating with prognosis; serum signatures included oncogenic/tumor-suppressive miRNAs, suggesting roles in disease progression. They performed uniplex analysis in microfluid chip Biomark System, which transforms PCR studies: it uses chips and counts with low input requirements, providing analysis sensitivity/yield necessary for PCR. Fluid integrated circuits permit discovery of one cell, with capacity of studying hundreds of individual genes via TaqMan and DNA-binding dye analyses; modified PCR showed that trespassing of multiplex PCR initiators presents harmful effects on miRNA quantification.

The PSA/free PSA present limited specificity to detect clinical, curable prostate cancer leading to unnecessary biopsy and indolent tumor detection/treatment; specificity to detect clinical disease improves by [–2] pro-PSA. Catalona et al. evaluated [2]pro-PSA, frcc PSA and PSA, via formula [–2]pro-PSA/free PSA × $PSA^{1/2}$, to enhance specificity to detect overall/high-grade prostate cancer.[158] They enrolled 892 men with no disease history, normal DRE, PSA 2–10 μM and Gleason ≥6 in prospective multi-institutional trial; they examined relation of serum PSA, free-to-total

PSA and prostate health index with biopsy. Primary end points were specific via index to detect overall/Gleason ≥7 compared with free-to-total PSA; in PSA 2–10 μM at 80%–95% sensitivity, index specificity exceeded PSA/free-to-total PSA. They associated increasing index with 4.7-fold increased risk of prostate cancer and 1.61-fold augmented possibility of Gleason ≥7; index exceeded free-to-total PSA to discriminate prostate cancer with score ≥7 from lower-grade/negative biopsy; it was not connected with age and prostate volume. Screening test is useful for patients with normal prostate examination in which PSA is 2–10 μM, which results diagnostic grey zone because most men with higher levels present gland cancer and most patients with lower levels do not; the higher pro-PSA/total PSA and lower free PSA, the more probable that men present aggressive disease. In PSA 2–4 μM, pro-PSA/free PSA ratio detected 90% cancers; index is useful in disease screening to decrease unnecessary biopsy, in men ages ≥50 with PSA 2–10 μM and negative DRE, with minimal sensitivity loss.

Sufficient-specificity/sensitivity lack among conventional cancer markers, for example, PSA for prostate disease, was widely recognized. Autoantibodies (autoAbs) are investigated as markers but remain elusive: obstacle is lack of sensitive/multiplex approach for quantifying autoAbs versus clinically relevant tumor-associated antigens (TAAs). In order to circumvent phage lysate preparation and recombinant protein purification, Xie et al. identified B-cell epitopes from prostate cancer-associated antigens (PCAAs).[159] They conjugated peptide epitopes from cancer/testis antigen 1B (CTAG1B) NY-ESO-1, X-antigen family member 1b (XAGE-1b), synovial sarcoma X-breakpoints-2/4 (SSX-2,4), prostate cancer overexpressed antigen α-methylacyl coenzyme-A (CoA) (AMACR), p90 autoantigen and lens epithelium-derived growth factor (LEDGF) p75 with seroMAP microspheres, allowing autoAbs multiplex measurement in serum. Simultaneous quantification of autoAbs + total PSA was achieved in one reaction and termed *A+PSA* assay. Peptide epitopes from six PCAAs were identified and confirmed that autoAbs versus epitopes reacted specifically with full-length protein. The A+PSA study used pre-surgery sera from 131 biopsy-confirmed prostate cancer and 121 BPH/prostatitis patients. A logistic regression-based A+PSA index enhanced sensitivities/specificities over PSA, distinguishing prostate cancer from nonmalignant cases; A+PSA reduced false positive rate and improved area under receiver operating characteristic (ROC) (AUROC) curve (AUC).

Prostasomes are microvesicles (diameter, 150 nM) that are produced/secreted by prostate acinar cells; it was hypothesized that invasive growth of malignant gland cells causes microvesicles, normally released into seminal fluid, to appear in interstitial space and peripheral circulation; prostasome suitability as blood marker in patients with gland cancer was tested via expanded proximity ligation assay (PLA). Tavoosidana et al. developed sensitive/specific assay (4PLA) for detection of complex target structures, for example, microvesicles, in which target is captured via immobilized antibody and detected via four other antibodies with attached DNA strands.[160] Requirement for coincident binding by five antibodies, to generate amplifiable reporter, results in increased specificity/sensitivity; assay detected elevated prostasome levels in blood samples from patients with prostate cancer before radical prostatectomy, compared with controls and benign-biopsy men. Medians for prostasome levels, in blood plasma of prostate cancer patients, were 2.5–sevenfold higher compared with control samples in two studies, and assay distinguished patients with high/medium prostatectomy Gleason (8-9/7) from low score (≤6), reflecting disease aggressiveness; approach that enables prostasome detection, in peripheral blood, results useful for diagnosis/prognosis assessment in organ-confined prostate cancer.

4.3.5 ANTENNAS AND APPLICATION TO OPTICAL MANIPULATION AND SENSING

Advances in nano-optics/fabrication made possible extending antenna concepts to optical frequencies, opening light–matter interaction control at nanometre. Quidant and co-workers reviewed optical antenna designs and discussed abilities to control light fields at nanometre[161,162] and direct nearby emitters emission.[163] They focused on antenna application to biosciences, for example, advances in plasmon-based nano-optical trapping[164] and localized surface plasmon resonance (LSPR) biosensing.[165] They discussed mixing both functionalities into integrated analytical platform for cancer detection. The LSPR early detection of circulation (SPEDOC) heat-shock proteins (HSPs) combines nano-optics, optical manipulation and microfluidics with HSPs to develop precursor of individualized cancer diagnosis/treatment devices. It, integrated in microfluidic environment, will exploit LSPRs supported by micro/nanoAu nanostructures: (1) to track 70 kDa HSPs (HSP70) in peripheral blood and (2) monitor its overexpression at cancer cell surface. It treats cancer patients at earlier stage and lower doses with consequent AE decrease.

4.3.6 EFFECT OF ZN^{2+} ON PROSTATIC CELL CYTOTOXICITY CAUSED BY TRICHOMONAS VAGINALIS

Vazquez-Carrillo et al. studied *Trichomonas vaginalis* proteome, morphology and cytotoxicity during interactions with prostatic DU-145 cells.[166] They showed that 37 proteins are expressed in Zn^{2+} presence, which downregulates protein/transcriptional levels of *T. vaginalis* cysteine (Cys) proteinase 65 kDa (TvCP65). Result is negative effect on trichomonal cytotoxicity. They identified differentially expressed proteins by MS.

4.3.7 METABOLISM PLATFORM TECHNOLOGY

Geno–phenotype union builds reference library (cf. Fig. 4.8).[167]

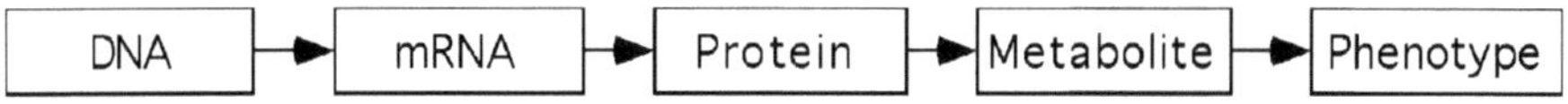

FIGURE 4.8 The union of genotype and phenotype to build a reference library.

Information flows for disease mechanism (cf. Fig. 4.9), for example, metabolites (glucose, cholesterol, L-threosine, etc.) for an illness (prostate neoplasia).[168]

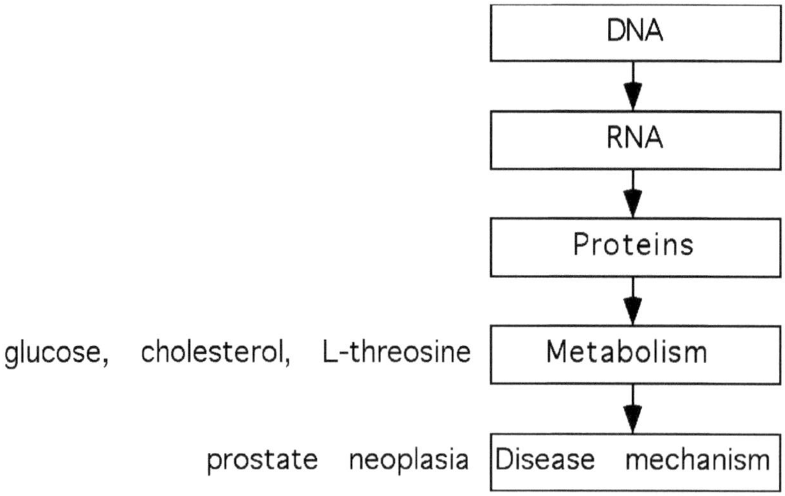

FIGURE 4.9 Information flow for disease mechanism: metabolites (glucose, etc.) for illness (neoplasia).

Biology advances via understanding metabolism (cf. Fig. 4.10); for example, prostate neoplasia aggressiveness was classified into benign, localized and metastatic via study of hundreds of thousands ion features; metabolism software showed that sarcosine modulates invasiveness.

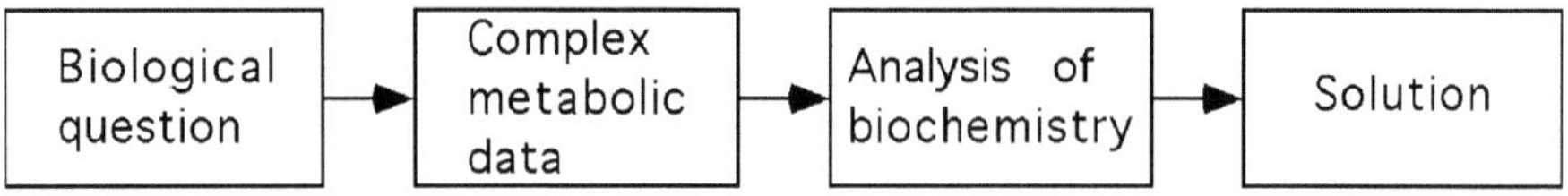

FIGURE 4.10 Advancing biology via understanding metabolism.

4.4 FINAL REMARKS

From the present results and discussion, the following final remarks can be drawn.

1. Commons exist between disparate complex systems arising in non-related areas of nature, society or technology; when complex networks represent systems, some features are well documented in literature. In order to discover the universal features, one needs to perform cross-comparative analysis of complex systems/behaviors via appropriate mathematical tools and physical concepts. Challenges exist facing metabolomics: compound identification, etc. Metabolomics solves routine problems: unknown drug mechanism of action and toxicology, etc. However, if studies cannot be executed quickly and with highest quality a limitation exists.
2. Successful complexity modeling/simulation were made possible via systems-biology predictive *in silico* approach, with objective of unravelling disease mechanisms. Reductionism resulted in promising drug candidates failing in final phases of clinical testing, which was because of knowledge lack about disease pathways at molecular level, which led to unforeseen adverse effects and unacceptable toxicity profiles. Mitochondrial deoxyribonucleic acid mutations are not only associated with predisposition to neuromuscular disease, but also to cancer.
3. Not only will be unique protein but protein set that via urine helps practitioner to decide without leaving aside present diagnostic marker PSA. Blood analysis can help to predict if men, who take medicine to reduce prostate size, will increase probability of contracting

aggressive cancer. Assay A+PSA represents platform that integrates autoAb signatures with conventional cancer marker, which aids in prostate disease and others diagnosis/prognosis.

4. Genotype–phenotype union resulted valid to elucidate disease mechanisms. A number of metabolic pathways were altered by diet. Prostate neoplasia aggressiveness was classified into benign, localized and metastatic via hundreds of thousands ion features. Metabolism software showed that sarcosine modulates invasiveness.
5. Zn^{2+} negatively affects growth, cytotoxicity and expression of *T. vaginalis* proteinases related to prostatic cell interaction.

Further work will deal with intracellular proteolysis, neoplasia, ubiquitin-proteasome system, cell cycle, cancer, vitamin-D, analogues, drugs, their relevance to cancer inhibition, prostate disease, stem cells and their connection with cancer, prostate disease, MEK/ERK pathway overactivation in liver tumors and resistance to transforming growth factor-β-induced death.

ACKNOWLEDGMENTS

The authors acknowledge support from Generalitat Valenciana (Project No. PROMETEO/2016/094) and Universidad Católica de Valencia *San Vicente Mártir* (Project No. 2019-217-001).

KEYWORDS

- **prostate neoplasia**
- **molecular test**
- **diagnostic**
- **gene expression marker**
- **prostate-specific antigen**
- **metabolism platform technology**

REFERENCES

1. Dunn, W. B. Current Trends and Future Requirements for the Mass Spectrometric Investigation of Microbial, Mammalian and Plant Metabolomes. Phys. Biol. **2008,** 5, 011001–1–24.

2. Lu, W.; Bennett, B. D.; Rabinowitz, J. D. Analytical Strategies for LC–MS-based Targeted Metabolomics. *J. Chromatogr., B* **2008,** *871*, 236–242.
3. Sabatine, M. S.; Liu, E.; Morrow, D. A.; Heller, E.; McCarroll, R.; Wiegand, R.; Berriz, G. F.; Roth, F. P.; Gerszten, R. E. Metabolomic Identification of Novel Biomarkers of Myocardial Ischemia. *Circulation* **2005,** *112*, 3868–3875.
4. Sawada, Y.; Akiyama, K.; Sakata, A.; Kuwahara, A.; Otsuki, H.; Sakurai, T.; Saito, K.; Hirai, M. Y. Widely Targeted Metabolomics Based on Large-scale MS/MS Data for Elucidating Metabolite Accumulation Patterns in Plants. *Plant Cell Physiol.* **2009,** *50*, 37–47.
5. Goodacre, R.; Vaidyanathan, S.; Dunn, W. B.; Harrigan, G. G.; Kell, D. B. Metabolomics by Numbers: Acquiring and Understanding Global Metabolite Data. *Trends Biotechnol.* **2004,** *22*, 245–252.
6. Fiehn, O. Metabolomics—the Link Between Genotypes and Phenotypes. *Plant Mol. Biol.* **2002,** *48*, 155–171.
7. Werner, E.; Heilier, J. F.; Ducruix, C.; Ezan, E.; Junot, C.; Tabet, J. C. Mass Spectrometry for the Identification of the Discriminating Signals from Metabolomics: Current Status and Future Trends. *J. Chromatogr., B* **2008,** *871*, 143–163.
8. Nordstrom, A.; Want, E.; Northen, T.; Lehtio, J.; Siuzdak, G. Multiple Ionization Mass Spectrometry Strategy Used to Reveal the Complexity of Metabolomics. *Anal. Chem.* **2008,** *80*, 421–429.
9. Patterson, A. D.; Li, H.; Eichler, G. S.; Krausz, K. W.; Weinstein, J. N.; Fornace, A. J., Jr.; Gonzalez, F. J.; Idle, J. R. UPLC-ESI-TOFMS-based Metabolomics and Gene Expression Dynamics Inspector Self-organizing Metabolomic Maps as Tools for Understanding the Cellular Response to Ionizing Radiation. *Anal. Chem.* **2008,** *80*, 665–674.
10. Wilson, I. D.; Nicholson, J. K.; Castro-Perez, J.; Granger, J. H.; Johnson, K. A.; Smith, B. W.; Plumb, R. S. High Resolution *Ultra Performance* Liquid Chromatography Coupled to oa-TOF Mass Spectrometry as a Tool for Differential Metabolic Pathway Profiling in Functional Genomic Studies. *J. Proteome Res.* **2005,** *4*, 591–598.
11. Want, E. J.; Nordström, A.; Morita, H.; Siuzdak, G. From Exogenous to Endogenous: The Inevitable Imprint of Mass Spectrometry in Metabolomics. *J. Proteome Res.* **2007,** *6*, 459–468.
12. Wilson, I. D.; Plumb, R.; Granger, J.; Major, H.; Williams, R.; Lenz, E. M. HPLC-MS-based Methods for the Study of Metabonomics. *J. Chromatogr., B* **2005,** *817*, 67–76.
13. Timischl, B.; Dettmer, K.; Kaspar, H.; Thieme, M.; Oefner, P. J. Development of a Quantitative, Validated Capillary Electrophoresis-time of Flight-mass Spectrometry Method with Integrated High- Confidence Analyte Identification for Metabolomics. *Electrophoresis* **2008,** *29*, 2203–2214.
14. Schauer, N.; Steinhauser, D.; Strelkov, S.; Schomburg, D.; Allison, G.; Moritz, T.; Lundgren, K.; Roessner-Tunali, U.; Forbes, M. G.; Willmitzer, L.; Fernie, A. R.; Kopka, J. GC-MS Libraries for the Rapid Identification of Metabolites in Complex Biological Samples. *FEBS Lett.* **2005,** *579*, 1332–1337.
15. Issaq, H. J.; Abbott, E.; Veenstra, T. D. Utility of Separation Science in Metabolomic Studies. *J. Sep. Sci.* **2008,** *31*, 1936–1947.
16. Want, E. J.; Cravatt, B. F.; Siuzdak, G. The Expanding Role of Mass Spectrometry in Metabolite Profiling and Characterization. *ChemBioChem* **2005,** *6*, 1941–1951.

17. Brindle, J. T.; Antti, H.; Holmes, E.; Tranter, G.; Nicholson, J. K.; Bethell, H. W.; Clarke, S.; Schofield, P. M.; McKilligin, E.; Mosedale, D. E.; Grainger, D. J. Rapid and Noninvasive Diagnosis of the Presence and Severity of Coronary Heart Disease Using ^{1}H-NMR-based Metabonomics. *Nat. Med.* **2002,** *8*, 1439–1444.
18. Clayton, T. A.; Lindon, J. C.; Cloarec, O.; Antti, H.; Charuel, C.; Hanton, G.; Provost, J. P.; Le Net, J. L.; Baker, D.; Walley, R. J.; Everett, J. R.; Nicholson, J. K. Pharmaco-metabonomic Phenotyping and Personalized Drug Treatment. *Nature (London)* **2006,** *440*, 1073–1077.
19. Griffin, J. L.; Schockcor, J. P. Metabolic Profiles of Cancer Cells. *Nat. Rev. Cancer.* **2004,** *4*, 551–561.
20. Holmes, E.; Tsang, T. M.; Huang, J. T. J.; Leweke, F. M.; Koethe, D.; Gerth, C. W.; Nolden, B. M.; Gross, S.; Schreiber, D.; Nicholson, J. K.; Bahn, S. Metabolic Profiling of CSF: Evidence that Early Intervention May Impact on Disease Progression and Outcome in Schizophrenia. *PLoS Med.* **2006,** *3*, 1420–1428.
21. Tolstikov, V.; Fiehn, O. Analysis of Highly Polar Compounds of Plant Origin: Combination of Hydrophilic Interaction Chromatography and Electrospray Ion Trap Mass Spectrometry. *Anal. Biochem.* **2002,** *301*, 298–307.
22. Van der Werf, M. J.; Overkamp, K. M.; Muilwijk, B.; Coulier, L.; Hankemeier, T. Microbial Metabolomics: Toward a Platform with Full Metabolome Coverage. *Anal. Biochem.* **2007,** *370*, 17–25.
23. Nordstrom, A.; O'Maille, G.; Qin, C.; Siuzdak, G. Nonlinear Data Alignment for UPLC-MS and HPLC-MS Based Metabolomics: Quantitative Analysis of Endogenous and Exogenous Metabolites in Human Serum. *Anal. Chem.* **2006,** *78*, 3289–3295.
24. Plumb, R.; Castro-Perez, J.; Granger, J.; Beattie, I.; Joncour, K.; Wright, A. Ultra-Performance Liquid Chromatography Coupled to Quadrupole-orthogonal Time-of-flight Mass Spectrometry. *Rapid Commun. Mass Spectrom.* **2004,** *18*, 2331–2337.
25. Kitteringham, N. R.; Jenkins, R. E.; Lane, C. S.; Elliott, V. L.; Park, B. K. Multiple Reaction Monitoring for Quantitative Biomarker Analysis in Proteomics and Metabolomics. *J. Chromatogr. B* **2009,** *877*, 1229–1239.
26. Woo, H. M.; Kim, K. M.; Choi, M. H.; Jung, B. H.; Lee, J.; Kong, G.; Nam, S. J.; Kim, S.; Bai, S. W.; Chung, B. C. Mass Spectrometry Based Metabolomic Approaches in Urinary Biomarker Study of Women's Cancers. *Clin. Chim. Acta* **2009,** *400*, 63–69.
27. Katajamaa, M.; Oresic, M. Data Processing for Mass Spectrometry-based Metabolomics. *J. Chromatogr. A* **2007,** *1158*, 318–328.
28. Broadhurst, D. I.; Kell, D. B. Statistical Strategies for Avoiding False Discoveries in Metabolomicsand Related Experiments. *Metabolomics* **2006,** *2*, 171–196.
29. Chen, J.; Zhao, X. J.; Fritsche, J.; Yin, P. Y.; Schmitt-Kopplin, P.; Wang, W. Z.; Lu, X.; Haring, H. U.; Schleicher, E. D.; Lenmann, R.; Xu, G. W. Practical Approach for the Identification and Isomer Elucidation of Biomarkers Detected in a Metabonomic Study for the Discovery of Individuals at Risk for Diabetes by Integrating the Chromatographic and Mass Spectrometric Information. *Anal. Chem.* **2008,** *80*, 1280–1289.
30. Kind, T.; Fiehn, O. Seven Golden Rules for Heuristic Filtering of Molecular Formulas Obtained by Accurate Mass Spectrometry. *BMC Bioinf.* **2007,** *8*, 105–1-20.
31. Kind, T.; Fiehn, O. Metabolomic Database Annotations *via* Query of Elemental Compositions: Mass Accuracy Is Insufficient Even at Less than 1 ppm. *BMC Bioinf.* **2006,** *7*, 234–1-10.

32. Gugisch, R.; Kerber, A.; Laue, R.; Meringer, M.; Rücker, C. History and Progress of the Generation of Structural Formulae in Chemistry and Its Applications. *MATCH Commun. Math. Comput. Chem.* **2007,** *58,* 239–280.
33. Evans, A. M.; DeHaven, C. D.; Barrett, T.; Mitchell, M.; Milgram, E. Integrated, Nontargeted Ultrahigh Performance Liquid Chromatography/Electrospray Ionization Tandem Mass Spectrometry Platform for the Identification and Relative Quantification of the Small-molecule Complement of Biological Systems. *Anal. Chem.* **2009,** *81,* 6656–6667.
34. Liu, J.; Litt, L.; Segal, M. R.; Kelly, M. J. S.; Pelton, J. G.; Kim, M. Metabolomics of Oxidative Stress in Recent Studies of Endogenous and Exogenously Administered Intermediate Metabolites. *Int. J. Mol. Sci.* **2011,** *12,* 6469–6501.
35. Laronga, C.; Drake, R. R. Proteomic approach to breast cancer. *Cancer Control* **2007,** *14,* 360–368.
36. Choo, J. R.; Nielsen, T. O. Biomarkers for Basal-like Breast Cancer. *Cancers* **2010,** *2,* 1040–1065.
37. Drake, R. R.; Cazares, L. H.; Jones, E. E.; Fuller, T. W.; Semmes, O. J.; Laronga, C. Challenges to Developing Proteomic-based Breast Cancer Diagnostics. *OMICS* **2011,** *15,* 251–259.
38. Torrens, F; Castellano, G. Modelling of Complex Multicellular Systems: Tumour–Immune Cells Competition. *Chem. Central J.* **2009,** *3* (Suppl. I), 75–1-1.
39. Torrens, F.; Castellano, G. Information Theoretic Entropy for Molecular Classification: Oxadiazolamines as Potential Therapeutic Agents. *Curr. Comput.-Aided Drug Des.* **2013,** *9,* 241–253.
40. Torrens, F.; Castellano, G. Molecular Classification of 5-amino-2-aroylquinolines and 4-aroyl-6,7,8-trimethoxyquinolines as Highly Potent Tubulin Polymerization Inhibitors. *Int. J. Chemoinf. Chem. Eng.* **2013,** *3* (2), 1–26.
41. Estrela, J. M.; Mena, S.; Obrador, E.; Benlloch, M.; Castellano, G.; Salvador, R.; Dellinger, R. W. Polyphenolic Phytochemicals in Cancer Prevention and Therapy: Bioavailability *versus* Bioefficacy. *J. Med. Chem.* **2017,** *60,* 9413–9436.
42. Torrens, F.; Castellano, G. Molecular Classification of Antitubulin Agents with Indole Ring Binding at Colchicine-Binding Site. In *Molecular Insight of Drug Design*; Parikesit, A. A., Ed.; InTechOpen: Vienna, 2018; pp 47–67.
43. Torrens, F.; Castellano, G. Molecular Classification of 2-Phenylindole-3-carbaldehydes as Potential Antimitotic Agents in Human Breast Cancer Cells. In *Theoretical Models and Experimental Approaches in Physical Chemistry: Research Methodology and Practical Methods*; Haghi, A. K., Thomas, S., Praveen, K. M., Pai, A. R., Eds.; Apple Academic–CRC: Waretown, NJ, in Press.
44. Torrens, F.; Castellano, G. *Cancer and Hypotheses on Cancer.* In *Molecular Chemistry and Biomolecular Engineering: Integrating Theory and Research with Practice*; Pogliani, L., Torrens, F., Haghi, A. K., Eds.; Apple Academic–CRC: Waretown, NJ, in Press.
45. Torrens, F.; Castellano, G. Precision Personalized Medicine from Theory to Practice: Cancer. In *Green Chemistry and Biodiversity: Principles, Techniques, and Correlations*; Aguilar, C. N., Ameta, S. C., Haghi, A. K., Eds.; Apple Academic–CRC: Waretown. NJ, in Press.
46. Torrens, F.; Castellano, G. AIDS Destroys Immune Defences: Hypothesis. *New Front. Chem.* **2014,** *23,* 11–20.

47. Torrens-Zaragozá, F.; Castellano-Estornell, G. Emergence, Spread and Uncontrolled Ebola Outbreak. *Basic Clin. Pharmacol. Toxicol.* **2015,** *117* (Suppl. 2), 38–38.
48. Torrens, F.; Castellano, G. 2014 Spread/Uncontrolled Ebola Outbreak. *New Front. Chem.* 2015, *24*, 81–91.
49. Torrens, F.; Castellano, G. Ebola Virus Disease: Questions, Ideas, Hypotheses and Models. *Pharmaceuticals* **2016,** *9*, 14–6-6.
50. Torrens, F.; Castellano, G. Clinical Translational Research: Cancer, Diabetes, and Cardiovascular Disease. In *Physical Biochemistry, Biophysics, and Molecular Chemistry: Applied Research and Interactions*; Torrens, F., Mahapatra, D. K., Haghi, A. K., Eds.; Apple Academic–CRC: Waretown, NJ, in Press.
51. Torrens, F.; Castellano, G. Primary Health, Smoking Habits and Immunotherapy in Cancer. In *Environmental Technology and Engineering Techniques: Basic Concepts and Health Interventions*; Khalaf, M. N., Smirnov, M. O., Kannan, P., Haghi, A. K., Eds.; Apple Academic–CRC: Waretown, NJ, in Press.
52. Torrens, F.; Castellano, G. Epigenomics, Epithelial Plasticity, Clinical Genetics and Rare Diseases. In *Environmental Technology and Engineering Techniques: Basic Concepts and Health Interventions*; Khalaf, M. N., Smirnov, M. O., Kannan, P., Haghi, A. K., Eds.; Apple Academic–CRC: Waretown, NJ, in Press.
53. Estrada, E., Fox, M., Higham, D. J., Oppo G. L., Eds., *Network Science: Complexity in Nature and Technology*; Springer: London, UK, 2010.
54. Sharan, R.; Ulitsky, I.; Shamir, R. Network-based Prediction of Protein Function. *Mol. Syst. Biol.* **2007,** *3*, 1–13.
55. Sharan, R.; Ideker, T. Protein Networks in Disease. *Genome Res.* **2008,** *18*, 644–652.
56. Jeong, H.; Mason, S. P.; Barabási, A. L.; Oltvai, Z. L. Lethality and Centrality in Protein Networks. *Nature (London)* **2001,** *411*, 41–42.
57. Yu, H.; Braun, P.; Yildirim, M. A.; Lemmens, I.; Venkatesan, K.; Sahalie, J.; Hirozane-Kishikawa, T.; Gebreab, F.; Li, N.; Simonis, N.; Hao, T.; Rual, J. F.; Dricot, A.; Vazquez, A.; Murray, R. R.; Simon, C.; Tardivo, L.; Tam, S.; Svrzikapa, N.; Fan, C.; de Smet, A. S.; Motyl, A.; Hudson, M. E.; Park, J.; Xin, X.; Cusick, M. E.; Moore, T.; Boone, C.; Snyder, M.; Roth, F. P.; Barabási, A. L.; Tavernier, J.; Hill, D. E.; Vidal, M. High-quality Binary Protein Interaction Map of the Yeast Interactome Network. *Science* **2008,** *322*, 104–110.
58. Ratmann, O.; Wiuf, C.; Pinney, J. W. From Evidence to Inference: Probing the Evolution of Protein Interaction Networks. *HFSP J.* **2009,** *3*, 290–306.
59. Przulj, N.; Wigle, D.; Jurisica, I. Functional Topology in a Network of Protein Interactions. *Bioinformatics* **2004,** *20*, 340–348.
60. Chua, H.; Sung, W.; Wong, L. Exploting Indirect Neighbors and Topological Weight to Predict Protein Function from Protein–protein Interactions. *Bioinformatics* **2006,** *22*, 1623–1630.
61. Schwikowski, B.; Uetz, P.; Fields, A. A Network of Protein–protein Interactions in Yeast. *Nat. Biotechnol.* **2000,** *18*, 1257–1261.
62. Vazquez, A.; Flammini, A.; Maritan, A.; Vespignani, A. Global Protein Function Prediction from Protein–protein Interaction Networks. *Nat. Biotechnol.* **2003,** *21*, 697–700.
63. Nabieva, E.; Jim, K.; Agarwal, A.; Chazelle, B.; Singh, M. Whole-proteome Prediction of Protein Function *via* Graph-theoretic Analysis of Interaction M1aps. *Bioinformatics* **2005,** *21*, 302–310.

64. Bader, G. D.; Hogue, C. W. V. An automatic method for finding molecular complexes in large protein interaction networks. *BMC Bioinform.* **2003,** *4,* 2–1-27.
65. King, A. D.; Przulj, N.; Jurisica, I. Protein Complex Prediction *via* Cost-based Clustering. *Bioinformatics* **2004,** *20,* 3013–3020.
66. Krogan, N.; Cagney, G.; Yu, H.; Zhong, G.; Guo, X.; Ignatchenko, A.; Li, J.; Pu, S.; Datta, N.; Tikuisis, A.; Punna, T.; Peregrín-Alvarez, J.; Shales, M.; Zhang, X.; Davey, M.; Robinson, M.; Paccanaro, A.; Bray, J.; Sheung, A. Beattie, B.; Richards, D.; Canadien, V.; Lalev, A.; Mena, F.; Wong, P.; Starostine, A.; Canete, M.; Vlasblom, J.; Wu, S.; Orsi, C.; Collins, S.; Chandran, S.; Haw, R.; Rilstone, J.; Gandi, K.; Thompson, N.; Musso, G.; St Onge, P.; Ghanny, S.; Lam, M.; Butland, G.; Altaf-Ul, A.; Kanaya, S.; Shilatifard, A.; O'Shea, E.; Weissman, J.; Ingles, C.; Hughes, T.; Parkinson, J.; Gerstein, M.; Wodak, S.; Emili, A.; Greenblatt, J. Global Landscape of Protein Complexes in the Yeast *Saccharomyces cerevisiae*. *Nature (London)* **2006,** *440,* 637–643.
67. Sharan, R.; Ideker, T.; Kelley, B. P.; Shamir, R.; Karp, R. M. Identification of Protein Complexes by Comparative Analysis of Yeast and Bacterial Protein Interaction Data. In *Proceedings of the Eighth Annual International Conference on Computational Molecular Biology (RECOMB'04)*; ACM: New York, NY, 2004; pp 282–289.
68. Berger, S. I.; Iyengar, R. Network Analyses in Systems Pharmacology. *Bioinformatics* **2009,** *25,* 2466–2472.
69. Yildirim, M. A.; Goh, K. I.; Cusick, M. E.; Barabási, A. L.; Vidal, M. Drug-target Network. *Nat. Biotechnol.* **2007,** *25,* 1119–1126.
70. Milenkovic, T.; Przulj, N. Uncovering Biological Network Function *via* Graphlet Degree Signatures. *Cancer Inform.* **2008,** *6,* 257–273.
71. Milenkovic, T.; Memisevic, V.; Ganesan, A. K.; Przulj, N. Systems-level Cancer Gene Identification from Protein Interaction Network Topology Applied to Melanogenesis-related Interaction Networks. *J. R. Soc. Interface* **2010,** *7,* 423–437.
72. Przulj, N. Biological Network Comparison Using Graphlet Degree Distribution. *Bioinformatics* **2007,** *23,* 177–183.
73. Ganesan, A. K.; Ho, H.; Bodemann, B.; Petersen, S.; Aruri, J.; Koshy, S.; Richardson, Z.; Le, L. Q.; Krasieva, T.; Roth, M. G.; Farmer, P.; White, M. A. Genome-wide siRNA-based Functional Genomics of Pigmentation Identifies Novel Genes and Pathways that Impact Melanogenesis in Human Cells. *PLoS Genet.* **2008,** *4,* e1000298–1-12.
74. Barabási, A. L.; Albert, R. Emergence of Scaling in Random Networks. *Science* **1999,** *286,* 509–512.
75. Bonchev, D. Complexity Analysis of Yeast Proteome Network. *Chem. Biodiv.* **2004,** *1,* 312–326.
76. Witten, T. M. (*M,R*)-systems, (*P,M,C*)-nets, Hierarchical Decay, and Biological Aging: Reminiscences of *Robert Rosen*. *Chem. Biodiv.* **2007,** *4,* 2332–2344.
77. Coma, M.; Naval Chamosa, J. Systems Biology Upends Drug Reprofiling. *Gen. Eng. Biotechnol. News* **2010,** *30* (18), 1–1.
78. Reyes, R.; Garrido, J.; Jaime, R. A.; Córdova, V.; Triana, J.; Villar, L.; Castro, J. C.; Fernúndez de Córdoba, P.; Urchueguía, J. F.; Navarro, E.; Montagud, A. Desarrollo de una plataforma computacional para el modelado metabólico de microorganismos. *Nereis* **2001,** *2011* (3), 25–31.
79. Kola, I.; Landis, J. Can the Pharmaceutical Industry Reduce Attrition Rates? *Nat. Rev. Drug Discov.* **2004,** *3,* 711–715.

80. Andersen, T.; Kragelund, C.; Reibel, J.; Nauntofte, B. Oral Adverse Drug Reactions to Cardiovascular Drugs. *Crit. Rev. Oral Biol. Med.* **2004,** *15*, 28–46.
81. Campillos, M.; Kuhn, M.; Gavin, A. C.; Jensen, L. J.; Bork, P. Drug Target Identification Using Side-effect Similarity. *Science* **2008,** *321*, 263–266.
82. Wishart, D. S.; Knox, C.; Guo, A. C.; Cheng, D.; Shrivastava, S.; Tzur, D.; Gautam, B.; Hassanali, M. DrugBank: A Knowledgebase for Drugs, Drug Actions and Drug Targets. *Nucleic Acids Res.* **2008,** *36*, D901–D906.
83. Kanehisa, M.; Goto, S.; Hattori, M.; Aoki-Kinoshita, K. F.; Itoh, M.; Kawashima, S.; Katayama, T.; Araki, M.; Hirakawa, M. From Genomics to Chemical Genomics: New Developments in KEGG. *Nucleic Acids Res.* **2006,** *34*, D354–D357.
84. Scheiber, J.; Chen, B.; Milik, M.; Sukuru, S. C. K.; Bender, A.; Mikhailov, D.; Whitebread, S.; Hamon, J.; Azzaoui, K.; Urban, L.; Glick, M.; Davies, J. W.; Jenkins, J. L. Gaining Insight Into Off-target Mediated Effects of Drug Candidates with a Comprehensive Systems Chemical Biology Analysis. *J. Chem. Inf. Model.* **2009,** *49*, 308–317.
85. Pache, R. A.; Zanzoni, A.; Naval, J.; Mas, J. M.; Aloy, P. Towards a Molecular Characterisation of Pathological Pathways. *FEBS Lett.* **2008,** *582*, 1259–1265.
86. Pujol, A.; Mosca, R.; Farrés, J.; Aloy, P. Unveiling the Role of Network and Systems Biology in Drug Discovery. *Trends Pharmacol. Sci.* **2010,** *31*, 115–123.
87. Che, D.; Liu, Q.; Rasheed, K.; Tao, X. Decision Tree and Ensemble Learning Algorithms with Their Applications in Bioinformatics. In *Software Tools and Algorithms for Biological Systems*; Arabnia, H. R., Tran, Q. N., Eds.; Advances in Experimental Medicine and Biology No. 696, Springer: New York, NY, 2011; pp 191–199.
88. Qu, Y.; Adam, B. L.; Yasui, Y.; Ward, M. D.; Cazares, L. H.; Schellhammer, P. F.; Feng, Z.; Semmes, O. J.; Wright, G. L., Jr. Boosted Decision Tree Analysis of Surface-enhanced Laser Desorption/ionization Mass Spectral Serum Profiles Discriminates Prostate Cancer from Noncancer Patients. *Clin. Chem.* **2002,** *48*, 1835–1843.
89. Sandstrom, P. A.; Mannie, M. D.; Buttke, T. M. Inhibition of Activation-induced Death in T Cell Hybridomas by Thiol Antioxidants: Oxidative Stress as Mediator of Apoptosis. *J. Leukocyte Biol.* **1994,** *55*, 221–226.
90. Sato, N.; Iwata, S.; Nakamura, K.; Hori, T.; Mori, K.; Yodoi, J. Thiol-mediated Redox Regulation of Apoptosis. *J. Immunol.* **1995,** *154*, 3194–3203.
91. Kohno, T.; Yamada, Y.; Hata, T.; Mori, H.; Yamamura, M.; Tomonaga, M.; Urata, Y.; Goto, S.; Kondo, T. Relation of Oxidative Stress and Glutathione Synthesis to CD95 (Fas/APO-1)-mediated Apoptosis of Adult T Cell Leukaemia Cells. *J. Immunol.* **1996,** *156*, 4722–4728.
92. Slater, A. F.; Nobel, C. S.; Maellaro, E.; Bustamante, J. Kimland, M.; Orrenius, S. Nitrone Spin Traps and a Nitroxide Inhibit a Common Pathway of Thymocyte Apoptosis. *Biochem. J.* **1995,** *306* (Pt. 3), 771–778.
93. Torres-Roca, J. F.; Lecoeur, H.; Amatore, C.; Gougeon, M. L. The Early Intracellular Production of a Reactive Oxygen Intermediate Mediates Apoptosis in Dexamethasone-treated Thymocytes. *Cell Death Differentiation* **1995,** *2*, 309–319.
94. Lane, D. P. A Death in the Life of p53. *Nature (London)* **1993,** *362*, 786–787.
95. Beaver, J. P.; Waring, P. A Decrease in Intracellular Glutathione Concentration Precedes the Onset of Apoptosis in Murine Thymocytes. *Eur. J. Cell Biol.* **1995,** *68*, 47–54.

96. Amatore, C.; Arbault, S.; Guille. M.; Lemaître, F. Electrochemical Monitoring of Single Cell Secretion: Vesicular Exocytosis and Oxidative Stress. *Chem. Rev.* **2008,** *108*, 2585–2621.
97. Hu, R.; Guille, M.; Arbault, S.; Lin, C. J.; Amatore, C. *In situ* electrochemical monitoring of Reactive Oxygen and Nitrogen Species Released by Single MG63 Osteosarcoma Cell Submitted to a Mechanical Stress. *Phys. Chem. Chem. Phys.* **2010,** *12*, 10048–10054.
98. Halliwell, B.; Gutteridge, J. M. C. Oxygen Toxicity, Oxygen Radicals, Transition Metals and Disease. *Biochem. J.* **1984,** *219*, 1–14.
99. Morris, J. G. Oxygen and the Obligate Anaerobe. *J. Appl. Bacteriol.* **1976,** *40*, 229–244.
100. Balentine, J. D. *Pathology of Oxygen Toxicity*; Academic: New York: NY, 1982.
101. Gilbert, D. L., Ed. *Oxygen and Living Processes: An Interdisciplinary Approach*; Springer: New York, NY, 1981.
102. Fridovich, I. Superoxide Dismutases. *Annu. Rev. Biochem.* **1975,** *44*, 147–159.
103. Fridovich, I. The biology of oxigen radicals. *Science* **1978,** *201*, 875–880.
104. Fridovich, I. Superoxide Radical: An Endogenous Toxicant. *Annu. Rev. Pharmacol. Toxicol.* **1983,** *23*, 239–257.
105. Itoh, S.; Kim, H. W.; Nakagawa, O.; Ozumi, K.; Lessner, S. M.; Aoki, H.; Akram, K.; McKinney, R. D.; Ushio-Fukai, M.; Fukai, T. Novel Role of Antioxidant-1 (Atox1) as a Copper-dependent Transcription Factor Involved in Cell Proliferation. *J. Biol. Chem.* **2008,** *283*, 9157–9167.
106. Keunen, E.; Remans, T.; Bohler, S.; Vangronsveld, J.; Cuypers, A. Metal-induced Oxidative Stress and Plant Mitochondria. *Int. J. Mol. Sci.* **2011,** *12*, 6894–6918.
107. Peñarrubia, L.; Andrés-Colás, N.; Moreno, J.; Puig, S. Regulation of Copper Transport in *Arabidopsis thaliana*: A Biochemical Oscillator? *J. Biol. Inorg. Chem.* **2010,** *15*, 29–36.
108. Andrés-Colás, N.; Perea-García, A.; Puig, S.; Peñarrubia, L. Deregulated Copper Transport Affects Arabidopsis Development Especially in the Absence of Environmental Cycles. *Plant Physiol.* **2010,** *153*, 170–184.
109. Perea-García, A.; Andrés-Colás, N.; Peñarrubia, L. Copper Homeostasis Influences the Circadian Clock in Arabidopsis. *Plant Signal. Behav.* **2010,** *5*, 1237–1240.
110. Borjigin, J.; Payne, A. S.; Deng, J.; Li, X.; Wang, M. M.; Ovodenko, B.; Gitlin, J. D.; Snyder, S. H. A Novel Pineal Night-specific ATPase Encoded by the Wilson Disease Gene. *J. Neurosci.* **1999,** *19*, 1018–1026.
111. Tobler, I.; Gaus, S. E.; Deboer, T.; Achermann, P.; Fischer, M.; Rulicke, T.: Moser, M.; Oesch, B.; McBride, P. A.; Manson, J. C. Altered Circadian Activity Rhythms and Sleep in Mice Devoid of Prion Protein. *Nature (London)* **1996,** *380*, 639–642.
112. Park, S. H.; Ozden, O.; Jiang, H.; Cha, Y. I.; Pennington, J. D.; Aykin-Burns, N.; Spitz, D. R.; Gius, D.; Kim, H. S. Sirt3, Mitochondrial ROS, Ageing, and Carcinogenesis. *Int. J. Mol. Sci.* **2011,** *12*, 6226–6239.
113. Przedborski, S.; Vila, M.; Jackson-Lewis, V. Neurodegeneration: What Is It and Where Are We? *J. Clin. Invest.* **2003,** *111*, 3–10.
114. Senderek, J.; Bergmann, C.; Ramaekers, V. T.; Nelis, E.; Bernert, G.; Makowski, A.; Züchner, S.; de Jonghe, P.; Rudnik-Schöneborn, S.; Zerres, K.; Schröder, J. M. Mutations in the *Ganglioside-induced Differentiation-associated Protein-1* (*GDAP1*) Gene in Intermediate Type Autosomal Recessive Charcot–Marie–Tooth Neuropathy. *Brain* **2003,** *126*, 642–649.

115. Takasaki, S. Mitochondrial Haplogroups Associated with Japanese Centenarians, Alzheimer's Patients, Parkinson's Patients, type 2 Diabetes Patients, Healthy Non-obese Young Males, and Obese Young Males. *J. Proteomics Bioinform.* **2011,** *4*, 106–112.
116. Koeppen, A. H.; Morral, J. A.; Davis, A. N.; Qian, J.; Petrocine, S. V.; Knutson, M. D.; Gibson, W. M.; Cusack M. J.; Li, D. The Dorsal Root Ganglion in Friedreich's Ataxia. *Acta Neuropathol.* **2009,** *118*, 763–776.
117. González-Cabo, P.; Bolinches-Amorós, A.; Cabello, J.; Ros, S.; Moreno, S.; Baylis, H. A.; Palau, F.; Vázquez-Manrique, R. P. Disruption of the ATP-binding cassette B7 (ABTM-1/ABCB7) Induces Oxidative Stress and Premature Cell Death in *Caenorhabditis elegans*. *J. Biol. Chem.* **2011,** *286*, 21304–21314.
118. Lotharius, J.; Brundin, P. Pathogenesis of Parkinson's Disease: Dopamine, Vesicles and α-synuclein. *Nat. Rev. Neurosci.* **2002,** *3*, 932–942.
119. Casella, L. Personal Communication.
120. Duvnjak, M.; Lerotic , I.; Barsic, N.; Tomasic, V.; Jukic, L. V.; Velagic, V. Pathogenesis and Management Issues for Non-alcoholic Fatty Liver Disease. *World J. Gastroenterol.* **2007,** *13*, 4539–4550.
121. He, Y.; Wu, J.; Dressman, D. C.; Iacobuzio-Donahue, C.; Markowitz, S. D.; Velculescu, V. E.; Diaz, L. A., Jr.; Kinzler, K. W.; Vogelstein, B.; Papadopoulos, N. Heteroplasmic Mitochondrial DNA Mutations in Normal and Tumour Cells. *Nature (London)* **2010,** *464*, 610–614.
122. Wallace, D. C. A Mitochondrial Paradigm of Metabolic and Degenerative Diseases, Aging, and Cancer: A Dawn for Evolutionary Medicine. *Annu. Rev. Genet.* **2005,** *39*, 359–407.
123. Petros, J. A.; Baumann, A. K.; Ruiz-Pesini, E.; Amin, M. B.; Sun, C. Q.; Hall, J.; Lim, S.; Issa, M. M.; Flanders, W. D.; Hosseini, S. H.; Marshall F. F.; Wallace, D. C. mtDNA Mutations Increase Tumorigenicity in Prostate Cancer. *Proc. Natl. Acad. Sci. U. S. A.* 2005, *102*, 719–724.
124. Koblyakov, V. A. Free Radicals and Inflammation. *Biochemistry (Moscow)* **2001,** *66*, 937–938.
125. Loeb, L. A.; Wallace D. C.; Martin, G. M. The Mitochondrial Theory of Aging and Its Relationship to Reactive Oxygen Species Damage and Somatic mtDNA Mutations. *Proc. Natl. Acad. Sci. USA* **2005,** *102*, 18769–18770.
126. Schumacker, P. T. Reactive Oxygen Species in Cancer Cells: Live by the Sword, Die by the Sword. *Cancer Cell* **2006,** *10*, 175–176.
127. Kannan, K.; Jain, S. K. Oxidative Stress and Apoptosis. *Pathophysiology* **2000,** *7*, 153–163.
128. Jezek, P.; Hlavata, L. Mitochondria in Homeostasis of Reactive Oxygen Species in Cell, Tissues, and Organism. *Int. J. Biochem. Cell Biol.* **2005,** *37*, 2478–2503.
129. Green, D. R.; Reed, J. C. Mitochondria and Apoptosis. *Science* **1998,** *281*, 1309–1312.
130. Li, J.; Huang, C. Y.; Zheng, R. L.; Cui, K. R.; Li, J. F. Hydrogen Peroxide Induces Apoptosis in Human Hepatoma Cells and Alters Cell Redox Status. *Cell. Biol. Int.* **2000,** *24*, 9–23.
131. Park, C.; So, H. S.; Shin, C. H.; Baek, S. H.; Moon, B. S.; Shin, S. H.; Lee, H. S.; Lee, D. W.; Park, R. Quercetin Protects the Hydrogen Peroxide-induced Apoptosis *via* Inhibition of Mitochondrial Dysfunction in H9c2 Cardiomyoblast Cells. *Biochem. Pharmacol.* **2003,** *66*, 1287–1295.

132. Adams, J. M.; Cory, S. The BCl-2 Protein Family: Arbiters of Cell Survival. *Science* **1998,** *281*, 1322–1326.
133. Choi, W. Y.; Choi, B. T.; Lee, W. H.; Choi, Y. H. Sulforaphane Generates Reactive Oxygen Species Leading to Mitochondrial Perturbation for Apoptosis in Human Leukemia U937 Cells. *Biomed. Pharmacother.* **2008,** *62*, 637–644.
134. Zhang, R.; Chae, S.; Kang, K. A.; Piao, M. J.; Ko, D. O.; Wang, Z. H.; Park, D. B.; Park, J. W.; You, H. J.; Hyun, J. W. Protective Effect of Butin Against Hydrogen Peroxide-induced Apoptosis by Scavenging Reactive Oxygen Species and Activating Antioxidant Enzymes. *Mol. Cell. Biochem.* **2008,** *318*, 33–42.
135. Zhang, R.; Lee, I. K.; Piao, M. J.; Kim, K. C.; Kim, A. D.; Kim, H. S.; Chae, S.; Kim, H. S.; Hyun, J. W. Butin (7,3',4'-trihydroxydihydroflavone) Reduces Oxidative Stress-induced Cell Death *via* Inhibition of the Mitochondria-dependent Apoptotic Pathway. *Int. J. Mol. Sci.* **2011,** *12*, 3871–3887.
136. Doll, A.; Abal, M.; Rigau, M.; Monge, M.; Gonzalez, M.; Demajo, S.; Colás, E.; Llauradó, M.; Alazzouzi, H.; Planagumá, J.; Lohmann, M. A.; Garcia, J.; Castellvi, S.; Ramon y Cajal, J.; Gil-Moreno, A.; Xercavins, J.; Alameda F.; Reventós, J. Novel Molecular Profiles of Endometrial Cancer—New Light through Old Windows. *J. Steroid Biochem. Mol. Biol.* **2008,** *108*, 221–229.
137. Rigau Resina, M. Un dilema diagnòstic: Buscant els biomarcadors del càncer de pròstata. *Mètode* **2011,** *2011* (69), 16–21.
138. Reynolds, M. A.; Kastury, K.; Groskopf, J.; Schalken, J. A.; Rittenhouse, H. Molecular Markers for Prostate Cancer. *Cancer Lett.* **2007,** *249*, 5–13.
139. Reynolds, M. A. Molecular Alterations in Prostate Cancer. *Cancer Lett.* **2008,** *270*, 13–24.
140. Day, J. R.; Jost, M.; Reynolds, M. A.; Groskopf J.; Rittenhouse, H. PCA3: From Basic Molecular Science to the Clinical Lab. *Cancer Lett.* **2011,** *301*, 1–6.
141. Van Kasteren, S. I.; Kramer, H. B.; Jensen, H. H.; Campbell, S. J.; Kirkpatrick, J.; Oldham, N. J.; Anthony D. C.; Davis, B. G. Expanding the Diversity of Chemical Protein Modification Allows Post-translational Mimicry. *Nature (London)* **2007,** *446*, 1105–1109.
142. Kumagai, J.; Fujimura, T.; Takahashi, S.; Urano, T.; Ogushi, T.; Horie-Inoue, K.; Ouchi, Y.; Kitamura, T.; Muramatsu, M.; Blumberg, B.; Inoue, S. Cytochrome P450 2B6 Is a Growth-Inhibitory and Prognostic Factor for Prostate Cancer. *Prostate* **2007,** *67*, 1029–1037.
143. Andriole, G.; Bostwick, D.; Brawley, O.; Gomella, L.; Marberger, M.; Tindall, D.; Breed, S.; Somerville, M.; Rittmaster, R.; REDUCE Study Group. Chemoprevention of Prostate Cancer in Men at High Risk: Rationale and Design of the Reduction by Dutasteride of Prostate Cancer Events (REDUCE) Trial. *J. Urol.* **2004,** *172*, 1314–1317.
144. Andriole, G. L.; Bostwick, D. G.; Brawley, O. W.; Gomella, L. G.; Marberger, M.; Montorsi, F.; Pettaway, C. A.; Tammela, T. L.; Teloken, C.; Tindall, D. J.; Somerville, M. C.; Wilson, T. H.; Fowler, I. L.; Rittmaster, R. S.; REDUCE Study Group. Effect of Dutasteride on the Risk of Prostate Cancer. *N. Engl. J. Med.* **2010,** *362*, 1192–1202.
145. Andreoiu, M.; Cheng, L. Multifocal Prostate Cancer: Biologic, Prognostic, and Therapeutic Implications. *Hum. Pathol.* **2010,** *41*, 781–793.
146. Ding, Z.; Wu, C. J.; Chu, G. C.; Xiao, Y.; Ho, D.; Zhang, J.; Perry, S. R.; Labrot, E. S.; Wu, X.; Lis, R.; Hoshida, Y.; Hiller, D.; Hu, B.; Jiang, S.; Zheng, H.; Stegh, A. H.; Scott, K. L.; Signoretti, S.; Bardeesy, N.; Wang, Y. A.; Hill, D. E.; Golub, T. R.; Stampfer, M. J.;

Wong, W. H.; Loda, M.; Mucci, L.; Chin, L.; DePinho, R. A. SMAD4-dependent Barrier Constrains Prostate Cancer Growth and Metastatic Progression. *Nature (London)* **2011,** *470*, 269–273.
147. Morgan, R.; Boxall, A.; Bhatt, A.; Bailey, M.; Hindley, R.; Langley, S.; Whitaker, H. C.; Neal, D. E.; Ismail, M.; Whitaker, H.; Annels, N.; Michael A.; Pandha, H. Engrailed-2 (EN2): A Tumor Specific Urinary Biomarker for the Early Diagnosis of Prostate Cancer. *Clin. Cancer Res.* **2011,** *17*, 1090–1098.
148. Crick, F. H. C. On Protein Synthesis. Symposia of the Society for Experimental Biology: No. 12, The Biological Replication of Macromolecules; Cambridge University: Cambridge, UK, 1958; pp 138–163.
149. Crick, F. Central Dogma of Molecular Biology. *Nature (London)* **1970,** *227*, 561–563.
150. Crick, F. DNA today. *Perspect. Biol. Med.* **1982,** *25*, 512–517.
151. Lopian, L.; Nussbaum-Shochat, A.; O'Day-Kerstein, K.; Wright, A.; Amster-Choder, O. The BglF Sensor Recruts the BglG Transcription Regulator to the Membrane and Releases It on Stimulation. *Proc. Natl. Acad. Sci. U. S. A.* **2003,** *100*, 7099–7104.
152. Raveh, H.; Lopian, L.; Nussbaum-Shochat, A.; Wright, A.; Amster-Choder, O. Modulation of Transcription Antitermination in the *bgl* Operon of *Escherichia coli* by the PTS. *Proc. Natl. Acad. Sci. U. S. A.* **2009,** *106*, 13523–13528.
153. Lopian, L.; Elisha, Y.; Nussbaum-Shochat, A.; Amster-Choder, O. Spatial and Temporal Organization of the *E. coli* PTS Components. *EMBO J.* **2010,** *29*, 3630–3645.
154. Nevo-Dinur, K.; Nussbaum-Shochat, A.; Ben-Yehuda, S.; Amster-Choder, O. Translation-Independent Localization of mRNA in *E. coli. Science* **2011,** *331*, 1081–1084.
155. Van der Lelij, P.; Chrzanowska, K. H.; Godthelp, B. C.; Rooimans, M. A.; Oostra, A. B.; Stumm, M.; Zdzienicka, M. Z.; Joenje, H.; de Winter, J. P. Warsaw Breakage Syndrome, a Cohesinopathy Associated with Mutations in the XPD Helicase Family Member DDX11/ChlR1, *Am. J. Hum. Genet.* **2010,** *86*, 262–266.
156. Farràs, R.; Baldin, V.; Gallach, S.; Acquaviva, C.; Bossis, G.; Jariel-Encontre, I.; Piechaczyk, M. JunB Breakdown in Mid-/late G_2 Is Required for Down-regulation of Cyclin A2 Levels and Proper Mitosis. *Mol. Cell. Biol.* **2008,** *28*, 4173–4187.
157. Moltzahn, F.; Olshen, A. B.; Baehner, L.; Peek, A.; Fong, L.; Stöppler, H.; Simko, J.; Hilton, J. F.; Carroll, P.; Blelloch, R. Microfluidic-based Multiplex qRT-PCR Identifies Diagnostic and Prognostic MicroRNA Signatures in the Sera of Prostate Cancer Patients. *Cancer Res.* **2011,** *71*, 550–560.
158. Catalona, W. J.; Partin, A. W.; Sanda, M. G.; Wei, J. T.; Klee, G. G.; Bangma, C. H.; Slawin, K. M.; Marks, L. S.; Loeb, S.; Broyles, D. L.; Shin, S. S.; Cruz, A. B.; Chan, D. W.; Sokoll, L. J.; Roberts, W. L.; van Schaik, R. H. N.; Mizrahi, I. A. A Multicenter Study of [–2]pro-prostate Specific Antigen Combined with Prostate Specific Antigen and Free Prostate Specific Antigen for Prostate Cancer Detection in the 2.0 to 10.0 ng/mL Prostate Specific Antigen Range. *J. Urol.* **2011,** *185*, 1650–1655.
159. Xie, C.; Kim, H. J.; Haw, J. G.; Kalbasi, A.; Gardner, B. K.; Li, G.; Rao, J.; Chia, D.; Liong, M.; Punzalan, R. R.; Marks, L. S.; Pantuck, A. J.; de la Taille, A.; Wang, G.; Mukouyama H.; Zeng, G. A Novel Multiplex Assay Combining Autoantibodies Plus PSA Has Potential Implications for Classification of Prostate Cancer from Non-malignant Cases. *J. Transl. Med.* **2011,** *9*, 43-1-11.
160. Tavoosidana, G.; Ronquist, G.; Darmanis, S.; Yan, J.; Carlsson, L.; Wu, D.; Conze, T.; Ek, P.; Semjonow, A.; Eltze, E.; Larsson, A.; Landegren, U. D.; Kamali-Moghaddam,

M. Multiple Recognition Assay Reveals Prostasomes as Promising Plasma Biomarkers for Prostate Cancer. *Proc. Natl. Acad. Sci. U. S. A.* **2011,** *108*, 8809–8814.

161. Volpe, G.; Molina-Terriza G.; Quidant, R. Deterministic Subwavelength Control of Light Confinement in Nanostructures. *Phys. Rev. Lett.* **2010,** *105*, 216802–1-4.
162. Volpe, G.; Volpe G.; Quidant, R. Fractal plasmonics: Subdiffraction Focusing and Broadband Spectral Response by a Sierpinski Nanocarpet. *Opt. Express* **2011,** *19*, 3612–3618.
163. Curto, A. G.; Volpe, G.; Taminiau, T. H.; Kreuzer, M. P.; Quidant, R.; van Hulst, N. F. Unidirectional Emission of a Quantum Dot Coupled to a Nanoantenna. *Science* **2010,** *329*, 930–933.
164. Juan, M. L.; Righini, M.; Quidant, R. Plasmon Nano-optical Tweezers. *Nat. Photon.* **2011,** *5*, 349–356.
165. Acimovic, S. S.; Kreuzer, M. P.; González, M. U.; Quidant, R. Plasmon Near-field Coupling in Metal Dimers as a Step Toward Single-molecule Sensing. *ACS Nano* **2009,** *3*, 1231–1237.
166. Vazquez-Carrillo, L. I.; Quintas-Granados, L. I.; Arroyo, R.; Hernández, G. M.; González-Robles, A.; Carvajal-Gamez B. I.; Álvarez-Sánchez, M. E. The Effect of Zn^{2+} on Prostatic Cell Cytotoxicity Caused by *Trichomonas vaginalis*. *J. Integrated Omics* **2011,** *1*, 198–210.
167. Nilsson, T.; Mann, M.; Aebersold, R.; Yates, J. R., III; Bairoch, A.; Bergeron, J. J. M. Mass Spectrometry in High-throughput Proteomics: Ready for the Big Time. *Nat. Methods* **2010,** *7*, 681–685.
168. Ryals, J. Personal Communication.

CHAPTER 5

Intracellular Proteolysis and Neoplasia: Ubiquitin-Proteasome System, Cell Cycle, and Cancer

FRANCISCO TORRENS[1*] and GLORIA CASTELLANO[2]

[1]*Institut Universitari de Ciència Molecular, Universitat de València, Edifici d'Instituts de Paterna, P. O. Box 22085, E-46071 València, Spain*

[2]*Departamento de Ciencias Experimentales y Matemáticas, Facultad de Veterinaria y Ciencias Experimentales, Universidad Católica de Valencia San Vicente Mártir, Guillem de Castro-94, E-46001 València, Spain*

Corresponding author. E-mail: torrens@uv.es

ABSTRACT

Affinity may include the ease of traverse via chreodes to the effector. Lag may be because of the time or concentration needed for a drug to displace transmitter molecules from the chreodes. Molecules unfit for the system may be excluded from the effector by not fitting the chreode patterns. Persistence may be because of the lingering of molecule in the chreodes after washout. Enzyme catalytic products leave the active site at a faster rate via chreodes, minimizing the retarding effect of diffusion-controlled rate limitation. Protein kinase-C-dependent ubiquitylation promotes non-proteolytic inactivation of suppressor Fbw7α which, although expressed in mature *Xenopus* eggs arrested in metaphase II, is nonfunctional explaining why cyclin-E can be stockpiled in mitotic-like phase.

5.1 INTRODUCTION

Activating protein-1 (AP-1) is a set of dimeric transcription factors, which bind to AP1/TRE–CRE deoxyribonucleic acid (DNA) motifs; best studied parts are Fos (c-Fos, FosB, fra-1 and fra-2)/jun (c-jun, jun-B and jun-D) families, which bind to DNA owing to DNA-binding domain (DBD) and dimerize via adjacent leucine (Leu) zipper (LZ) one. Its dimers act posi/negatively on transcription depending on composition, target gene, cell context, and environmental signals.[1–7] Its TRE/CRE motifs are found in genes. It regulates cell processes, for example, proliferation, differentiation, apoptosis, and responses to stress and is essential for physiological organism functions; it is implicated in pathologies, for example, tumorigenesis, via effects on cell fate. Although its proteins can be oncogenic, major contribution is effector of upstream oncogenic events; for example, Fos/jun members expression is altered by mutated rat sarcoma (Ras) protein, which is instrumental for cell transformation. Deregulated Fos/jun protein expression is associated with neoplasias.[8] Although AP-1 is tumorigenesis promoter, components display oncosuppressor activity, for example, under c-Fos/jun-B.

In AP-1cell-division control, Fos/jun proteins regulate expression of cell-cycle controllers, for example, cyclins/cyclin-dependent kinase inhibitors (cki), and transcriptional control of the latter genes depends on changes in Fos/jun proteins levels. Depending on expression condition and extracellular cues, Fos/jun proteins manifest opposite functions: for example, when quiescent cells are stimulated by mitogens, c-Fos/jun exert positive effects on cell division via cyclin-D1 gene induction in gap $(G)_1$;[9–11] however, they act as apoptosis effectors otherwise. In cell-cycle control, jun-B exerts dual function: though it is cell-division inhibitor,[12] senescence inducer and tumor suppressor at least in myeloid lineage,[13–15] it results cell-division promoter. Its expression, which is low in quiescent cells, is transiently induced by mitogenic stimuli during $G_{0\rightarrow 1}$ transition before returning to an intermediate level,[16–18] with both being instrumental for progression towards S-phase[19] in which rapid advance depends on jun-B, which expression rises at $G_1 \rightarrow S$ transition, to regulate positively cyclin-A2 gene transcription.[20] However, contrasting with proliferation-stimulation functions, sustained jun-B accumulation throughout G_1, in response to antiproliferative signals, leads to cell-cycle arrest via $p16^{INK4\alpha}$cki gene induction and downregulation of that of cyclin-D1. Cells are blocked before entering S-phase, which is followed by senescence. Levels of jun-B result low in mitotic and early G_1; in contrast c-jun levels remain constant during period with, however, an increase in transcriptional activity for early G_1. As jun-B represses and c-jun activates cyclin-D1 promoter, low former levels in mitotic and early G_1 lead progression via G_1 toward S-phase, which increases cyclin-D1 transcription.

In earlier publications it was informed trichomonacidal drugs,[21–25] the modeling of complex multicellular systems and tumor–immune cells competition,[26] information theoretic entropy for molecular classification of oxadiazolamines as potential therapeutic agents,[27] molecular classification of 5-amino-2-aroylquinolines and 4-aroyl-6,7,8-trimethoxyquinolines as highly potent tubulin polymerization inhibitors,[28] polyphenolic phytochemicals in cancer prevention, therapy, bioavailability versus bioefficacy,[29] molecular classification of antitubulin agents with indole ring binding at colchicine-binding site,[30] molecular classification of 2-phenylindole-3-carbaldehydes as potential antimitotic agents in human breast cancer cells,[31] cancer, its hypotheses,[32] precision personalized medicine (PPM) from theory to practice and cancer.[33] It was reported how human immunodeficiency virus/acquired immunodeficiency syndrome destroy immune defences, hypothesis,[34] 2014 emergence, spread, uncontrolled Ebola outbreak,[35,36] Ebola virus disease, questions, ideas, hypotheses, models,[37] clinical translational research, cancer, diabetes, cardiovascular disease,[38] primary health, smoking habits, immunotherapy in cancer,[39] epigenomics, epithelial plasticity, clinical genetics, rare diseases,[40] predictive *in silico* global metabolism analysis and disease mechanism.[41] The present report reviews intracellular proteolysis, neoplasia, ubiquitin-proteasome system (UPS), cell cycle and cancer. Protein destruction determines cellular functions in central nervous system (CNS): dysfunctions cause pathologies, for example, Alzheimer (AD), Parkinson and Huntington's diseases. Topics are related to proteolytic role in normal/degeneration neuronal physiology, for example, protein autophagic and UPS pathways destructions in AD, etc.; intracellular proteolysis is a complex process, which regulates proteins levels to keep cell homeostasis/physiology. Alterations in molecular mechanisms that control process cause multiple pathologies, for example, cancer; new drugs were developed orientated directly towards activity control of molecular machinery involved in intracellular proteolysis, increasing impact to human health. Present report reviews state of the art on research in proteolytic enzymes, inhibitors and substrates, from viewpoints of basic, clinical and pharmaceutical industry investigations: studies will help to define therapeutic strategies in oncology and orientate research in proteolysis, cancer, etc. Main objective is to provide information on growth of educational, scientific and industry investigations among chemists/pharmacists and chemic/pharmaceutical engineers, and medium for mutual communication between international academia/industry. Reporting methods/applications in medical informatics and including latest coverage of chemical databases, aim is to present research in intracellular proteolysis.

5.2 COMPUTATIONAL METHOD

Diffusion via bulk H_2O is compared versus across effector landscape; proximal relation exists among metabolic events; catalysis rate is biased directed diffusion; microviscosity is important. Smoluchowski introduced[42,43] and Debye enriched[44] concept of diffusion-controlled reaction; early view invoked three-dimensional (3D) random walk of ligand via bulk H_2O to effector, which would be too slow to accomplish effector complex tasks/communications. Welch, Adam and Eigen agreed that two-dimensional (2D) surface diffusion to active site enhances diffusion-controlled reaction rate. Hasinoff concluded that fast enzyme reactions are limited by ligand diffusion to effector. Berg and Purcell described model of ligand skipping over cell surface with encountering-effector chance; cell membrane surface supports 2D diffusion but contains no variety to influence direction preference.[45,46] Rhodes, Sarmiento and Herbette invoked protein as general target followed by lateral diffusion to effector across/via membrane. Blum described possible ligand guidance to effector via folds created by Bonnet transformation. Chen and Zhou explained model in which effector-outside protein plays major role as *promoter*, giving rise to fast ligand flow to effector. Sweet-tasting molecule administration produced response; washout created persistence. Birch postulated queue formation on receptor protein surface; exiting from queue after washout provided persistence until queue emptied; concept invoked: (1) washout-resistant residence on protein surface, (2) favored location, and (3) directional influence.

H_2O around hydrophobic solute presents greater attraction for itself than for solute; there is greater bent for H_2O to H-bond to nearby H_2O than to hydrate solute, which molecules may/not bond to each other depending on electrostriction: hydrophobic solutes tend to occupy cavities while hydrophilic ones show avidity for H_2O and hydrate. H_2O on protein surface is *ordered* because of hydrophobic/electrostriction influences generated by side chains; ordering may extend out from protein surface, several H_2O layers; H^+ exchange between two H_2O occurs at 10^{-14}s: landscape is hydrodynamic; ordered H_2O on surface may facilitate ligand diffusion to active site, enhancing catalysis rates or receptor binding. Kier et al. built ligand diffusion models over protein surfaces.[47] Ligands target protein by diffusion from bulk H_2O; they establish residence on protein surface via long-range forces; they follow chreodes (probability pathways) created by side-chain hydrophobic effect; they approach effectors at rates faster than random motion; queues form in chreodes; reaction products may leave effector via chreodes; there exists side-chain patterns on each protein that form code creating chreodes specific for existing effectors. Waddington analyzed chreode on epigenetic landscape.[48] Measured drug affinity for receptor may include presence in nearby chreodes; molecules not system part may be excluded from receptor contact because they do not fit into chreodes. *Lag* may be because of time/concentration needed for drug molecule to displace normal transmitter in its chreodes; persistence may be because of molecule lingering in chreodes time after washout. Enzyme reactions may be fast because substrates diffuse to receptor and products diffuse away from active site rapidly, via chreodes.

More can be done: to develop information about surface residues on proteins, attempt to decipher codes creating chreodes, examine effector-binding data in broader target light and consider drug design and quantitative structure–activity relationship (QSAR) as possible consequence of broader target presented by trajectories via chreodes. Models of complex ligand/side chain/H_2O invoke ingredients: (1) ligand diffusion is via H_2O cavities, (2) which pattern is defined by side chains, (3) which produce hydrophobic effect. Affinity includes traverse ease via chreodes to effector; lag may be because of time/concentration needed for drug to displace transmitter molecules from chreodes; molecules unfit for system are excluded from effector not fitting chreode patterns; persistence results because of molecule lingering in chreodes after washout; enzyme catalytic products leave active site at faster rate via chreodes, minimizing delay of diffusion-controlled rate limitation.[49–52] Several H_2O molecules/nanostructures were reported.[53] Carcinogenesis shows identifying-biomarker opportunities (cf. Fig. 5.1).[54] Santilli devised cancer elimination disrupting intermolecular magnecular bond.[55] Special issue of *Accounts of Chemical Research* is devoted to theranostic nanomedicine.[56]

5.3 RESULTS: CHROMOSOME INSTABILITY AND CANCER

5.3.1 GENOMIC INSTABILITY AND CANCER

Tumorigenesis is imbalance between cell-cycle control mechanisms and mutation rates within genes.[57] Genomic instability is classified into microsatellite instability (MIN), associated with mutator phenotype, and chromosome instability (CIN) recognized by gross chromosomal abnormalities. Three intracellular mechanisms are involved in DNA damage repair, which leads to mutator phenotype: nucleotide (NER)/base excision repairs (BER) and mismatch repair (MMR). Pathway of CIN is associated with mutation accumulation in tumor suppression genes and oncogenes. Defects in DNA MMR/CIN pathways are responsible for congenital cancer predisposition syndromes, for example, hereditary non-polyposis colorectal carcinoma (HNPCC), Bloom syndrome (BS), ataxia telangiectasia (A-T) and Fanconi anaemia (FA). While there are genetic contributors to CIN/MIN, there exist epigenetic factors that emerged damaging to cell-cycle control: hypermethylation of tumor suppressor and DNA MMR gene promoter regions results epigenetic mechanism of gene silencing, which contributes to tumorigenesis. Telomere shortening increases mouse genetic instability and tumor formation, underscoring telomere length and telomerase activity in maintaining genomic integrity. Murine models provided insights for discovering pathways, in progression to cancer, and to elucidate cross talk among different pathways. Charames and Bapat reviewed genomic instability mechanisms and relevance to cancer.[58]

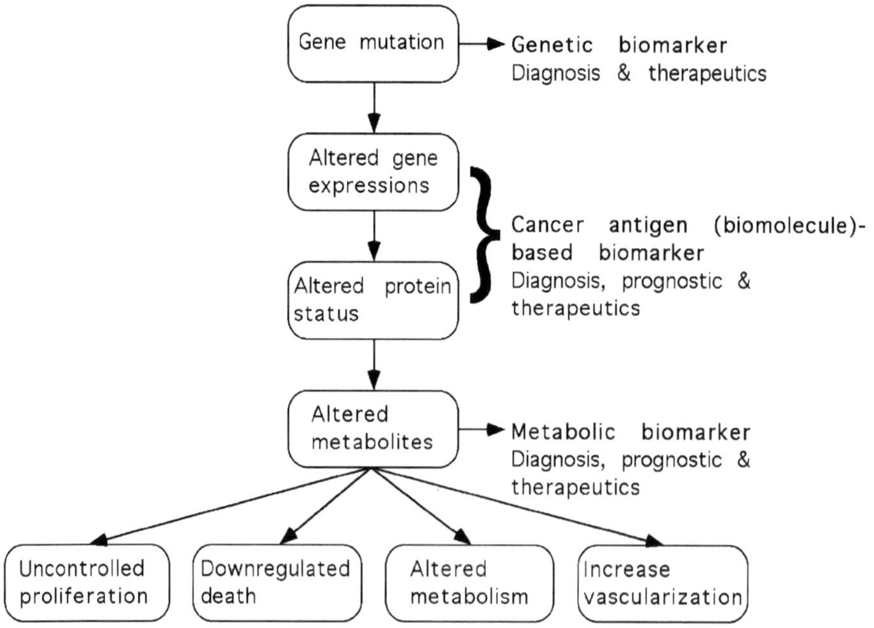

FIGURE 5.1 The process of carcinogenesis, showing opportunities of identifying biomarker.

5.3.2 DNA DAMAGE RESPONSE AS CANDIDATE ANTICANCER IN EARLY TUMORIGENESIS

During cancer evolution, incipient tumor experiences *oncogenic stress*, which evokes counter-response to eliminate hazardous cells. However, stress nature and inducible anticancer barrier remain elusive, which elicits cell-growth arrest/death. Two independent pathways lead to apoptosis in thymocytes: not/tumor protein 53 (p53) mediated (cf. Fig. 5.2).[59] Kalimuthu and Se-Kwon reviewed cell survival and apoptosis signalling as therapeutic target for cancer.[60] Li et al. revised the yin-yang of DNA damage response (DDR).[61]

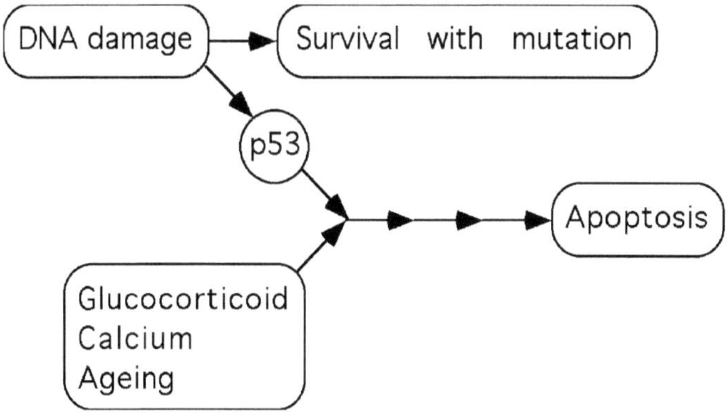

FIGURE 5.2 Two independent pathways lead to apoptosis in thymocytes: not/p53 mediated.

Bartkova et al. showed that in clinical specimens from different stages of human urinary bladder, breast, lung and colon tumors, early precursor lesions commonly express markers of activated DDR (cf. Fig. 5.3), for example, phosphorylated kinases A-T mutated (ATM) and checkpoint homolog-2 (Chk2), and phosphorylated hystone H2AX and p53.[62] Similar checkpoints were induced in cultured cells upon different-oncogene expression, which deregulate DNA replication. Together with genetic analyses, for example, genome-wide assessment of allelic imbalances, they showed that early in tumorigenesis, human cells activate A-T/Rad3-related (ATR)-ATM-regulated DDR protein/protein interaction network (PIN), which delays/prevents cancer. Mutations compromising checkpoint, for example, defects in ATM-Chk2-p53 pathway, allow cell proliferation, survival, increased genomic instability and tumor progression.

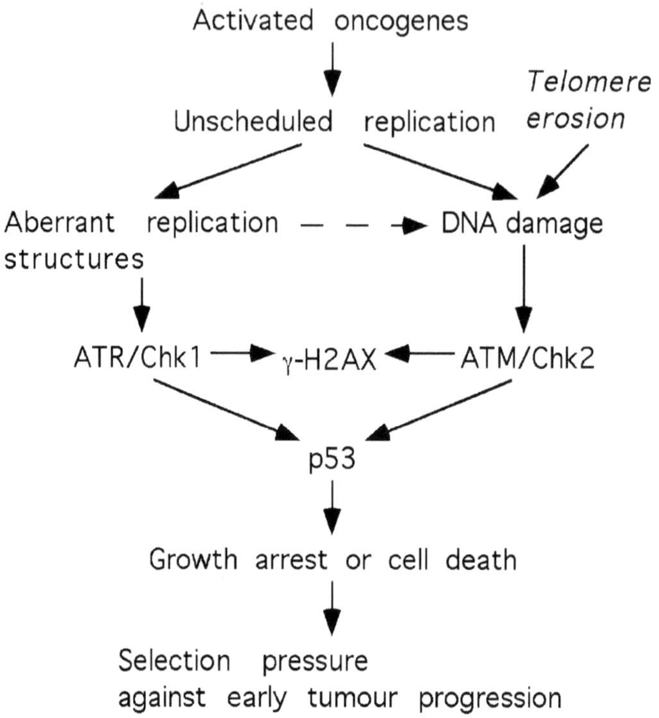

FIGURE 5.3 Model of the deoxyribonucleic acid (DNA) damage response (DDR) in tumorigenesis.

5.3.3 ACTIVATION OF DNA DAMAGE CHECKPOINT: GENOMIC INSTABILITY IN PRECANCER

Genes of DNA damage checkpoint, for example, *p53*, are frequently mutated in human cancer but selective pressure for their inactivation remains elusive. Gorgoulis et al. analyzed human lung hyperplasias, which retained wild-type (WT) *p53* genes and presented no gross CHI, and found DDR signs, for example, H2AX/Chk2 phosphorylation, p53 accumulation, p53-binding protein-1 (53BP1) focal staining and apoptosis.[63] Progression to carcinoma was associated with *p53/53BP1* inactivation and decreased apoptosis. A DDR was observed in dysplastic nevi and human skin xenografts, in which hyperplasia was induced by growth factor overexpression. Lung/experimentally induced skin hyperplasias showed allelic imbalance, at loci that are prone to DNA double-strand break (DSB) formation, when DNA replication is compromised. They proposed that from its earliest stages, cancer development is associated with DNA replication stress, which leads to DNA DSBs, genomic instability and selective pressure for *p53* mutations.

5.3.4 IN VIVO FORMATION/REPAIR OF DNA DOUBLE-STRAND BREAKS AFTER TOMOGRAPHY

Ionizing radiation (IR) leads to deleterious effects in humans importantly to cancer induction. The DSBs are major genetic lesions introduced by IR that can initiate carcinogenesis. Löbrich et al. enumerated γ-H2AX foci as DSB measure in lymphocytes, from individuals undergoing thorax/abdomen computed tomography (CT) examination.[64] Number of CT-induced DSBs depends linearly on dose–length product, radiodiagnostic (RD) unit that is proportional to delivered local dose and exposed-body length. Lymphocyte analysis, sampled up to one-day postirradiation (PI), provided kinetics for in vivo loss of γ-H2AX foci that correlated with DSB repair. In contrast to results obtained in vitro, normal individuals repair DSBs to background levels. Patient, who previously showed adverse effects after radiotherapy (RT), displayed γ-H2AX focus levels at various sampling times PI that were several times higher than normal individuals; γ-H2AX/pulsed-field gel electrophoresis (PFGE) analyses of fibroblasts, obtained from patient, confirmed DSB repair defect. Fibroblasts showed significant in vitro radioselectivity (RS). Data showed that in vivo DSB induction/repair can be assessed in individuals exposed to low radiation doses (RDs), adding further dimension to DSB repair studies, and providing opportunity to identify repair-compromised individuals after diagnostic irradiation.

5.3.5 ATM/ATR SUBSTRATE ANALYSES REVEAL PROTEIN NETS RESPONSIVE TO DNA DAMAGE

Cellular responses to DNA damage are mediated by kinases, for example, ATM/ATR. Outlines of signal transduction portion of pathway are known but little is recognized about DDR physiological scope. Matsuoka et al. performed large-scale proteomic analysis of proteins phosphorylated in response to DNA damage, on consensus sites recognized by ATM/ATR, and identified >900 regulated phosphorylation sites encompassing >700 proteins.[65] Functional analysis of data subset indicated that list is highly enriched for DDR-involved proteins. Protein set is highly interconnected, and they identified PINs not previously DDR-linked. Database paints broader landscape for DDR than was appreciated and opens research into DDRs in mammals.

5.3.6 PROTEOMICS REVEALS APOPTOSIS-ASSOCIATED PROTEINS IN DEFICIENT LUNG CELLS

Formation of reactive oxygen species (ROS) plays role in toxicities induced by most potent dioxin 2,3,7,8-tetrachlorodibenzo-*p*-dioxin (TCDD), in mammalian cells, since it promotes cell proliferation, growth arrest and apoptosis. Lin group reported if TCDD induce oxidative stress (OS)/DNA damage in human ERα(+)/MCF-7 and ERα(−)/MDA-MB-231 breast cancer cells, and if this be accompanied by DNA repair initiation.[66] Results indicated that MCF-7/MDA-MB-231 cell viability was concentration/time-dependently reduced by TCDD. They observed increases in ROS formation and decays in intracellular glutathione (GSH) in both after TCDD treatment. Cell death extent was greater in MCF-7 than in MDA-MB-231, whereas ROS formation and GSH depletion was greater in MDA-MB-231 than in MCF-7. They observed that at non-cytotoxic concentration, TCDD decreased intracellular nicotinamide adenine dinucleotide (NAD) (phosphate) hydride [NAD(P)H]/NAD^+ in MCF-7/MDA-MB-231. Decreases were blocked by three types of poly[adenosine diphosphate (ADP)-ribose] polymerase-1 (PARP-1) inhibitors. Catalytic activation of PARP-1 in TCDD-treated cells was confirmed by detection of polymers of ADP-ribose-modified PARP-1 using Western blotting. They showed increases in DNA strand breaks in MCF-7/MDA-MB-231 cells exposed to TCDD, as measured by single-cell gel electrophoresis (Comet) assay, which confirms that TCDD induces decays in intracellular NAD(P)H/NAD^+ via PARP-1 activation via formation of DNA strand breaks. They showed that OS/DNA damage was greater in MDA-MB-231 than in MCF-7, with strong correlation to oestrogen receptor (OR) status. The ROS is determinant in mediating induction of oxidative DNA damage/repair in human breast cancer cells exposed to TCDD, and that TCDD-induced OS and DNA damage contribute to TCDD-induced carcinogenesis. The TCDD binds to aryl hydrocarbon receptor (AhR), ligand-activated transcription factor, which subsequently alters gene expression. Epidemiological studies suggested that TCDD exposure enhanced pulmonary disease risk. They adopted comparative proteomic approach, using two-dimensional gel electrophoresis (2DE), to identify protein differential expression between TCDD-treated control and AhR-deficient lung adenocarcinoma cells.[67] They elucidated 15 proteins from 17 differential protein spots that four of which were downregulated and 11, upregulated. Western blotting confirmed differential expression for five proteins: GRB10, YWHAG, HSP27, LGALS1 and ACTB. Analysis of PIN illustrated 12 proteins linked to network containing cysteinyl aspartate-specific protease (caspase)-3, p53, MDM2 and Akt/PKB, which are associated with apoptosis. The GRB10 plays role to regulate apoptosis development in human lung cells by TCDD exposure. It is expected to lead to insights into apoptosis in lung adenocarcinoma by TCDD exposure.

5.3.7 PROTEIN PHOSPHORYLATION IN RESPONSE TO DNA DAMAGE VIA YEAST ARRAYS

Two collections of chromosomally tagged yeast *Saccharomyces cerevisiae* strains were designed to detect protein/protein interactions, via cross-and-capture system. Suter *et at.* screened collections for proteins that are phoshorylated in response to DNA damage, by electrophoretic shift analysis.[68] Modification of proteins that are known targets for checkpoint kinases was confirmed. They identified MMR protein postmeiotic segregation increased-1 (Pms1) as target for phosphorylation, in response to DNA damage and replication fork arrest. Genetic analysis revealed that phosphorylation is dependent on checkpoint activation, by ATM/ATR (yeast Mec1p/Tel1p) kinases. They showed that cross-and-capture was efficiently used to detect posttranslational modifications (PTMs), which modulate/control protein function in eukaryotic cells. Ljungman reviewed targeting DDR in cancer.[69]

5.3.8 CELL-BASED SCREEN ATR INHIBITOR WITH SYNTHETIC LETHAL PROPERTIES: CANCER

Oncogene activation generates replication-born DNA damage (replicative stress). Primary responder to replicative stress is not ATM but ATR, which is kinase involved in signal transduction in DDR, which mutation is related to little frequent syndromes with cerebral ataxia manifestations. An ATR-study limitation is potent-inhibitor lack. Toledo et al. described cell-based screening strategy that identified ATR inhibitors in nanomolar range.[70] Pharmacological ATR inhibition generates replicative stress, leading to chromosomal breakage in presence of conditions that stall replication forks. Inhibition of ATR is poisonous for *p53*-deficient cells, toxicity being exacerbated by replicative stress-generating conditions, for example, cyclin-E overexpression. They searched for ATR inhibitors that may be efficient versus cells that present *p53* mutations, that is, anticancerous. They developed two antitumoral molecules that act inhibiting ATR, which is activated by DNA damages, and which target is gene-*p53* mutations frequent in various cancer types. One is dual phosphatidylinositol-3-OH kinase (PI3K) and mammalian target of rapamycin (mTOR) inhibitor NVP-BEZ235, which is tested for cancer chemotherapy, but that they showed, is potent against ATM, ATR and catalytic subunit of DNA-dependent protein kinase (DNA-PKcs). They described its antitumoral mechanism. Study opens way to oppose tumor response loss to usual chemotherapy.

5.3.9 THE MAMMALIAN TARGET OF RAPAMYCIN SIGNALLING PATHWAY IN HUMAN CANCER

Conserved serine/threonine (Ser/Thr) kinase mTOR, downstream effector of pathway PI3K/AKT, forms multiprotein complexes: mTOR complex (mTORC)1/2. The mTORC1 is sensitive to rapamycin, activates S6K1/4EBP1, which are involved in messenger ribonucleic acid (RNA) (mRNA) translation; it is activated by stimuli, for example, growth factors, nutrients, energy and stress signals, and essential signalling pathways, for example, PI3K, MAPK and AMPK, in order to control cell growth, proliferation and survival. The mTORC2 is resistant to rapamycin and insensitive to nutrients/energy signals; it activates PKC-α/AKT and regulates actin cytoskeleton. Element deregulation of pathway mTOR (*PI3K* amplification/mutation, *PTEN* function loss, AKT overexpression and S6K1/4EBP1/eIF4E overexpression) was reported in many cancer types, particularly melanoma, where alterations in pathway components showed effects on tumor progression. The mTOR is appealing therapeutic target, and its inhibitors, for example, rapamycin analogues deforolimus, everolimus and temsirolimus, are submitted to clinical trials for treating multiple cancers, alone or in combination with other-pathway inhibitors. Temsirolimus/everolimus were approved by FDA for treatment of renal cell carcinoma, PNET and giant-cell astrocytoma. Small molecules, which inhibit mTOR kinase activity, and dual PI3K-mTOR inhibitors are being developed. Pópulo et al. reviewed signalling pathway mTOR in human cancer with aim to survey research, signalling molecular mechanisms, for example, upstream activation and downstream effectors, and mTOR role in cancer, mainly melanoma.[71]

5.3.10 WEE1 CONTROLS GENOMIC STABILITY IN REPLICATION REGULATING ENDONUCLEASE

Genome replication and integrity protection are essential for cell survival. Mechanism loss that protect DNA frees development of diseases, for example, cancer. The H2AX presents phosphorylation role in DNA repairing complexes. Domínguez-Kelly et al. discovered function mechanism of genomic stability regulator contributing to fight cancer.[72] In high-throughput screen (HTS) studying H2AX phosphorylation, they identified Wee1 [nuclear kinase of checkpoint (cell-cycle regulator) family] as genomic-stability regulator (cf. Fig. 5.4), which downregulation not only induced H2AX phosphorylation, but also triggered general DDR and caused block in DNA replication, resulting in cell accumulation in S-phase. The Wee1-defiecient cells showed decay in replication fork speed indicating Wee1 involvement in DNA replication. Inhibiting Wee1 in cells treated with short hydroxyurea handling enhanced DDR, which suggests that Wee1 specifically protects stability of stalled replication forks. The DDR induced by Wee1 depletion depends on Mus81-Eme1 endonuclease (which cuts DNA fragments residual from replication), and they found that Mus81/Wee1 co-depletion abrogated S-phase delay. The Wee1/Mus81 interact in vivo suggesting direct regulation. Results show Wee1 role in controlling Mus81/DNA replication in human cells.

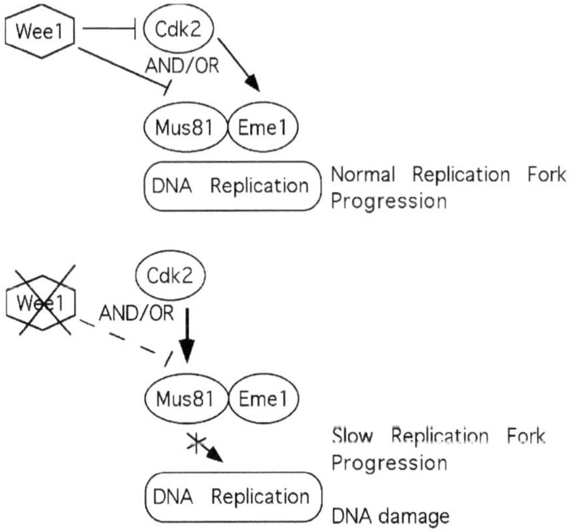

FIGURE 5.4 Model for Wee1 regulating deoxyribonucleic acid (DNA) replication by controlling Mus81.

5.3.11 CHROMOSOME INSTABILITY SYNDROMES

The CIN syndromes A-T, FA and BS were known for years.[73] Nijmegen breakage syndrome (NBS) and A-T-like disorder (ATLD) were identified. The A-T, ATLD and NBS form disorder group, which show similar cellular features, which result from consequences of increased sensitivity to IR.[74] They share clinical features particularly A-T/ATLD and show immunodeficiency. The A-T/NBS present predisposition to lymphoid tumors. The FA is caused by mutations in eight genes although most are accounted for by *FANCA*/*FANCC*; rare BS is caused by mutation in gene *BLM*. Common important feature results that identified genes are involved in aspects of recombination repair of DNA damage.

5.3.12 MITOTIC CHROMOSOMAL INSTABILITY AND CANCER: MURINE MODELING OF DISEASE

Stepwise progression, from early dysplastic lesion to full-blown metastatic malignancy, is associated with genomic instability rises. Mitotic CIN, inability to faithfully segregate equal chromosome complements to both daughter cells during mitosis, is widespread in solid tumors, which serves as fuel for tumorigenic progression. How CIN arises in tumors and what consequences it causes are debated. Schvartzman et al. reviewed models that recapitulate observations from human cancer.[75]

5.3.13 ROLE OF NEURO-ASTRO-VASCULAR UNIT IN ETIOLOGY OF ATAXIA TELANGIECTASIA

Recognition that brain pathologies do not affect neurones only but rather are, to a large extent, pathologies of glial cells and vasculature opens perspectives in understanding CNS genetic disorders. In order to validate neuro-glial-vascular unit role in etiology of genome-instability disorders, Meshulam et al. reported cell death and morphological aspects of neuroglia nets and associated vasculature in murine model of A-T.[76] They found that A-T-mutated protein deficiency was consistent with aberrant astrocytic morphology and vasculature alterations, accompanied by reactive gliosis. Similar findings could be reported in case of other genetic disorders. Observations bolster notion that astrocyte-specific pathologies hampered vascularization and astrocyte–endothelium interactions in CNS play role in etiology of genome-instability brain disorders and underlie degeneration.

5.4 RESULTS: INTRACELLULAR PROTEOLYSIS AND ITS RELATIONSHIP TO CANCER

5.4.1 BALANCING LIFE/DEATH DECISIONS: CDK2-DEPENDENT FOXO1 PHOSPHORYLATION

Transcription-factor phosphorylation is crucial to if DNA damage in cell results in death or repair/survival.[77] Usually cyclin-dependent kinase 2 (Cdk2) function is abolished after DNA damage (cf. Fig. 5.5); its inhibition plays central role in DNA injury-induced cell-cycle arrest/repair; however, if it influences cell survival under genotoxic stress is unknown.[78] Forkhead box-O (FOXO) transcription factors are emerging as cell survival regulators. The Cdk2 specifically phosphorylated FOXO1 at Ser[249] in vitro/vivo, which phosphorylation resulted in FOXO1 cytoplasmic localization/inhibition, which was abrogated upon DNA damage via cell-cycle checkpoint pathway, which is Chk1/2 dependent. Silencing of FOXO1, by small interfering RNA (siRNA), diminished DNA damage-induced death in p53-de/proficient cells; effect was reversed by restored FOXO1 expression depending on Ser[249] phosphorylation. Functional Cdk2/FOXO1 interaction provides mechanism that regulates apoptotic cell death after DSB.

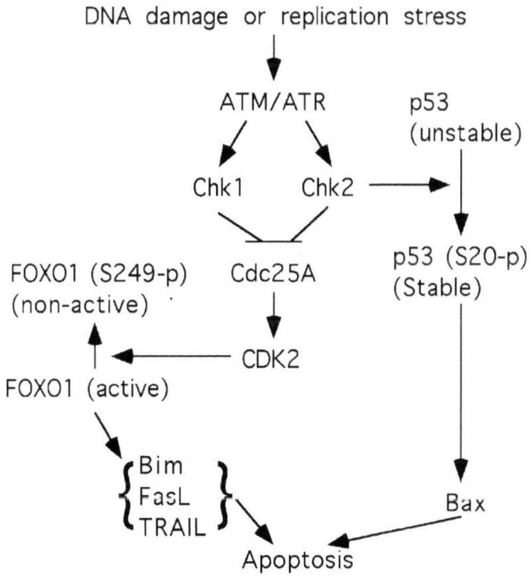

FIGURE 5.5 Role model for FOXO1 Cdk2-mediated phosphorylation/regulation in apoptotic DDR.

Yeung and Durocher reviewed engineering DDR without DNA damage.[79] Work achieved feat of activating DNA injury checkpoint in DNA-spoil absence, revealing protein/chromatin association importance for DDR activation, amplification and maintenance. Mullor reported functional assays in vertebrates to evaluate antioxidant properties.[80]

5.4.2 WARSAW BREAKAGE SYNDROME: COHESINOPATHY ASSOCIATED WITH MUTATIONS

Fe/S-containing DNA helicases XPD, FA group-J (FANCJ), probable adenosine triphosphate (ATP)-dependent RNA DDX11 and RTEL represent small subclass of superfamily-2; XPD/FANCJ were connected to genetic instability syndromes xeroderma pigmentosum (Xp)/FA. Van der Lelij et al. reported man with biallelic mutations in *DDX11*.[81] Defective DDX11 was associated with unique cellular phenotype, in which FA (drug-induced chromosomal breakage) and Roberts syndrome (sister chromatid cohesion defects) features co-existed; DDX11-deficient patient represented another cohesinopathy, besides Cornelia de Lange and Roberts syndromes, and showed that DDX11 functions at DNA repair/sister chromatid cohesion interface.

5.4.3 MOTIF INVOLVED IN CONTROL OF RAPID PROTEASOMAL DEGRADATION OF C-FOS

Proto-oncoprotein c-Fos is rapid/transiently expressed in cells undergoing $G_0 \rightarrow S$, in response to stimulation for growth by serum. Under conditions, rapid protein decay occurring after induction is accounted for by efficient proteasome recognition/degradation. Sequences rich in proline (Pro, P), glutamic acid (Glu, E), Ser (S) and Thr (T) (PEST) constitute protein instability determinants; c-Fos contains three such motifs, one of which comprises C-terminal 20 amino acids (AAs) and is major determinant of c-Fos instability. Using site-directed mutagenesis and expression system reproducing c-*fos* gene transient expression in transfected cells, Acquaviva et al. analyzed turnover of c-Fos mutants deleted of PESTs in synchronized murine embryo fibroblasts.[82] They showed no role for both internal PESTs in c-Fos instability; however, C-terminal PEST deletion led to only twofold protein stabilization. They reported that c-Fos instability, during $G_0 \rightarrow S$, is governed by major non-PEST destabilizer and C-terminal degradation-accelerating element.

Dissection of c-Fos C-terminal region showed that degradation-accelerating effect is not contributed by whole PEST, but by P-T-Leu (L) (PTL) tripeptide, which is not PEST motif and acts in PEST-environment absence. The PTL is conserved in members of *fos* multigene family; nevertheless, contribution to protein instability is restricted to c-Fos, suggesting that mechanisms whereby Fos proteins are broken down are different. Phosphorylation mediated by MAP kinases of two Ser's close to PTL, which are phosphorylated all over $G_0 \rightarrow S$, stabilizes c-Fos; however, they informed that PTL does not exert effect counteracting phosphorylation stabilization under conditions.

5.4.4 PROTEASOME DEGRADES C-FOS PROTO-ONCOPROTEIN INDEPENDENTLY

Prior ubiquitinylation of unstable c-Fos proto-oncoprotein is required for proteasome recognition/degradation. In contrast, Bossis et al. reported that although c-Fos forms conjugates with ubiquitin in vivo, nonubiquitinylatable c-Fos mutants showed regulated degradation identical to WT protein in living cells, under two classical conditions: transient *c-fos* gene expression during $G_{0\rightarrow1}$ upon mitogen stimulation and constitutive expression throughout asynchronous growth.[83] Destruction of c-Fos in $G_{0\rightarrow1}$ is unusual because it depends on two distinct but cumulative mechanisms; they reported that one mechanism involved C-terminal destabilizer, which did not need active ubiquitin cycle, whereas the other concerned N-terminal underminer dependent on ubiquitinylation of upstream c-Fos breakdown effector. In addition to providing insights into mechanisms of c-Fos protein destruction, consequence is that ubiquitinylation-dependent proteasomal degradation claimed for proteins should be reassessed.

5.4.5 PKC-DEPENDENT UBIQUITYLATION PROMOTES SUPPRESSOR Fbw7α INACTIVATION

The Fbw7 (hCdc4) is substrate-recruiting subunit of SKP1/CUL1/F-box protein (SCF) ubiquitin ligase complex, responsible for degradation of several cell-cycle regulators/proto-oncogenes, for example, cyclin-E, c-Myc, Aurora-A, Notch, c-jun and Mc11. Cyclin-E, Cdk2 activator, accumulates in dividing cells in late G_1, activates Cdk2 at $G_1 \rightarrow S$ and is degraded as cells progress via S-phase; Fbw7-function loss, for example, in many cancers, leads to deregulation of cyclin-E expression, which is no more restricted to $G_1 \rightarrow S$, and cyclin-E accumulation in mitosis. Coux's group observed that Fbw7α, although expressed in mature frog *Xenopus* eggs arrested in metaphase II, is nonfunctional explaining why cyclin-E can be stockpiled in mitotic-like phase.[84] They found that in *Xenopus* eggs and early embryos, Fbw7α inactivation correlates with its protein kinase-C (PKC)-dependent polyubiquitylation, which is maintained until end of early rapid cleavage cycles where cyclin-E undergoes abruptly degraded. Negative PKC-dependent regulation of Fbw7α is conserved during human somatic cell cycles, resulting into cyclin-E periodic expression. Findings reveal mechanism critical for Fbw7α temporal regulation and that functions are altered by PKC deregulation, which occurs in tumor types.

5.4.6 F-BOX PROTEIN FBXO11 TARGETS BCL6 FOR DEGRADATION AND IS MUTATED

B-cell lymphoma (BCL)6 protein is proto-oncogene product implicated in pathogenesis of human BCLs. Binding specific DNA sequences, BCL6 controls transcription of genes involved in B-cell development, differentiation and activation. All functions are critical for mature B-cells to form germinal centers, sites within lymph nodes that are key to humoral immune response. Critical BCL6 function is repression of genes involved in DDR and cell-cycle checkpoints, allowing B-cells to tolerate DNA breaks generated by somatic hypermutation and class switch recombination in immunoglobulin (Ig) genes. Tolerance of DNA damage plays role in BCL6-mediated lymphomagenesis. The BCL6 results overexpressed in most patients with aggressive diffuse large B-cell lymphoma (DLBCL), most common one in adulthood, and transgenic mice constitutively expressing BCL6 in B-cells develop DLBCL similar to human. In many DLBCL patients, BCL6 overexpression is achieved via promoter translocation (40%)/hypermutation (15%); however, many other DLBCLs overexpress BCL6 via unknown mechanism. Pagano's group showed that BCL6 is targeted for ubiquitylation/proteasomal degradation by SCF ubiquitin ligase complex, which contains orphan F-box only protein 11 (FBXO11).[85] Gene encoding FBXO11 resulted deleted/mutated in 19.3% DLBCL cells, and aberration in *FBXO11* correlated with increased BCL6 levels/stability. Similarly *FBXO11* was deleted/mutated in 12.7% human primary DLBCLs. Tumor-derived FBXO11 mutants displayed impaired abilities to induce BCL6 degradation in cell system. Expression of FBXO11 in DLBCL cells lacking gene inhibited cell proliferation and induced cell death. They provided molecular mechanism controlling BCL6 stability and showed that FBXO11 mutations/deletions, observed in human DLBCLs, contribute to lymphomagenesis via BCL6 stabilization.

5.4.7 SCF^{FBXO9}-MEDIATED DEGRADATION OF TEL2 AND TTI1 CONTROLS PI3K-RELATED KINASES

The PI3K-related kinases (PIKKs) mTOR, ATM, ATR, DNA-PKcs, SMG-1 and TRRAP control cell growth, proliferation, survival, and genome surveillance in mammalian cells. Evolutionarily conserved telomere maintenance-2 (Tel2)/Tel2-interacting protein-1 (Tti1) proteins regulate mammalian-PIKK abundance and affect functions; however, mechanism of how Tel2/Tti1 are regulated to direct individually PIKK functions remained unknown. Fernández-Sáiz's group reported that Tel2/Tti1 undergo ubiquitylated/ processed for proteasomal degradation, via previous orphan SCF^{Fbxo9} ubiquitin ligase, in response to mitogen deprivation by which mTORC1/2-complexes activity is regulated to control cell size, protein translation and apoptosis.[86] They investigated Fbxo9 role in pathogenesis of multiple myeloma (MM), second most common hematological malignancy, characterized by high/ progressive levels of genomic instability. Increasing evidence, from different comparative genomic hybridization (CGH) arrays, showed copy number gains and Fbxo9 overexpression in MM patient samples. Proteasomal inhibition was successfully introduced into MM therapy, indicating presence of disease-specific deregulated ubiquitylation. They found, using quantitative polymerase chain reaction (qPCR), Fbxo9 overexpression in MM cell lines/ patient samples. Short hairpin RNA (shRNA)-mediated inactivation of Fbxo9 resulted in apoptotic response in MM cells with high Fbxo9 expression. They identified Fbxo9-Tel2/Tti1 axis as PIKK regulatory hub, whose deregulation contributes to MM development/progression.

5.4.8 JUN-B BREAKDOWN IN G_2 REQUIRED FOR DOWNREGULATION OF CYCLIN-A2 LEVELS

The jun-B is best known as cell proliferation inhibitor, senescence inducer and tumor suppressor although it presents cell-division-promoting activity. Cell-division cycle effects were studied mostly in G_1/S phases whereas G_2/M result elusive. Using cell synchronization experiments, Farràs et al. showed that jun-B levels, which were high in S-phase, dropped during middle-to-late

G_2 because of accelerated phosphorylation-dependent degradation by proteasome.[87] Forced expression of ectopic jun-B protein, in late G_2, indicated that its decay is necessary for subsequent cyclin-A2 reduction in prometaphase, which is essential for proper mitosis. Abnormal jun-B expression in late G_2 entailed mitotic defect variety. As aberrations cause genetic instability, their findings contrasted with acknowledged tumor suppressor activity of jun-B and revealed mechanism, by which its deregulation contributes to tumorigenesis.

5.4.9 JUN-B DEGRADATION IN MITOSIS AND DEREGULATED MECHANISMS OF CONTROL

Cell-cycle regulation consists of three layers. Ubiquitin ligase E3 confers specificity to UPS, which is involved in cell-cycle regulation; for example, UPS deregulation in breast, prostate, etc. cancers and human papiloma virus (HPV) infection constitute high risk of cervical neoplasia; UPS pathway is emerging drug target in cancer therapy. Farràs et al. reported jun-B degradation in mitosis and aberrant consequences of deregulated control mechanisms. They analyzed jun-B case, which is overexpressed in cancer types. Is jun-B tumor suppressor/oncogene? Murine jun-B knockout (KO) acts as tumor suppressor; however, it is not clear in humans. In cell cycle, jun-B decays in middle-to-late G_2. Is jun-B degradation dependent on proteasome? Is degradation necessary for mitosis? The jun-B contains phosphodegron motif; E3 SCF^{Fbw7} recognizes consensus phosphodegron. It interacts with Fbw7 E3; it accumulates in mitotic DLD1 $Fbw7^{-/-}$ cells; mutations in phosphodegron increase its half-life compared to WT. It undergoes sequential phosphorylation; kinase GSK3 inhibition reduces its phosphorylation on T255/S251. Its degradation correlations and pharmacological inhibition are analyzed. There is GSK3β genetic inactivation in late G_2. Mutant AAA-jun-B is not ubiquilated and accumulates as cell reaches mitosis. The jun-B is degraded by GSK3-SCF^{Fbw7} in G_2: SCF^{Fbw7} appears mutated in several cancers; its overexpression alters mitosis duration, gene expression and cytokinesis; its degradation in G_2 is necessary to repress cyclin-A2. Other transcriptional jun-B was examined. Gene expression analysis of cell-cycle regulators was performed in cells expressing non-degradable AAA-jun-B mutant. The DDX11 is repressed in AAA-jun-B expression, AAA-jun-B and $Fbw7^{-/-}$ cells. The jun-B binds to DDX11 promoter. The DDX11 jun-B-dependent downregulation causes chromatid cohesion defects.

5.4.10 JUN-B APPLICATIONS IN CANCER: ANAPLASTIC LARGE-CELL LYMPHOMA AND KINASE

The jun-B was overexpressed in anaplastic large-cell lymphoma (ALCL); anaplastic lymphoma kinase (ALK) ALCL produced survival/proliferation; GSK3β-dependent jun-B degradation was impaired in ALCL; DDX11-dependent jun-B repression was ALCL. Chromatid cohesion was analyzed. Farràs et al. proposed working model for jun-B protein regulation during cell cycle. Primer kinase is not identified. Is this an Aurore kinase? Look at phosphorylation: degradation is fast; there are three monophosphorylations that are homologous but there are particular monophosphorylation sites, which are stable/unable to degradation. There is no inhibitor specific for jun-B, which structure is unknown; only dimerization domain is accepted. If instead of removing jun-B one inhibits it, whether it would result the same phenotype, one would show that jun-B is transcription factor. Intention is not to form complex. Is jun-B/DDX11 axis deregulated in other cancers? Does deregulated jun-B/DDX11 cause effects?

5.4.11 GSK3-SCF^{FBW7} TARGETS JUN-B FOR DEGRADATION IN G_2 TO PRESERVE CHROMATID

The jun-B acts either as tumor suppressor or oncogene depending on cell context; it is upregulated in ALK-positive ALCL where it enhances cell proliferation. Although its overexpression is linked to lymphomagenesis, mechanisms are obscure; jun-B undergoes co-ordinated phosphorylation-dependent ubiquitylation during G_2. Pérez-Benavente's group characterized critical consensus phosphodegron, which controls jun-B turnover, and identified GSK3/SCF^{FBXW7} as kinase/E3 ubiquitin ligase responsible for its degradation in G_2.[88] Pharmacological/genetic inactivation of GSK3-FBXW7-jun-B axis induced jun-B accumulation, in G_2/M, and entailed transcriptional repression of DNA helicase *DDX11*, leading to premature sister chromatid separation. Abnormal phenotype, because of GSK3β/jun-B/DDX11 pathway deregulation, is phenocopied in ALK-positive ALCLs. Results reveal mechanism by which mitosis progression and chromatid cohesion are regulated, via GSK3/SCF^{FBXW7}-mediated jun-B proteolysis, essential in maintaining genetic fidelity during mitosis.

5.4.12 P53 INDUCES TRANSCRIPTION AND TRANSLATION TO SUPRESS CELL PROLIFERATION

Cell growth and proliferation are tightly connected to ensure that appropriately sized daughter cells are generated following mitosis. Energy stress blocks cell growth and proliferation, a critical response for survival under extreme conditions. Excessive oncogenic stress leads to p53 activation and the induction of senescence, an irreversible state of cell-cycle arrest and a critical component in tumorigensis suppression. Nutrient-sensing and mitogenic cues converge on a major signalling node, which regulates mTOR activity. Although transcriptional responses to energy and oncogenic stresses were examined by gene expression experiments, a global exploration of the modulation of mRNA translation in response to these conditions is lacking. Loayza-Puch et al. combined RNA sequencing and ribosomal profiling analyses to delineate systematically modes of transcriptional/translational regulation induced in response to conditions of limited energy, oncogenic stress and cellular transformation.[89] They detected a key role for mTOR/p53 in these distinct physiological states, and provided the first genome-wide association study (GWAS) demonstration that p53 activation results in mTOR inhibition and a consequent global repression of protein translation. They confirmed the role of the direct p53 target genes Sestrin1/2 in this response, as part of the broad modulation of gene expression induced by p53 activation.

5.4.13 AGEING/CANCER: REPROGRAMING VIA MI-2/NURD/MTOT KINASE/METABOLISM MODULATION

From non-conventional angles, Zhang argued *Don't put all eggs in one basket* and natural *self-organization* are among keys to unveil ageing and cancer.[90] The former is behind phenomena, for example, asymmetry division, sharing interaction, chaotic and oscillatory gene expressions, sex, stemness, tumor and ageing damage heterogeneities; the latter is a core of system biology-based cell attractors theory, and evolution in this vision and reflections. He examined ageing and cancer inevitabilities, and postulated that the sharing in modern-life origin and multicellularity commencement is essential for cell normality, and the carcinogenesis is speciation. As for the nucleosome remodeling and histone deacetylase (Mi-2/NuRD), he explained the carcinogenesis alongside diseased multicellularity. He extended it to mTOR protein kinase and metabolism in ageing and cancer (cf. Fig. 5.6). He proposed the modulation on them to reprogram ageing and cancer.

Developmental program → Ageing program → Cancer stem-like program

FIGURE 5.6 Reprograming of Mi-2/NuRD, mTOR and metabolism (3M) in ageing and cancer.

5.4.14 LONG NON-CODING RNAS AND P53 REGULATION

Advent of high-throughput sequencing technologies allowed genome sequencing at unprecedented depth. Most transcribed RNAs were classified as non-coding RNAs (ncRNAs), among which long ncRNAs (lncRNAs) emerge as regulators in many bioprocesses. Baldassarre and Masotti reviewed role of lncRNAs that are under p53 control or able to regulate its activity, because of central role of p53 pathway in many conditions.[91] They revised strategies/computational tools need to unravel lncRNAs multifaceted roles and pave way to development of diagnos/therapeutic applications based on lncRNAs.

5.4.15 SOD1 LOSS INDUCES INTRINSIC SUPEROXIDE ACCUMULATION: GROWTH ARREST

Oxidative damages induced by a redox disequilibrium cause age-related changes in cells and tissues. Superoxide dismutase (SOD) enzymes play a major role in the antioxidant system and catalyze superoxide radicals $O_2^{\bullet-}$. Since cytoplasmic-SOD (SOD1) loss resulted in aging-like phenotypes in several types of murine tissue, SOD1 is essential for tissue-homeostasis maintenance. In order to clarify SOD1 cellular function, Watanabe et al. investigated the cellular phenotypes of *Sod1*-deficient fibroblasts.[92] They showed that *Sod1* deficiency impaired proliferation and induced apoptosis associated with $O_2^{\bullet-}$ accumulation in the cytoplasm and mitochondria in fibroblasts. The *Sod1* loss also decreased the mitochondrial membrane potential and led to DNA damage-mediated p53 activation. Antioxidant treatments effectively improved the cellular phenotypes via suppression of intracellular $O_2^{\bullet-}$ accumulation and p53 activation in *Sod1*-deficient fibroblasts. In vivo experiments revealed that transdermal treatment with a vitamin-C (Vit-C) derivative significantly reversed the skin thinning commonly associated with the upregulated p53 action in the skin. Their findings revealed that intrinsic $O_2^{\bullet-}$ accumulation promoted p53-mediated growth arrest/apoptosis and mitochondrial dysfunction in the fibroblasts.

5.4.16 NEIL BENCE: MANIPULATING DEGRADATION

Bence group found that protein aggregates were tied to impairment of the ubiquitin–proteasome system.[93] They ruled out the possibility that aggregates lead to the sequestering of the ubiquitin complex, making it unavailable.[94] They discovered new inhibitors of the ubiquitin-activating enzyme.[95]

5.4.17 EFFECT OF ZN^{2+} ON PROSTATIC CELL CYTOTOXICITY CAUSED BY TRICHOMONAS VAGINALIS

Vazquez-Carrillo et al. studied *Trichomonas vaginalis* proteome, morphology and cytotoxicity during interactions with prostatic DU-145 cells.[96] They showed that 37 proteins are expressed in Zn^{2+} presence, which downregulates protein/transcriptional levels of *T. vaginalis* cysteine (Cys) proteinase 65kDa (TvCP65). Result is negative effect on trichomonal cytotoxicity. They identified differentially expressed proteins by mass spectrometry (MS).

5.5 FINAL REMARKS

From the present results and discussion, the following final remarks can be drawn:

1. The DDX11 shares sequence with FANCJ and DEAH-box helicases XPD/RTEL, which contain one Fe/S cluster between domains IA/II; FANCJ/XPD connect to genetic-instability syndromes Fanconi anaemia and xeroderma pigmentosum, and inactivation increases cancer risk. Inactivation of DDX11 causes genetic instability combining Fanconi anaemia/Roberts syndrome; given DDX11-deficient cell hypersensitivity for mitomycin-C/campothecin, which interfere with DNA replication, DDX11 functions at replication-coupled acid repair/sister chromatid cohesion interface.
2. In *Xenopus* eggs/embryos, Fbw7α inactivation correlates with protein kinase-C-dependent polyubiquitylation, which is maintained

until end of early rapid cleavage cycles where cyclin-E undergoes degraded. Kinase-C-dependent negative regulation of Fbw7α is conserved during human somatic cell cycles, resulting in cyclin periodic expression. In mechanism for Fbw7α regulation, functions change by kinase deregulation, which occurs in tumor types.

3. In mechanism controlling B-cell lymphoma-6 protein stability, FBXO11 mutations/deletions in human diffuse large B-cell lymphomas contribute to lymphomagenesis, via 6-protein stabilization.
4. The Fbxo9-telomere maintenance-2/interacting protein-1 axis is phosphatidylinositol-3 kinase-related kinase regulatory hub, whose deregulation contributes to multiple-myeloma development/ progression.
5. Cancer jun-B implication is context dependent: besides cell proliferation inhibition/senescence, it suppresses tumor; its murine overexpression inhibits B-cell transformation by oncogene v-abl. It contributes to tumor phenotype; as disrupted passage via mitosis leads to chromosome missegregation and aneuploid progeny production, it favors that jun-B overexpression in late G_2 represent oncogenic mechanism.
6. Pharmacological/genetic inactivation of GSK3-FBXW7-jun-B axis induced jun-B accumulation, in G_2/M, and entailed *DDX11* transcriptional repression leading to premature sister chromatid separation. Abnormal phenotype, because of GSK3β/jun-B/DDX11 pathway deregulation, is phenocopied in anaplastic lymphoma kinase-positive anaplastic large-cell lymphomas. Mitosis progression and chromatid cohesion are regulated via GSK3/SCF^{FBXW7}-mediated jun-B proteolysis, essential in maintaining genetic fidelity during mitosis.
7. A bimodal tumor-suppressive regulatory program activated by p53 was delineated, in which cell-cycle arrest is imposed mainly at the transcriptional level, whereas cell growth inhibition is enforced by global repression of the translation machinery.
8. Gaps exist but self-organization insights in system biology point of ageing/cancer suggest that both are inevitable, and reversible or preventable. Breaking *Don't put all eggs in one basket* and *self-organization* promote ageing and cancer. Cancer may hijack host system with/out switching their original functions. Inhibitors for Mi-2/NuRD and its functional partner chromatin remodeling system, mTOR could separately reprogram and reverse ageing/cancer. However, it is tempting to find out effective combination rapamycin/

mTOR inhibitors, vitamin-C and histone deacetylase inhibitors of low-dosage and lesser side effect in such reversals. Simplified genome-wide expression partners could be developed as ensemble markers for ageing-related diseases or cancer.

9. Zn^{2+} negatively affects growth, cytotoxicity and expression of *T. vaginalis* proteinases related to prostatic cell interaction.

Further work will deal with vitamin-D, analogues, drugs, their relevance to cancer inhibition, prostate disease, stem cells and their connection with cancer, prostate disease, MEK/ERK pathway overactivation in liver tumors and resistance to transforming growth factor-β-induced death.

ACKNOWLEDGMENTS

The authors thank support from Generalitat Valenciana (Project No. PROMETEO/2016/094) and Universidad Católica de Valencia San Vicente Mártir (Project No. 2019-217-001).

KEYWORDS

- **Jun-B**
- **c-Fos**
- **oncogene**
- **protein degradation**
- **proteasome**
- **PEST motif**
- **chromosome instability**

REFERENCES

1. Chinenov, Y.; Kerppola, T. K. Close Encounters of Many Kinds: Fos–Jun Interactions that Mediate Transcription Regulatory Specificity. *Oncogene* **2001,** *20*, 2438–2452.
2. Eferl, R.; Wagner, E. F. AP-1: A Double-edged Sword in Tumorigenesis. *Nat. Rev. Cancer* **2003,** *3*, 859–868.
3. Hess, J.; Angel, P.; Schorpp-Kistner, M. AP-1 Subunits: Quarrel and Harmony Among Siblings. *J. Cell Sci.* **2004,** *117*, 5965–5973.

4. Jochum, W.; Passegue, E.; Wagner, E. F. AP-1 in Mouse Development and Tumorigenesis. *Oncogene* **2001,** *20*, 2401–2412.
5. Mechta-Grigoriou, F.; Gerald, D.; Yaniv, M. The Mammalian Jun Proteins: Redundancy and Specificity. *Oncogene* **2001,** *20*, 2378–2389.
6. Shaulian, E.; Karin, M. AP-1 as a Regulator of Cell Life and Death. *Nat. Cell Biol.* **2002,** *4*, E131–E136.
7. Shaulian, E.; Karin, M. AP-1 in Cell Proliferation and Survival. *Oncogene* **2001,** *20*, 2390–2400.
8. Milde-Langosch, K. The Fos Family of Transcription Factors and Their Role in Tumourigenesis. *Eur. J. Cancer* **2005,** *41*, 2449–2461.
9. Albanese, C.; Johnson, J.; Watanabe, G.; Eklund, N.; Vu, D.; Arnold, A.; Pestell, R. G. Transforming p21ras Mutants and c-Ets-2 Activate the Cyclin D1 Promoter through Distinguishable Regions. *J. Biol. Chem.* **1995,** *270*, 23589–23597.
10. Bakiri, L.; Lallemand, D.; Bossy-Wetzel, E.; Yaniv, M. Cell Cycle-dependent Variations in c-Jun and JunB Phosphorylation: A Role in the Control of Cyclin D1 Expression. *EMBO J.* **2000,** *19*, 2056–2068.
11. Herber, B.; Truss, M.; Beato M.; Muller, R. Inducible Regulatory Elements in the Human Cyclin D1 Promoter. *Oncogene* **1994,** *9*, 1295–1304.
12. Passegué, E.; Wagner, E. F. JunB Supresses Cell Proliferation by Transcriptional Activation of p16(INK4α) Expression. *EMBO J.* **2000,** *19*, 2969–2979.
13. Passegué, E.; Jochum, W.; Schorpp-Kistner, M.; Mohle-Steinlein, U.; Wagner, E. F. Chronic Myeloid Leukemia with Increased Granulocyte Progenitors in Mice Lacking junB Expression in the Myeloid Lineage. *Cell* **2001,** *104*, 21–32.
14. Passegué, E.; Wagner, E. F.; Weissman, I. L. JunB Deficiency Leads to a Myeloproliferative Disorder Arising from Hematopoietic Stem Cells. *Cell* **2004,** *119*, 431–443.
15. Steidel, U.; Rosenbauer, F.; Verhaak, R. G.; Gu, X.; Ebralidze, A.; Otu, H. H.; Klippel, S.; Steidel, C.; Bruns, I.; Costa, D. B.; Wagner, K.; Aivado, M.; Kobbe, G.; Valk, P. J.; Passegué, E.; Libermann, T. A.; Delwel, R.; Tenen, D. G. Essential Role of Jun Family Transcription Factors in PU.1 Knoxkdown-induced Leukemic Stem Cells. *Nat. Genet.* **2006,** *38*, 1269–1277.
16. Kovary K.; Bravo, R. Existence of Different Fos/Jun Complexes During the G_0-to-G_1 Transition and During Exponential Growth in Mouse Fibroblasts: Differential Role of Fos Proteins. *Mol. Cell. Biol.* **1992,** *12*, 5015–5023.
17. Kovary, K.; Bravo, R. Expression of Different Jun and Fos Proteins During the G_0-to-G_1 Transition in Mouse Fibroblasts: In Vitro and In Vivo Associations. *Mol. Cell. Biol.* **1991,** *11*, 2451–2459.
18. Lallemand, D.; Spyrou, G.; Yaniv, M.; Pfarr, C. M. Variations in Jun and Fos Protein Expression and AP-1 Activity in Cycling, Resting and Stimulated Fibroblasts. *Oncogene* **1997,** *14*, 819–830.
19. Kovary, K.; Bravo, R. The Jun and Fos Protein Families are Both Required for Cell Cycle Progression in Fibroblasts. *Mol. Cell. Biol.* **1991,** *11*, 4466–4472.
20. Andrecht, S.; Kolbus, A.; Hartenstein, B.; Angel, P.; Schorpp-Kistner, M. Cell Cycle Promoting Activity of JunB through Cyclin A Activation. *J. Biol. Chem.* **2002,** *277*, 35961–35968.
21. Marrero-Ponce, Y.; Machado-Tugores, Y.; Montero Pereira, D.; Escario, J. A.; Gómez Barrio, A.; Nogal-Ruiz, J. J.; Ochoa, C.; Arán, V. J.; Martínez-Fernández, A. R.; García Sánchez, R. N.; Montero-Torres, A.; Torrens, F.; Meneses-Marcel, A. A Computer-based

Approach to the Rational Discovery of New Trichomonacidal Drugs by Atom-type Linear Indices. *Curr. Drug Discov. Technol.* **2005,** *2*, 245–265.

22. Marrero-Ponce, Y.; Meneses-Marcel, A.; Castillo-Garit, J. A.; Machado-Tugores, Y.; Escario, J. A.; Gómez Barrio, A.; Montero Pereira, D.; Nogal-Ruiz, J. J.; Arán, V. J.; Martínez-Fernández, A. R.; Torrens, F.; Rotondo, R.; Ibarra-Velarde, F.; Alvarado, Y. J. Predicting Antitrichomonal Activity: A Computational Screening Using Atom-based Bilinear Indices and Experimental Proofs. *Bioorg. Med. Chem.* **2006,** *14*, 6502–6524.
23. Marrero-Ponce, Y.; Meneses-Marcel, A.; Rivera-Borroto, O. M.; García-Domenech, R.; de Julián-Ortiz, J. V.; Montero, A.; Escario, J. A.; Gómez Barrio, A.; Montero Pereira, D.; Nogal, J. J.; Grau, R.; Torrens, F.; Vogel, C.; Arán, V. J. Bond-based Linear Indices in QSAR: Computational Discovery of Novel Anti-trichomonal Compounds. *J. Comput.-Aided Mol. Design* **2008,** *22*, 523–540.
24. Meneses-Marcel, A.; Rivera-Borroto, O. M.; Marrero-Ponce, Y.; Montero, A.; Machado Tugores, Y.; Escario, J. A.; Gómez Barrio, A.; Montero Periera, D.; Nogal, J. J.; Kouznetsov, V. V.; Ochoa Puentes, C.; Bohórquez, A. R.; Grau, R.; Torrens, F.; Ibarra-Velarde, F.; Arán, V. J. New Antitrichomonal Drug-like Chemicals Selected by Bond (Edge)-based TOMOCOMD–CARDD Descriptors. *J. Biomol. Screen.* **2008,** *13*, 785–794.
25. Rivera-Borroto, O. M.; Marrero-Ponce, Y.; Meneses-Marcel, A.; Escario, J. A.; Gómez Barrio, A.; Arán, V. J.; Martins Alho, M. A.; Montero Pereira, D.; Nogal, J. J.; Torrens, F.; Ibarra-Velarde, F.; Vera Montenegro, Y.; Huesca-Guillén, A.; Rivera, N.; Vogel, C. Discovery of Novel Trichomonacidals Using LDA-driven QSAR Models and Bond-based Bilinear Indices as Molecular Descriptors. *QSAR Comb. Sci.* **2009,** *28*, 9–26.
26. Torrens, F; Castellano, G. Modelling of Complex Multicellular Systems: Tumour–Immune Cells Competition. *Chem. Central J.* **2009,** *3* (Suppl. I), 75–1-1.
27. Torrens, F.; Castellano, G. Information Theoretic Entropy for Molecular Classification: Oxadiazolamines as Potential Therapeutic Agents. *Curr. Comput.-Aided Drug Des.* **2013,** *9*, 241–253.
28. Torrens, F.; Castellano, G. Molecular Classification of 5-amino-2-aroylquinolines and 4-aroyl-6,7,8-trimethoxyquinolines as Highly Potent Tubulin Polymerization Inhibitors. *Int. J. Chemoinf. Chem. Eng.* **2013,** *3* (2), 1–26.
29. Estrela, J. M.; Mena, S.; Obrador, E.; Benlloch, M.; Castellano, G.; Salvador, R.; Dellinger, R. W. Polyphenolic Phytochemicals in Cancer Prevention and Therapy: Bioavailability versus Bioefficacy. *J. Med. Chem.* **2017,** *60*, 9413–9436.
30. Torrens, F.; Castellano, G. Molecular Classification of Antitubulin Agents with Indole Ring Binding at Colchicine-Binding Site. In *Molecular Insight of Drug Design*; Parikesit, A. A., Ed.; InTechOpen: Vienna, 2018; pp 47–67.
31. Torrens, F.; Castellano, G. Molecular Classification of 2-Phenylindole-3-carbaldehydes as Potential Antimitotic Agents in Human Breast Cancer Cells. In *Theoretical Models and Experimental Approaches in Physical Chemistry: Research Methodology and Practical Methods*; Haghi, A. K., Thomas, S., Praveen, K. M., Pai, A. R., Eds.; Apple Academic–CRC: Waretown, NJ, in Press.
32. Torrens, F.; Castellano, G. Cancer and Hypotheses on Cancer. In *Molecular Chemistry and Biomolecular Engineering: Integrating Theory and Research with Practice*; Pogliani, L., Torrens, F., Haghi, A. K., Eds.; Apple Academic–CRC: Waretown, NJ, in Press.
33. Torrens, F.; Castellano, G. Precision Personalized Medicine from Theory to Practice: Cancer. In *Green Chemistry and Biodiversity: Principles, Techniques, and Correlations*;

Aguilar, C. N., Ameta, S. C., Haghi, A. K., Eds.; Apple Academic–CRC: Waretown. NJ, in Press.
34. Torrens, F.; Castellano, G. AIDS Destroys Immune Defences: Hypothesis. *New Front. Chem.* **2014,** *23*, 11–20.
35. Torrens-Zaragozá, F.; Castellano-Estornell, G. Emergence, Spread and Uncontrolled Ebola Outbreak. *Basic Clin. Pharmacol. Toxicol.* **2015,** *117* (Suppl. 2), 38–38.
36. Torrens, F.; Castellano, G. 2014 Spread/Uncontrolled Ebola Outbreak. *New Front. Chem.* **2015,** *24*, 81–91.
37. Torrens, F.; Castellano, G. Ebola Virus Disease: Questions, Ideas, Hypotheses and Models. *Pharmaceuticals* **2016,** *9*, 14–6-6.
38. Torrens, F.; Castellano, G. Clinical Translational Research: Cancer, Diabetes, and Cardiovascular Disease. In *Physical Biochemistry, Biophysics, and Molecular Chemistry: Applied Research and Interactions*; Torrens, F., Mahapatra, D. K., Haghi, A. K., Eds.; Apple Academic–CRC: Waretown, NJ, in Press.
39. Torrens, F.; Castellano, G. Primary Health, Smoking Habits and Immunotherapy in Cancer. In *Environmental Technology and Engineering Techniques: Basic Concepts and Health Interventions*; Khalaf, M. N., Smirnov, M. O., Kannan, P., Haghi, A. K., Eds.; Apple Academic–CRC: Waretown, NJ, in Press.
40. Torrens, F.; Castellano, G. Epigenomics, Epithelial Plasticity, Clinical Genetics and Rare Diseases. In *Environmental Technology and Engineering Techniques: Basic Concepts and Health Interventions*; Khalaf, M. N., Smirnov, M. O., Kannan, P., Haghi, A. K., Eds.; Apple Academic–CRC: Waretown, NJ, in Press.
41. Torrens, F.; Castellano, G. Predictive in Silico Global Metabolism Analysis: Disease Mechanism. In *Processing and Characterization of Green Engineered and Sustainable Materials*; Haghi, A. K., Ed.; Apple Academic–CRC: Waretown, NJ, in Press.
42. Smoluchowski, M. Drei vorträge über Diffusion, Brownsche molekularbewegung und koagulation von kolloidteilchen. 1. *Phys. Z.* **1916,** *17*, 557–571.
43. Smoluchowski, M. Drei vorträge über diffusion, Brownsche molekularbewegung und koagulation von kolloidteilchen. 2. *Phys. Z.* **1916,** *17*, 585–599.
44. Debye, P. Reaction Rates in Ionic Solutions. *Trans. Electrochem. Soc.* **1942,** *82*, 265–272.
45. Berg, H. C.; Purcell, E. M. Physics of Chemoreception. *Biophys. J.* **1977,** *20*, 193–219.
46. Berezhkovskii, A. M.; Szabo, A. Effect of Ligand Diffusion on Occupancy Fluctuations of Cell-Surface Receptors. *J. Chem. Phys.* **2013,** *139*, 121910–1-7.
47. Kier, L. B.; Seybold, P. G.; Cheng, C. K. *Modeling Chemical Systems Using Cellular Automata*; Springer: Dordrecht, The Netherlands, 2005.
48. Waddington, C. H. *The Strategy of the Genes*; George Allen and Unwin: London, UK, 1957.
49. Gráf, L.; Jancsó, Á.; Szilágyi, L.; Hegyi, G.; Pintér, K.; Náray-Szabó, G.; Hepp, J.; Medzihradszky, K.; Rutter, W. J. Electrostatic Complementarity within the Substrate-binding Pocket of Trypsin. *Proc. Natl. Acad. Sci. USA* **1988,** *85*, 4961–4965.
50. Náray-Szabó, G. Quantitative Estimation of Activities of Mutant Enzymes. *Catal. Lett.* **1989,** *2*, 185–190.
51. Kirby, A. J. Molecular Recognition of Transition States. *Philos. Trans. R. Soc. London, Ser. A* **1993,** *345*, 67–76.
52. Torrens, F.; Castellano, G. Topological Charge-transfer Indices: From Small Molecules to Proteins *Curr. Proteomics* **2009,** *6*, 204–213.

53. Whitney, C. K. On the Several Molecules and Nanostructures of Water. *Int. J. Mol. Sci.* **2012,** *13*, 1066–1094.
54. Karley, D.; Gupta, D.; Tiwari, A. Biomarkers: The Future of Medical Science to Detect Cancer. *J. Mol. Biomark Diagn.* **2011,** *2*, 118–1-7.
55. Gandzha, I.; Kadeisvily, J. *New Sciences for a New Era: Mathematical, Physical and Chemical Discoveries of Ruggero Maria Santilli*; Sankata: Tebahal, Kathmandu, Nepal, 2010.
56. Chen, X.; Gambhir, S. S.; Cheon, J. Theranostic Nanomedicine. *Acc. Chem. Res.* **2011,** *44*, 841–841.
57. Perera. S.; Bapart, B. Genetic Instability in Cancer. *Atlas Genet. Cytogenet. Oncol. Haematol.* **2007,** *2007* (January), 1–1.
58. Charames, G. S.; Bapat, B. Genomic Instability and Cancer. *Curr. Mol. Med.* **2003,** *3*, 589–596.
59. Beaver, J. P.; Waring, P. A Decrease in Intracellular Glutathione Concentration Precedes the Onset of Apoptosis in Murine Thymocytes. *Eur. J. Cell Biol.* **1995,** *68*, 47–54.
60. Kalimuthu, S.; Se-Kwon, K. Cell Survival and Apoptosis Signaling as Therapeutic Target for Cancer: Marine Bioactive Compounds. *Int. J. Mol. Sci.* **2013,** *14*, 2334–2354.
61. Li, X.; Xu, H.; Xu, C.; Lin, M.; Song, X.; Yi, F.; Feng, Y.; Coughlan, K. A.; Cho, W. C. S.; Kim, S. S.; Cao, L. The Yin-Yang of DNA Damage Response: Roles in Tumorigenesis and Cellular Senescence. *Int. J. Mol. Sci.* **2013,** *14*, 2431–2448.
62. Bartkova, J.; Horesi, Z.; Koed, K.; Krämer, A.; Tort, F.; Zieger, K.; Guldberg, P.; Sehested, M.; Nesland, J. M.; Lukas, C.; Ørntoft, T.; Lukas, J.; Bartek, J. DNA Damage Response as a Candidate Anti-cancer Barrier in Early Human Tumorigenesis. *Nature (London)* **2005,** *434*, 864–870.
63. Gorgoulis, V. G.; Vassiliou, L. V. F.; Karakaidos, P.; Zacharatos, P.; Kotsinas, A.; Liloglou, T.; Venere, M.; DiTullio, R. A., Jr.; Kastrinakis, N. G.; Levy, B.; Kletsas, D.; Yoneta, A.; Herlyn, M.; Kittas, C.; Halazonetis, T. D. Activation of the DNA Damage Checkpoint and Genomic Instability in Human Precancerous Lesions. *Nature (London)* **2005,** *434*, 907–913.
64. Löbrich, M.; Rief, N.; Kühne, M.; Heckmann, M.; Fleckenstein, J.; Rübe, C.; Uder, M. In Vivo Formation and Repair of DNA Double-strand Breaks After Computed Tomography Examinations. *Proc. Natl. Acad. Sci. USA* **2005,** *102*, 8984–8989.
65. Matsuoka, S.; Ballif, B. A.; Smogorzewska, A.; McDonald, E. R., III; Hurov, K. E.; Luo, J.; Bakalarski, C. E.; Zhao, Z.; Solimini, N.; Lerenthal, Y.; Shiloh, Y.; Gygi, S. P.; Elledge, S. J. ATM and ATR Substrate Analysis Reveals Extensive Protein Networks Responsive to DNA Damage. *Science* **2007,** *316*, 1160–1166.
66. Lin, P. H.; Lin, C. H.; Huang, C. C.; Chuang, M. C.; Lin, P. 2,3,7,8-Tetrachlorodibenzo-*p*-dioxin (TCDD) Induces Oxidative Stress, DNA Strand Breaks, and Poly(ADP-ribose) Polymerase-1 Activation in Human Breast Carcinoma Cell Lines. *Toxicol. Lett.* **2007,** *172*, 146–158.
67. Hsiao, E. S. L.; Chang, Y. W.; Lin, P.; Liao, P. C. Proteomics Investigation Reveals Apoptosis-associated Proteins in Aryl Hydrocarbon Receptor-deficient Human Lung Cells Treated with 2,3,7,8-tetrachlorobenzo-*p*-dioxin. *J. Proteomics Bioinform.* **2012,** *5* (1), 15–23.
68. Suter, B.; Graham, C. I.; Stagljar, I. Exploring Protein Phosphorylation in Response to DNA Damage Using Differentially Tagged Yeast Arrays. *Biotechniques* **2008,** *45*, 581–584.

69. Ljungman, M. Targeting the DNA Damage Response in Cancer. *Chem. Rev.* **2009,** *109,* 2929–2950.
70. Toledo, L. I.; Murga, M.; Zur, R.; Soria, R.; Rodriguez, A.; Martinez, S.; Oyarzabal, J.; Pastor, J.; Bischoff, J. R.; Fernandez-Capetillo, O. A Cell-based Screen Identifies ATR Inhibitors with Synthetic Lethal Properties for Cancer-associated Mutations. *Nat. Struct. Mol. Biol.* **2011,** *18,* 721–727.
71. Pópulo, H.; Lopes, J. M.; Soares, P. The mTOR Signalling Pathway in Human Cancer. *Int. J. Mol. Sci.* **2012,** *13,* 1886–1918.
72. Domínguez-Kelly, R.; Martín, Y.; Koundrioukoff, S.; Tanenbaum, M. E.; Smits, V. A. J.; Medema, R. H.; Debatisse, M.; Freire, R. Wee1 Controls Genomic Stability During Replication by Regulating the Mus81-Eme1 Endonuclease. *J. Cell Biol.* **2011,** *194,* 567–579.
73. Gisselsson, D. Chromosomal Instability in Cancer: Causes and Consequences. *Atlas Genet. Cytogenet. Oncol. Haematol.* **2001,** *2001* (May), 1-1.
74. Taylor, A. M. R. Chromosome Instability Syndromes. *Best Pract. Res. Clin. Haematol.* **2001,** *14,* 631–644.
75. Schvartzman, J. M.; Sotillo, R.; Benezra, R. Mitotic Chromosomal Instability and Cancer: Mouse Modelling of the Human Disease. *Nat. Rev. Cancer* **2010,** *10,* 102–115.
76. Meshulam, L.; Galron, R.; Kanner, S.; de Pittà, M.; Bonofazi, P.; Goldin, M.; Frenkel, D.; Ben-Jacob, E.; Barzilai, A. The Role of the Neuro–astro–vascular Unit in the Etiology of Ataxia Telangiectasia, *Frontier Pharmacol.* **2012,** *3,* 151–1-9.
77. Bartek, J.; Lukas, J. Balancing Life-or-death Decisions. *Science* **2006,** *314,* 261–262.
78. Huang, H.; Regan, K. M.; Lou, Z.; Chen, J.; Tindall, D. J. CDK2-dependent Phosphorylation of FOXO1 as an Apoptotic Response to DNA Damage. *Science* **2006,** *314,* 294–297.
79. Yeung, M. T.; Durocher, D. Engineering a DNA Damage Response Without DNA Damage. *Gen. Biol.* **2008,** *9,* 227–1-4.
80. Mullor, J. L. Personal Communicaton.
81. Van der Lelij, P.; Chrzanowska, K. H.; Godthelp, B. C.; Rooimans, M. A.; Oostra, A. B.; Stumm, M.; Zdzienicka, M. Z.; Joenje, H.; de Winter, J. P. Warsaw Breakage Syndrome, a Cohesinopathy Associated with Mutations in the XPD Helicase Family Member DDX11/ChlR1. *Am. J. Hum. Genet.* **2010,** *86,* 262–266.
82. Acquaviva, C.; Brockly, F.; Ferrara, P.; Bossis, G.; Salvat, C.; Jariel-Encontre, I.; Piechaczyk, M. Identification of a C-terminal Tripeptide Motif Involved in the Control of Rapid Proteasomal Degradation of c-Fos Proto-oncoprotein During the G_0-to-S Phase Transition. *Oncogene* **2001,** *20,* 7563–7572.
83. Bossis, G.; Ferrara, P.; Acquaviva, C.; Jariel-Encontre, I.; Piechaczyk, M. c-Fos Proto-oncoprotein Is Degraded by the Proteasome Independently of Its Own Ubiquitinylation in vivo. *Mol. Cell. Biol.* **2003,** *23,* 7425–7436.
84. Coux, O. Personal Communication.
85. Pagano, M. Personal Communication.
86. Fernández-Sáiz, V. Personal Communication.
87. Farràs, R.; Baldin, V.; Gallach, S.; Acquaviva, C.; Bossis, G.; Jariel-Encontre, I.; Piechaczyk, M. JunB Breakdown in Mid-/late G_2 Is Required for Down-regulation of Cyclin A2 Levels and Proper Mitosis. *Mol. Cell. Biol.,* **2008,** *28,* 4173–4187.
88. Pérez-Benavente, B. Personal Communication.

89. Loayza-Puch, F.; Drost, J.; Rooijers, K.; Lopes, R.; Elkon, R.; Agami, R. p53 Induces Transcriptional and Translational Programs to Suppress Cell Proliferation and Growth. *Gen. Biol.* **2013,** *14*, R32–1-34.
90. Zhang, Y. Ageing and Cancer: Breaking the *Don't Put All Eggs in One Basket* and Natural *Self-organisation*, and their Potential Reprogramming via Modulation of Mi-2/NuRD, mTOR Kinase and Metabolism. *Enz. Eng.* **2013,** *2* (1), 109–1-8.
91. Baldassarre, A.; Masotti, A. Long Non-coding RNAs and p53 regulation. *Int. J. Mol. Sci.* **2012,** *13*, 16708–16717.
92. Watanabe, K.; Shibuya, S.; Koyama, H.; Ozawa, Y.; Toda, T.; Yokote, K.; Shimizu, T. *Sod1* Loss Induces Intrinsic Superoxide Accumulation Leading to p53-mediated Growth Arrest and Apoptosis. *Int. J. Mol. Sci.* **2013,** *14*, 10998–11010.
93. Bence, N. F.; Sampat, R. M.; Kopito, R. R. Impairment of the Ubiquitin–proteasome System by Protein Aggregation. *Science* **2001,** *292*, 1552–1555.
94. Bennett, E. J.; Bence, N. F.; Jayakumar, R.; Kopito, R. R. Global Impairment of the Ubiquitin–proteasome System by Nuclear or Cytoplasmic Protein Aggregates Precedes Inclusion Body Formation. *Mol. Cell* **2005,** *17*, 351–365.
95. Chen, J. J.; Tsu, C. A.; Gavin, J. M.; Milhollen, M.A; Bruzzese, F. J.; Mallender, W. D.; Sintchak, M. D.; Bump, N. J.; Yang, X.; Ma, J.; Loke, H.-K.; Xu, Q.; Li, P.; Bence, N. F.; Brownell, J. E.; Dick, L. R. Mechanistic Studies of Substrate-assisted Inhibition of Ubiquitin-activating Enzyme by Adenosine Sulfamate Analogues. *J. Biol. Chem.* **2011,** *286*, 40867–40877.
96. Vazquez-Carrillo, L. I.; Quintas-Granados, L. I.; Arroyo, R.; Hernández, G. M.; González-Robles, A.; Carvajal-Gamez, B. I.; Álvarez-Sánchez, M. E. The Effect of Zn^{2+} on Prostatic Cell Cytotoxicity Caused by *Trichomonas vaginalis*. *J. Integr. Omics* **2011,** *1*, 198–210.

CHAPTER 6

Vitamin D, Analogues, Drugs, and their Relevance to Cancer Inhibition

FRANCISCO TORRENS[1*] AND GLORIA CASTELLANO[2]

[1]*Institut Universitari de Ciència Molecular, Universitat de València, Edifici d'Instituts de Paterna, P. O. Box 22085, E-46071 València, Spain*

[2]*Departamento de Ciencias Experimentales y Matemáticas, Facultad de Veterinaria y Ciencias Experimentales, Universidad Católica de Valencia San Vicente Mártir, Guillem de Castro-94, E-46001 València, Spain*

[*]*Corresponding author. E-mail: torrens@uv.es*

ABSTRACT

When ultraviolet rays hit the skin, they initiate a series of reactions in the body that ultimately result in the synthesis of an active form of vitamin D_3, an essential hormone best known for promoting healthy bone formation by regulating blood concentrations of HPO_4^{2-} and Ca^{2+}. However, most if not all cells present receptors for vitamin D_3, suggesting that metabolites of the vitamin may play a role in myriad physiological processes. Indeed vitamin-D deficiencies, which could be caused by inadequate exposure to sunlight, were linked to a growing list of disorders, for example, osteoporosis, certain types of cancers, heart disease and susceptibility to infection, and autoimmune diseases, for example, multiple sclerosis, arthritis and type-1 diabetes.

6.1 INTRODUCTION

Molecular basis, for thinking that vitamin (Vit)-D (cf. Table 6.1) presents potential to prevent cancer, lies in its role in range of cellular mechanisms central to neoplasia development. From plants, Vit-D_2 presents lesser activity in human cells. Effects are mediated via Vit-D receptors (VDRs) expressed in cancer cells.

TABLE 6.1 Both Major Forms (Vitamers) of Vitamin (Vit)-D Exist.

Name	Chemical composition	Structure
Vit D_2	Ergocalciferol (made from ergosterol)	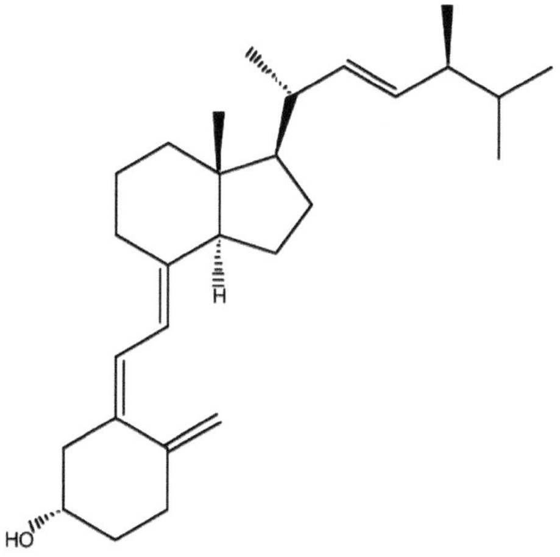
Vit D_3	Cholecalciferol (made from 7-dehydrocholesterol in the skin)	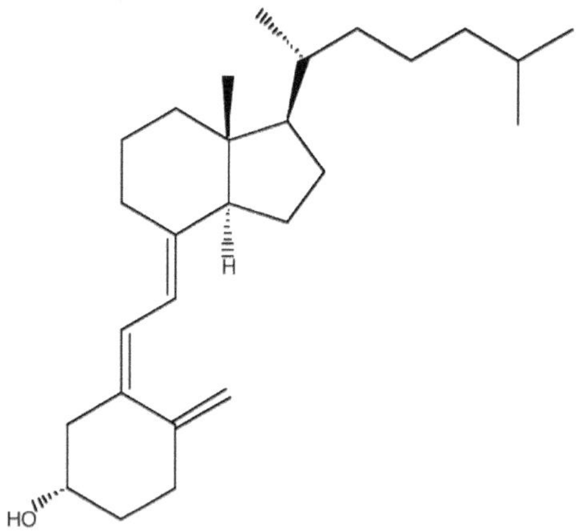

Genetic factors exist involved with cancer incidence/mortality, which are more common in northern latitudes (cf. Fig. 6.1).[1] The $1,25(OH)_2D_3$ is a hormone.

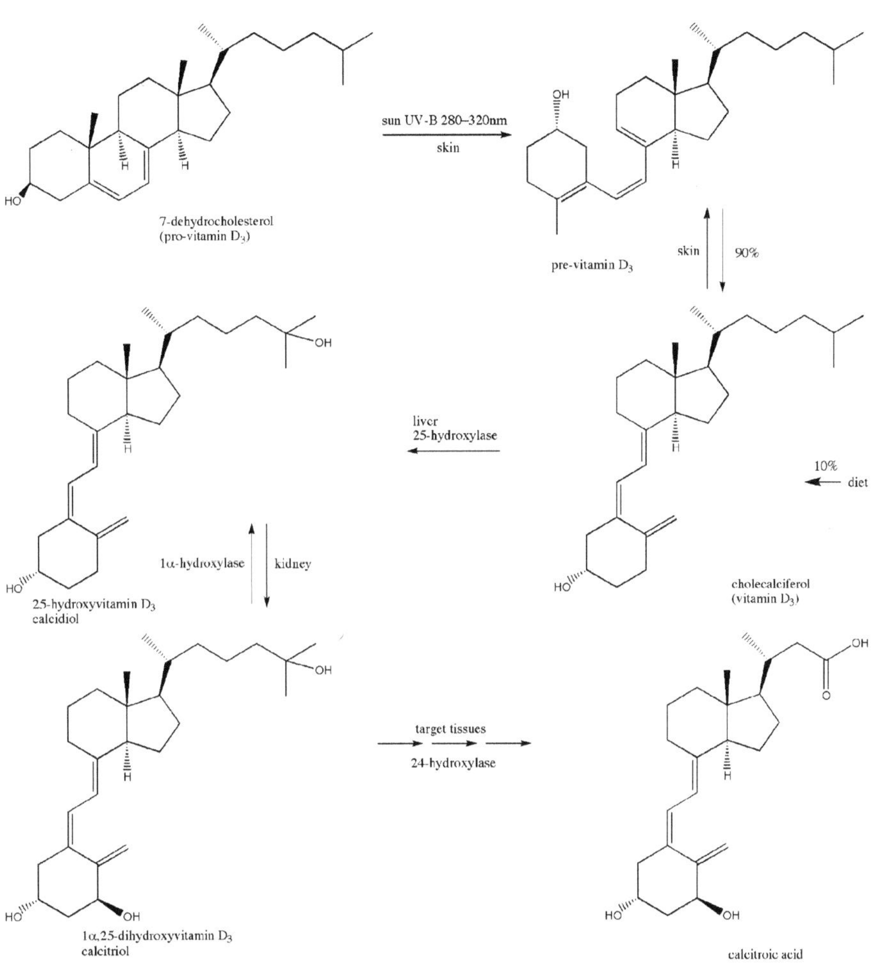

FIGURE 6.1 Vit D_3 synthesis, activation and catabolism; it is produced in skin by 7-dehydrocholesterol photolytic cleavage followed by thermal isomerization; it is transported to liver via serum Vit-D-binding protein where it is converted to 25(OH)D_3, major circulating metabolite of Vit D_3. Final activation, 1α-hydroxylation, occurs primarily in kidney forming 1,25$(OH)_2D_3$, its hormonal form. Catabolism inactivation is performed by 24-hydroxylase, which catalyzes oxidation series resulting in side-chain cleavage. Pro-Vit D_2 suffers similar process.

Vitamin D inhibits cell proliferation and promotes apoptosis in vitro, and several tissues locally produce its physiologically active form 1α,25-dihydroxyvitamin-D [1,25$(OH)_2$D], which presents anticarcinogenicity (cf. Fig. 6.2).[2] Tissues express VDR that enables 1,25$(OH)_2$D cellular action. Basic research strengthened credibility of hypothesis largely derived from ecological studies in US, by which higher Vit-D status is associated with lower cancer risk. 25-Hydroxyvitamin-D [25(OH)D] is precursor of physiologically active $1,25(OH)_2D$. Serum 25(OH)D level is result of skin exposure to sunlight, total Vit-D intake and other factors, for example, age/skin pigmentation. It varies with season with highest levels in summer/autumn.

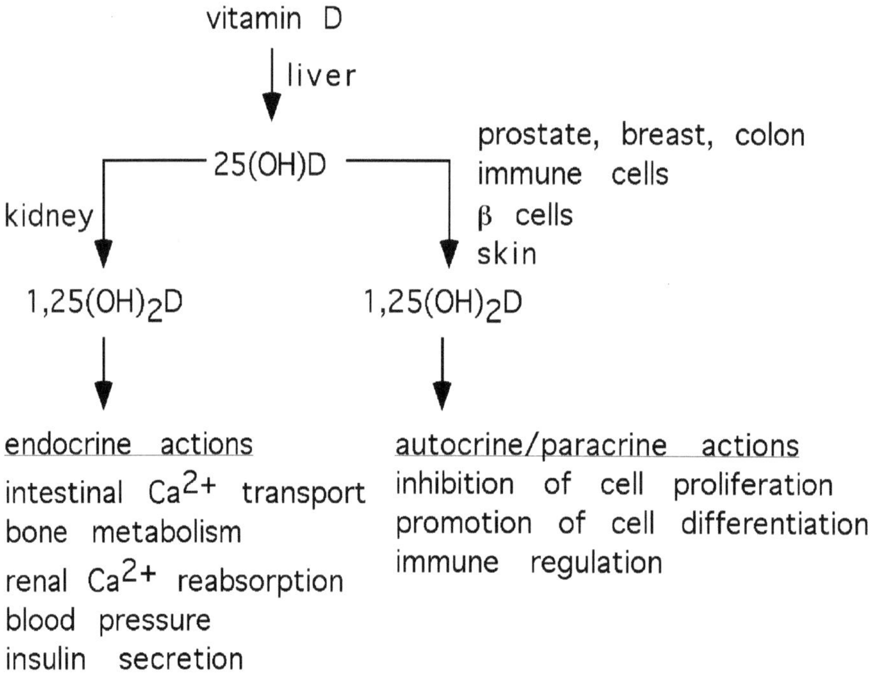

FIGURE 6.2 Renal and extrarenal $1,25(OH)_2D_3$ production serves endo, auto and paracrine functions.

The 25(OH)D shows half-life in circulatory system of 2–3 weeks[3]. In contrast, serum $1,25(OH)_2D$ is tightly biochemically regulated except in extreme-deficiency situations, in keeping with its role in Ca^{2+} homeostasis; it presents circulating half-time of 5–15 h and exhibits seasonal variability[4]. Serum 25(OH)D is considered as reflecting Vit-D status better than $1,25(OH)_2D$. First report of serum 25(OH)D/cancer inverse association was published for colorectal cancer (CRC) in US[5]. Observational studies examined serum 25(OH)D levels in relation to cancer risk. Meta-study analyzed 35 Vit-D/cancer studies. Researchers determined that 10 μM increase in serum Vit-D is associated with 15% lower risk of colon cancer. Analysis found

11% lower risk for breast cancer although finding is insignificant. Vitamin D was reviewed as a link between osteoporosis, obesity and diabetes.[6] Expression of VDR positively correlated with survival of urothelial bladder cancer patients.[7] Molecular and clinical studies were revised on role of Vit-D beyond the skeletal function.[8] Relationship between Vit-D structure and biological activity was reviewed.[9] A special issue on Vit-D and human health was published.[10]

In earlier publications, it was informed the modeling of complex multicellular systems and tumor–immune cells competition,[11] information theoretic entropy for molecular classification of oxadiazolamines as potential therapeutic agents,[12] molecular classification of 5-amino-2-aroylquinolines and 4-aroyl-6,7,8-trimethoxyquinolines as highly potent tubulin polymerization inhibitors,[13] polyphenolic phytochemicals in cancer prevention, therapy, bioavailability verses bioefficacy,[14] molecular classification of antitubulin agents with indole ring binding at colchicine-binding site,[15] molecular classification of 2-phenylindole-3-carbaldehydes as potential antimitotic agents in human breast cancer cells,[16] cancer, its hypotheses,[17] precision personalized medicine (PPM) from theory to practice and cancer.[18] It was reported how human immunodeficiency virus/acquired immunodeficiency syndrome destroy immune defences, hypothesis,[19] 2014 emergence, spread, uncontrolled Ebola outbreak,[20,21] Ebola virus disease, questions, ideas, hypotheses, models,[22] clinical translational research, cancer, diabetes, cardiovascular disease,[23] primary health, smoking habits, immunotherapy in cancer,[24] epigenomics, epithelial plasticity, clinical genetics, rare diseases,[25] predictive *in silico* global metabolism analysis, disease mechanism,[26] intracellular proteolysis, neoplasia, ubiquitin-proteasome system (UPS), cell cycle and cancer.[27] Present report reviews state of the art of Vit-D inhibition of prostate cancer.

6.2 COMPUTATIONAL METHOD

Human beings are made mostly of water and that perspective makes water a subject fitting for deep mathematical study. By *deep* Whitney meant looking at many possible ionic structures and investigating the possible macroscopic consequences thereof.[28] Water does a number of interesting things, not all of which one is intimately familiar with since one lives mostly within the circumstances of planet Earth's surface. However, there are interesting behaviors of some other molecules that are in deep ways similar to water. Therefore, she included some of those other molecules to ponder as analogues of water or vice versa, and guided to look for behaviors one already knows about from the other molecules. Therefore, she included not just water but also methane, other hydrocarbons and other atoms/molecules involved with water in various technolgies. The analytical tools used throughout is *algebraic chemistry* (AC), which is a quantitative model for predicting the energy budget for all sorts of changes between different ionization states of atoms that are involved in molecules or nanostructures undergoing chemical reactions or changes of physical state. She used the model to develop information about water: its several possible isomer molecules, the ways in which they might interact, their implications in regard to macroscopic physical states. Analogies between water and other simple molecules are highlighted.

Self-appointed pseudo-scientists generally dub as *semantic* basic advances without any serious study, because beyond their comprehension, which is also the case for Santilli's discoveries in biology, although by a rapidly decaying number of academicians.[29] The problem for pseudo-scientists is that all Santilli's discoveries present concrete applications, under development by industry and certainly not by academia with due exceptions because of novelty. Santilli introduced the notion of magnetoliquid, for the specific intent of initiating the transition from microwave devices exciting individual molecules of a liquid, for example, water, to a new generation of equipment that *disrupts the magnecular bond between molecules*, which development is evidently prohibited by the conventional *H-bridges* notion, because of their pure nomenclature character without quantitative treatment. By comparison, magnecular bonds in a liquid can indeed be treated quantitatively. Additionally all magnetic effects present a temperature at which they disappear (Curie temperature).

A new equipment that disrupts the magnecular bond between molecules can be attempted in a number of ways, for example, via microwaves, causing the magnecular Curie temperature at the *microscopic level of individual molecular couplings*, which equipment is currently under development by industry. One of the most important possible applications is cancer cure. Santilli devised a possible future elimination of cancer, which principle is the disruption of the magnecular bond between molecules whether within a cell, deoxyribonucleic acid (DNA) or other structures depending on the case at hand. In the event the procedure is possible, it is predicted not to require surgery, because the achievement of the Curie temperature for the disruption of magnecular bonds in DNA can be achieved by two microwaves, which are individually non-troublesome for human tissues but are disruptive solely at their intersection.

6.3 RESULTS: MOLECULAR BASIS FOR THINKING THAT VIT-D PREVENTS CANCERS

6.3.1 HLA-A2 GENE FREQUENCY/LATITUDE/OVARIAN/PROSTATE CANCER MORTALITY RATES

Molecular-target therapies are approaches to prostate/ovarian cancer treatment but, to ensure best response, careful patient selection based on immunological characteristics must be performed. De Petris et al. screened for human leukocyte antigen (HLA) type, 24 patients with advanced ovarian and 26 men with hormone-refractory prostate cancers, in order to be recruited to vaccine protocols.[30] Typing of HLA was performed with polymerase chain reaction (PCR) in ovarian and serological assay in prostate cancer patients. They extended results to population level, comparing HLA genotype frequencies in Europe with ovarian/prostate cancer mortality rates; they observed HLA-A2 phenotype overrepresentation in both compared to normal Swedish population. Higher phenotype frequency of allele found in Scandinavia decays as one moves south in Europe. Ovarian/prostate cancer mortality rates drop as demographic changes in HLA-A2. Observations must be confirmed by research to elucidate if HLA-A2 higher frequency is present, at diagnosis (risk factor), or is selected during disease course (prognostic factor). Fact suggests different strategies for specific immunotherapy in addition to first-line conventional treatments.

6.3.2 EPIDEMIOLOGY OF VITAMIN D, AND CANCER INCIDENCE AND MORTALITY

Giovannucci reviewed Vit-D epidemiology and cancer incidence/mortality.[31] In vitro/animal studies showed that Vit-D presents anticancer benefit, for example, versus progression/metastasis, against disease spectrum. Supporting Vit-D anticancer effect is cell ability to convert 25(OH)D primary circulating form into 1,25$(OH)_2$D, most active type (cf. Fig. 6.3). No epidemiologic study measured Vit-D concentrations/intakes on risk of total cancer incidence/mortality; however, higher rates of total disease mortality in regions with less ultraviolet (UV)-B, and among African-Americans and overweight/obese people, each associated with lower circulating Vit-D, are compatible with its benefit on mortality. Poorer survival from cancer, in individuals diagnosed in months when Vit-D levels are lowest, suggests its benefit against late carcinogenesis.

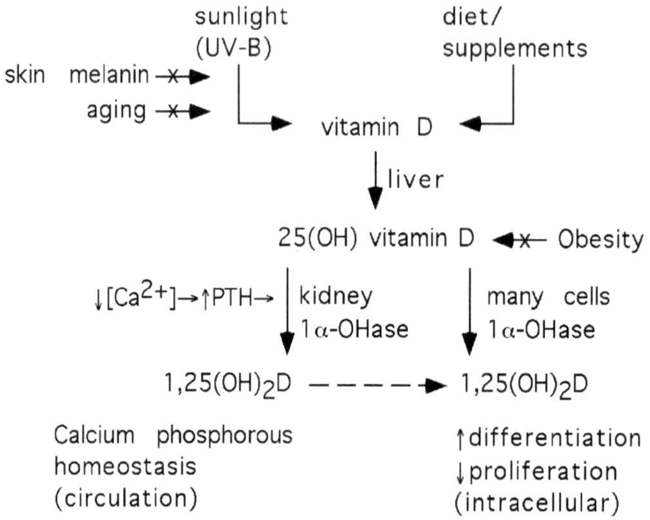

FIGURE 6.3 Vit-D/cancer mechanism. Main Vit-D sources are sunlight and diet/supplements. Darker skin, age and obesity are associated with lower 25(OH)D. Anti-cancer effects are due to conversion of 25(OH)D to 1,25$(OH)_2$D within cells, although circulating 1,25$(OH)_2$D contributes.

Individual cancer sites that were examined in relation to Vit-D status are colorectal, prostate (cf. Table 6.2) and breast diseases.

TABLE 6.2 Vitamin-D, Calcium and Cancer: Summary of Relationships from Epidemiologic Studies.

	Colorectal	Prostate
Ecologic (UV-B)	↓↓	↓ (weak)
Vitamin-D intake	↓↓	0
Circulating 25(OH)D		↓[a]
Circulating 1,25$(OH)_2$D	0	↓ (aggressive)
		0/↑ (non-aggressive)
Calcium/milk intake	↓↓	↑↑ (aggressive)
		0 (non-aggressive)

↓↓ = strong protection; ↓ = suggestive protection; ↑↑ = strong increased risk.
[a]Perhaps only for clinically low 25(OH)D levels.

For breast cancer, data are sparse to support conclusion. Evidence that higher 25(OH)D levels, via increased sunlight exposure or dietary/supplement intake, inhibit colorectal carcinogenesis is substantial (cf. Fig. 6.4a).

Bioevidence for 25(OH)D anticancer role is strong for prostate disease but epidemiologic data were not supportive (cf. Fig. 6.4b). Higher circulating $1,25(OH)_2D$ results more important than 25(OH)D for protection against aggressive, poorly differentiated prostate cancer. Unlike colorectal tumors, prostate cancers loose ability to hydroxylate 25(OH)D to $1,25(OH)_2D$ and rely on circulation as the latter main source; suppression of the latter circulating levels, by Ca^{2+} intake, explains why higher Ca^{2+}/milk ingestions increase risk of advanced prostate cancer. Given Vit-D potential benefit, research results priority.

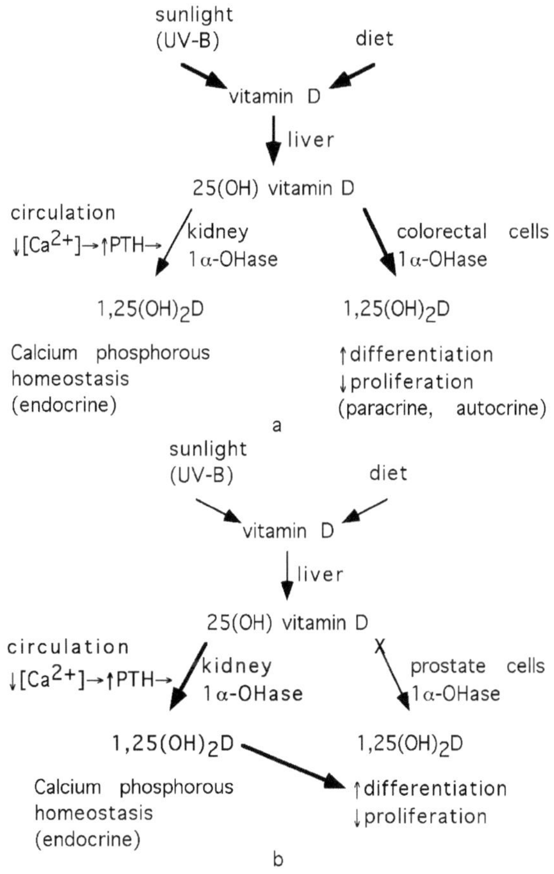

FIGURE 6.4 Mechanism for why Vit-D influences prostate/colorectal cancers differently: (a) Colorectal: cells maintain 1-α-hydroxylase activity, and main influence is from para/autocrine $1,25(OH)_2D$ produced intracellularly from 25(OH)D. (b) Prostate: 1-α-hydroxylase-activity loss reduces para/autocrine influence/reliance, on total circulating 25(OH)D, and increases circulating $1,25(OH)_2D$ power, which is tightly regulated by Ca^{2+}/P status.

6.3.3 MOLECULAR BASIS OF THE POTENTIAL OF VITAMIN-D TO PREVENT CANCER

Ingraham et al. reviewed molecular basis of Vit-D to prevent cancer, covering research findings in cell biology, epidemiology and clinical trials on its protection against breast, colon, prostate, lung and ovary cancer development.[32] They revised recommendations for optimal Vit-D, movement towards standards revision and reflections on healthy exposure to sunlight. They conducted literature search on databases Medline/Web of Knowledge for articles using terms: *vitamin D, calcitriol, cancer, chemoprevention, nuclear receptor, vitamin D receptor, apoptosis, cell cycle, epidemiology* and *cell adhesion molecule*. They selected articles, which focused on epidemiologic/clinical evidence for Vit-D's effects, and obtained additional reports from retrieved reference lists. Active Vit-D form presents protection against cancer development. Epidemiological studies show an inverse association between sun exposure, 25(OH)D serum levels and Vit-D intakes and developing/surviving cancer risk. Protective Vit-D effects result from its role as nuclear transcription factor, which regulates cell growth, differentiation, apoptosis and cellular mechanisms central to cancer development. Individuals present serum Vit-D levels lower than what appears to protect against cancer, and researchers are revising guidelines for optimal health, which will lead to improved public health policies and reduced disease risk. Research supports that improving Vit-D status would undergo protection versus neoplasia development. Researchers revise recommendations for optimal serum levels and sensible intensities of sun exposure to greater ones: 1000–4000 IU intakes will lead to more healthy level of serum 25(OH)D at 75 nM, which will defend against breast, colon, prostate, ovary, lungs and pancreas cancers. First randomized trial protected against breast cancer, and other clinical ones will lead to improved public health policies and fewer diseases.

6.3.4 META-ANALYSIS OF SERUM 25(OH)D LEVELS, CANCERS AND COLONIC ADENOMA

Epidemiological studies suggested reduced cancer risk associated with high Vit-D status. Gandini et al. reviewed with observational study meta-analyses of serum 25(OH)D level and colorectal, breast and prostate cancers and colonic adenoma.[33] They searched literature and did meta-regression analysis to compute dose-response effects. Because in case-control studies, serum 25(OH)D level is measured after cancer diagnosis, they did separate analyses for case-control/prospective studies. They identified 35 studies. The seven studies on colorectal adenomas were heterogeneous in terms of endpoint/control, for major confounding factors, and they did not perform data meta-analysis. Summary relative risk (SRR) and 95% confidence interval, for 10 μM increase in serum 25(OH)D, was 0.85±0.06 for colorectal (n = 2630 cases in N = 9 studies), 0.89±0.09 for breast (n = 6175, N = 10) and 0.99±0.04 for prostate (n = 3956, N = 11) cancers. For breast neoplasia, case-control studies (n = 3030) presented major limitations and obtained SRR = 0.83±0.04, whereas prospective studies (n = 3145) SRR = 0.97±0.06. For colorectal/breast cancers, case/control differences in season of blood draw, overweight/obesity or physical inactivity could not explain results. They found consistent inverse relationship between serum 25(OH)D levels and colorectal neoplasia; they discovered no breast/prostate cancers association.

6.3.5 VITAMIN D_3: PROTEASES, PROTEASE INHIBITORS AND CANCER

CRC is major human neoplasia. Epidemiologic/preclinical studies showed that 1,25$(OH)_2D_3$, most active metabolite of Vit D_3, is wide but not understood antitumor. Proteins are degraded by proteases whose activity results in turn controlled by endogenous inhibitors. Another critical component of protein degradation is UPS, which operates in nucleus/cytosol. Álvarez-Díaz group, using transcriptomic analyses of 1,25$(OH)_2D_3$ action in human CRC cells, revealed genes encoding proteases/inhibitors, for example, 1,25$(OH)_2D_3$ candidate target genes.[34] One gene is *CST5*, which encodes cystatin-D, inhibitor of several cysteine (Cys) proteases of family cathepsin, with more restricted pattern of tissue expression and narrower inhibitory profile than other cystatins. Their results showed that cystatin-D presents unexpected tumor suppressor effects in CRC, and its expression is lost during CRC progression. They characterized cystatin-D action mechanism in CRC and 1,25$(OH)_2D_3$ role controlling protein degradation. The 1,25$(OH)_2D_3$ increases cystatin-D protein levels in CRC cells. Induction is specific as no cystatin-SN initiation is found. Secreted-cystatin-D level is increased after 1,25$(OH)_2D_3$ treatment. Ectopic cystatin-D expression inhibits proliferation/migration in CRC cells. Using mutant cystatin-D proteins, they observed that Cys-protease inhibition is responsible for mediated migration inhibition but not proliferation inhibition. They studied effects of ectopic cystatin-D expression on bioparameters. Using cytokine arrays, they saw that expression is modulated by cystatin-D. By enzyme-linked immunosorbent assay (ELISA), they validated CX3CL1 (fractalkine) expression regulation by cystatin-D. Transcriptomic studies, using oligonucleotide microarrays, showed that ectopic cystatin-D expression regulates 69-genes ribonucleic acid (RNA): 23 (33%) result upregulated whereas 46 (67%), downregulated. Using Western blot/immunofluorescence, they validated E-cadherin (Cadh) regulation by cystatin-D overexpression. Microarray analysis of DNA showed that 1,25$(OH)_2D_3$ modifies UPS-genes expression, in SW480-ADH CRC cells.

6.3.6 COLON CANCER: VITAMIN-D AND WNT/β-CATENIN SIGNALING PATHWAY

Initial alteration of most CRCs is in genes adenomatous polyposis coli *(APC)*, *CTNNB1*/Cadh-associated protein β 1 (β-catenin, β-Cat) and *AXIN*.[35] Complex β-Cat/T-cell transcription factor 4 (TCF-4) imposes crypt progenitor phenotype on CRC cells (cf. Fig. 6.5).[36] Mutation of *APC*, *CTNNB1* or *AXIN* causes aberrant activation of wingless (Wg)/Int (Wnt)/β-Cat signaling pathway, which in some cancers is because of abnormally high expression of Wnt factors and/or inhibitors loss. Route Wnt/β-Cat is constitutively active in most colorectal adenomas and considered an initial/crucial event in human CRC.

Actions of 1,25$(OH)_2D_3$ are classic–endocrine [Ca^{2+}/inorganic phosphate (P_i) homeostasis and bony-biology regulation] and new–paracrine/autocrine (antiproliferative, pro-apoptotic/autophagia, pro-differentiating, anti-invasive, anti-angiogenic, anti-microbial and immunomodulation). Its use in anticancer therapy is hindered by hypercalcaemic effect at healing doses. It/analogues (deltanoids), designed to keep antitumor action and reduce hypercalcaemic one, are in pre-/clinical phases alone/in combination versus different neoplasias. It/VDR are distributed over organism indicating multiple effects/regulatory actions.[37] Expression of VDR is lost during colon cancer progression, which implies insensibility to 1,25$(OH)_2D_3$/analogues treatment. Rosen et al. reviewed Vit-D nonskeletal effects.[38] Muñoz group showed that Vit-D_3 promotes differentiation of colon carcinoma cells by E-Cadh induction and β-Cat-signaling inhibition.[39] They indicated that Ras homologue gene family (Rho), member A (RhoA)-Rho-associated coiled-coil containing protein kinase (ROCK)/p38 mitogen-activated protein kinase

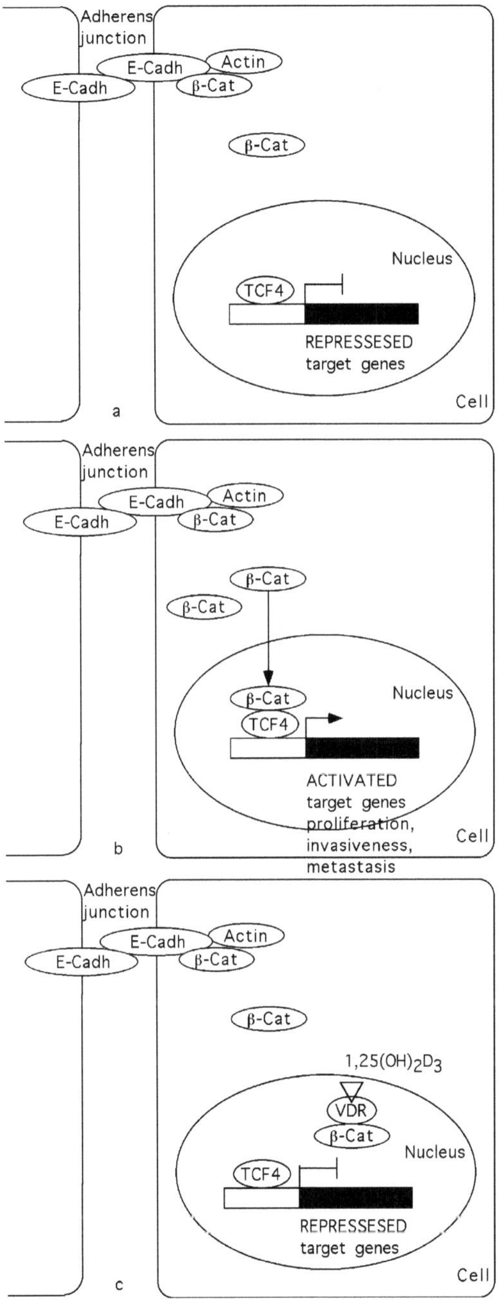

FIGURE 6.5 Wnt/β-Cat route: (a) normal epithelium, OFF; (b) colon cancer, ON; (c) $1,25(OH)_2D_3$, ON.

(p38MAPK)-mitogen- and stress-activated kinase 1 (MSK1) mediate Vit-D effects on gene expression, phenotype and Wnt pathway in colon cancer cells.[40] Cano et al. explained that transcription factor Snail controls epithelial–mesenchymal transitions by repressing E-Cadh expression.[41] Nieto reviewed superfamily Snail of Zn-finger transcription factors.[42] Muñoz group showed that SNAIL represses VDR expression/responsiveness in human colon cancer.[43] Bonilla group reported E-Cadh/VDR regulation by SNAIL/Zn finger E-box-binding homeobox (ZEB) 1, in colon cancer, and clinicopathological correlations.[44] Peinado et al. discussed Snail, ZEB and basic helix-loop-helix (bHLH) factors in tumor progression.[45] The Snail1/2 repress VDR expression and inhibit $1,25(OH)_2D_3$ action (cf. Fig. 6.6).

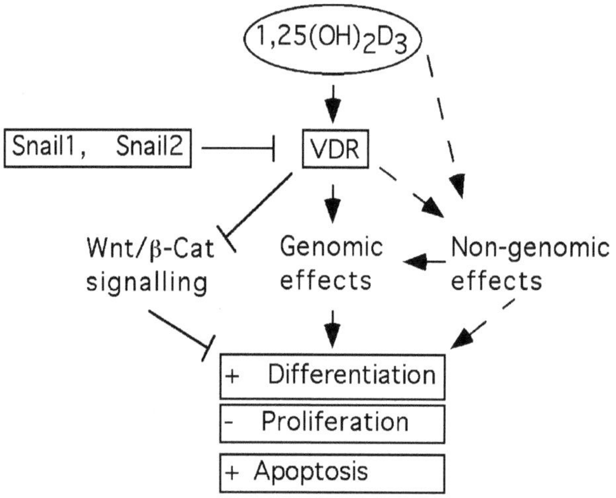

FIGURE 6.6 The Snail1 and Snail2 repress VDR expression and inhibit the action of $1,25(OH)_2D_3$.

Álvarez-Díaz et al. indicated that Cystatin-D is candidate tumor suppressor gene induced by Vit-D in human colon cancer cells.[46] Sprouty homologue-2 (*Drosophila*, SPRY2) expression is greater in invading-front cells (cf. Fig. 6.7). Casal group explained that proteomic analysis of $1,25(OH)_2D_3$ action on human colon cancer cells reveals link to splicing regulation.[47] They discussed colon cancer, Vit-D and Wnt/β-Cat route. It is not clear Vit-D route in nocturne (bat)/hair (mouse) mammals. Excess of Vit-D causes hypercalcaemia (Ca^{2+} excess in blood). Recommended blood Vit-D concentration is 20–30 μg/L[1]; lower levels in winter can be increased with supplements (without Ca^{2+}). Expression of VDR is induced by butyrate (from fiber digestion in gut), phitoestrogens (from soja, etc.) in colon epithelium via receptor type-β, epigallocatechin gallate (main polyphenol in green tea), etc. Vitamin-D is self-regulated: excessive UV-B degradates it; with ageing/obesity (it accumulates in fat), it works worse; Vit-A (cod liver oil, etc.) inhibits it.

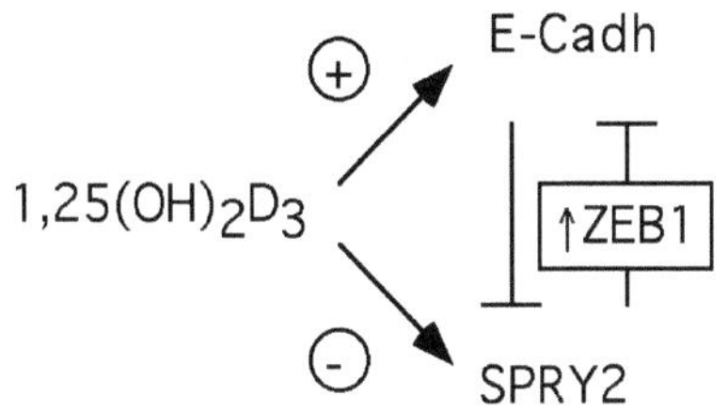

FIGURE 6.7 Expression of SPRY2 is greater in the invading-front cells.

6.3.7 VITAMIN D AND DEATH BY SUNSHINE

Sunlight exposure is major cause of skin cancer. The UV from the sun damages DNA by direct absorption and can cause skin cell death; it produces reactive oxygen species (ROSs) that interact with DNA to cause indirectly oxidative DNA damage (DD); it increases p53 accumulation in skin cells, which upregulates repair genes but promotes death of irreparably damaged cells. Sunlight benefit is Vit-D, which is formed following 7-dehydrocholesterol exposure in skin cells to UV. Relatively inert Vit-D is metabolized to bioactive compounds, for example, 1,25$(OH)_2D_3$. Its therapeutic uses proved beneficial in cancer types, and it prevents UV-induced cell death and DD in human skin cells. Dixon et al. reviewed its effects in skin cells that were exposed to UV, specifically signaling pathways involved in Vit-D-induced protection of skin cells from UV.[48]

6.4 RESULTS: VITAMIN-D, ANALOGUES AND PROSTATE CANCER PREVENTON/THERAPY

6.4.1 1α,25-DIHYDROXYVITAMIN D_3 RECEPTORS AND ACTIONS IN PROTEASE CANCER

Deficiency of Vit-D promotes prostate cancer with unknown mechanism. Skowronski et al. examined LNCaP, DU-145 and PC-3 human prostate neoplasic cell lines for presence of specific $1,25(OH)_2D_3$ (calcitriol, DCI) VDRs, and employed them to study hormone effects on cell proliferation/differentiation.[49] Ligand-binding experiments showed classical VDR in three lines, with apparent dissociation constant of 7.5, 5.4 and 6.3 × 10^{-11} M, respectively; corresponding binding capacity for all lines resulted 27, 31 and 78 fmol/mg protein. They confirmed VDR presence in three lines by immunocytochemistry; they identified one major 4.6 kb messenger RNA (mRNA) transcript, hybridizing with specific human VDR complementary DNA (cDNA) probe in all lines. Only DU-145/PC-3 exhibited $1,25(OH)_2D_3$-stimulated induction of 24-hydroxylase (cytochrome P450, CYP24, cf. Table 6.3) mRNA employed as Vit action marker. Physiological $1,25(OH)_2D_3$ levels inhibited LNCaP/PC-3 proliferation. However, in spite of high-affinity VDR presence, DU-145 proliferation was not inhibited by $1,25(OH)_2D_3$, which treatment caused dose-dependent stimulation of prostate-specific antigen (PSA) secretion by LNCaP. Three lines possess specific VDR and $1,25(OH)_2D_3$ treatment elicits antiproliferation/differentiation, which supports that Vit-D might exert benefit on prostate cancer risk.

Kumagai et al. reported CYP2B6 as growth-inhibitory/prognostic factor for prostate cancer.[50]

TABLE 6.3 Summary of Some CYP Families in Humans: Some of the Genes and Proteins they Encode.

Family	Function	Members	Names
CYP2	Drug and steroid metabolism	13 subfamilies, 16 genes, 16 pseudogenes	CYP2A6, CYP2A7, CYP2A13, CYP2B6, CYP2C8, CYP2C9, CYP2C18, CYP2C19, CYP2D6, CYP2E1, CYP2F1, CYP2J2, CYP2R1, CYP2S1, CYP2U1, CYP2W1
CYP24	Vit-D degradation	1 subfamily, 1 gene	CYP24A1
CYP27	*varied*	3 subfamilies, 3 genes	CYP27B1 (Vit-D_3 1α-hydroxylase, activates Vit-D_3), CYP27C1 (unknown function)

6.4.2 ACTIONS OF VITAMIN D_3, ANALOGUES ON HUMAN PROSTATE CANCER CELL LINES

Epidemiological studies suggest that Vit-D deficiency might promote prostate cancer, although mechanism is not understood. Skowronski et al. showed VDRs presence in LNCaP, PC-3 and DU-145 human prostate carcinoma cell lines, and in primary cultures of stromal/epithelial cells derived from normal/malignant gland tissues.[51] They showed that $1,25(OH)_2D_3$ elicits antiproliferation in cells; they compared $1,25(OH)_2D_3$ bioactions to natural Vit-D_3 metabolites and its several synthetic analogues, known to exhibit lesser hypercalcemic activity in vivo. In ligand-binding competition experiments, they showed following potency order in displacing $[3H]1,25(OH)_2D_3$ from VDR: EB-1089 > $1,25(OH)_2D_3$ > MC-903 > $1,24,25(OH)_3D_3$ > 22-oxacalcitriol (OCT) > 1α,25-dihydroxy-16-enecholecalciferol (Ro24-2637) > $25(OH)D_3$, with EB-1089 twofold more potent than native hormone. They found no competitive activity for 25-hydroxy-16,23-dienecholecalciferol. When compared for ability to inhibit LNCaP proliferation, MC-903, EB-1089, OCT and Ro24-2637 exhibited four, three and twofold greater inhibitory activity than $1,25(OH)_2D_3$. Although OCT/Ro24-2637 exhibit ×10/14 lower affinity for VDR than $1,25(OH)_2D_3$, they inhibited LNCaP proliferation with potency greater than native hormone. Relative potency of Vit-D_3 metabolites/analogues, to inhibit cell proliferation, correlated with compounds ability to stimulate PSA secretion, by LNCaP, and potency to induce $25(OH)D_3$-CYP24 mRNA transcript in PC-3. Synthetic Vit-D_3 analogues, which exhibit reduced calcemic activity, can elicit antiproliferative effects and other bioactions in LNCaP/PC-3. Although VDR binding is critical for $1,25(OH)_2D_3$ action, analogue data indicate that additional factors contribute to bioresponse magnitude. Antiproliferative effects of several synthetic analogues, which exhibit lesser calcemic activity than $1,25(OH)_2D_3$, suggest that compounds result useful as additional therapeutic option for prostate cancer treatment.

6.4.3 VITAMIN D AND PROSTATE CANCER

Feldman et al. showed VDR presence in various human prostate cancer cell lines and primary cultures derived form normal, benign prostatic hyperplasia (BPH) and gland cancer.[52] The $1,25(OH)_2D$ induced several bioresponses in cells, for example, growth inhibition and PSA stimulation. Based on examples in different malignant/prostate cells that Vit-D is antiproliferative and promotes cellular maturation, it results important cellular modulator of growth/differentiation and Ca^{2+} homeostasis regulator. It presents benefit on malignancies, for example, prostate cancer; its role in disease remains undetermined but $1,25(OH)_2D$ results useful in chemoprevention/differentiation therapy. They believe available data provide basis for its use to treat prostate cancer in patients, and research is warranted to define better its potential therapeutic utility.

6.4.4 VIT-D INHIBITION OF PROSTATE ADENOCARCINOMA GROWTH AND METASTASIS

Risk factors for prostate cancer-related mortality include old age, black race and residence in northern latitudes. Getzenberg et al. examined in vitro/vivo effects of 1,25$(OH)_2D_3$ and lesser-hypercalcemic analogues, on Dunning rat prostate adenocarcinoma model.[53] In order to evaluate 1,25$(OH)_2D_3$ effect on prostate cancer in vitro, they used highly metastatic Mat-lylu (MLL) and moderate R3327-AT-2 (AT-2) Dunning gland cell lines, and examined effects on growth, clonogenicity, differentiation and cell cycle. In vivo analysis included compound effects examination on tumor growth/metastasis. Using 3-day microculture tetrazolium (3-(4,5-dimethylthiazol)-2,5-diphenyltetrazolium bromide) (MTT) and 7-day clonogenic assays, 1,25$(OH)_2D_3$ showed growth inhibitory effect with concentration for 50% inhibition (IC_{50}) of 20 μM for MLL/AT-2. Cell-cycle analysis of treated MLL cells [10 μM 1,25$(OH)_2D_3$ for 48 h] presented 25% more cells in $G_{0/1}$ phases than did control ones. In order to examine in vivo effect of 1,25$(OH)_2D_3$ and lesser hypercalcemic analogue, Ro25-6760 (6760), on MLL prostate cancer growth and metastasis, tumors (5×10^5 cells) were implanted subcutaneously into Copenhagen rats flank, on day that treatment was initiated with 1,25$(OH)_2D_3$ (1 μg)/6760 (1–5 μg); rats received treatment 3 times a week. After 3 weeks, 1,25$(OH)_2D_3$ and 6760 (5 μg dosing) resulted in tumor volume inhibition and reduction in lung metastases number/size. Preclinical studies showed profound in vitro/vivo antiproliferative/differentiating 1,25$(OH)_2D_3$/6760 effects, on prostate cancer, and suggested that drugs undergo benefit in advanced-disease treatment.

6.4.5 TREATMENT OF EARLY RECURRENT PROSTATE CANCER: 1α,25-DIHYDROXYVITAMIN D_3

Experimental/epidemiological data indicate that 1,25$(OH)_2D_3$ presents anti-proliferation on human prostate cancer cells. Gross et al. performed open label, nonrandomized pilot trial to determine whether 1,25$(OH)_2D_3$ therapy is safe/efficacious for early recurrent prostate cancer.[54] They hypothesized that 1,25$(OH)_2D_3$ therapy slows PSA rise rate compared with pretreatment one. After primary treatment with radiation/surgery, recurrence was indicated by rising serum PSA levels documented on at least three occasions. Seven subjects completed 6–15 months of 1,25$(OH)_2D_3$ therapy, starting with 0.5 μg drug daily and slowly increasing to maximum dose of 2.5 μg daily, depending on individual calciuric/calcemic responses. Each subject served as his own control comparing PSA rise rate before/after 1,25$(OH)_2D_3$ treatment. As determined by multiple regression analysis, PSA rise rate during versus before 1,25$(OH)_2D_3$ therapy significantly decreased in 6/7 patients, while in remaining man deceleration in PSA rise rate did not reach significance. Overall decreased rate of PSA rise was significant (Wilcoxon signed-rank test). Dose-dependent hypercalciuria limited given maximal 1,25$(OH)_2D_3$ therapy (1.5–2.5 μg daily). Pilot study provided preliminary evidence that 1,25$(OH)_2D_3$ slows PSA rise rate in select cases, although dose-dependent calciuric adverse effects limit its clinical usefulness. Development of 1,25$(OH)_2D_3$ analogues with decreased calcemic side effects results promising, since they may be more effective for treating prostate cancer.

6.4.6 LIAROZOLE-1,25$(OH)_2D_3$ SYNERGESIS TO INHIBIT DU-145 PROSTATE CANCER GROWTH

The 1,25$(OH)_2D_3$ inhibits proliferation of many cancer cells in culture but not aggressive human prostate disease cell line DU-145. Feldman and co-workers postulated that 1,25$(OH)_2D_3$-resistant phenotype, in DU-145, results from high levels of 25(OH)D-CYP24 expression induced by 1,25$(OH)_2D_3$ treatment.[55] As CYP24 initiates 1,25$(OH)_2D_3$ catabolism, they presumed that high level of enzyme induction could limit Vit-antiproliferation effectiveness. In order to examine hypothesis, they explored combination therapy with liarozole fumarate (R85246), imidazole derivative in trials for prostate cancer therapy; as imidazole derivatives inhibit CYPs, they postulated that R85246 inhibits CYP24 increasing 1,25$(OH)_2D_3$ half-life enhancing its antiproliferation on DU-145. Cell growth was assessed by viable-cell measurement using 3-(4,5-dimethylthiazol-2-yl)-5-(3-carboxymethoxyphenyl)-2-(4-sulphophenyl)-2H-tetrazolium (MTS) assay. Used alone neither 1,25$(OH)_2D_3$ (1–10 nM) nor liarozole (1–10 μM) inhibited DU-145 growth; however, when added together 10 nM/1 μM inhibited growth 65% after four culture days. They used thin-layer chromatography (TLC), to assess CYP24 activity, and showed that liarozole (1–100 μM) inhibited it in dose-dependent manner. Liarozole treatment caused increase in $1,25(OH)_2D_3$ half-life from 11 to 31 h, which produced homologous VDR upregulation and, in drug presence, effect was amplified enhancing Vit. Western blot analyses demonstrated that DU-145, treated with $1,25(OH)_2D_3$/liarozole, presented greater VDR upregulation than cells treated with either. They showed that liarozole augments $1,25(OH)_2D_3$ ability to inhibit DU-145 growth. Mechanism results because of CYP24 activity inhibition, leading to increased $1,25(OH)_2D_3$ half-life and augmentation of VDR homologous upregulation. They raised possibility that combination therapy, using $1,25(OH)_2D_3$/liarozole or other CYP24 inhibitors in nontoxic doses, serves as prostate cancer treatment. They reviewed vit-D and prostate cancer.[56]

6.4.7 GLUCOCORTICOIDS PROMOTE ANDROGEN-INDEPENDENT PROSTATE-CANCER GROWTH

Androgen receptor (AR) is involved in prostate cancer development, growth and progression, which usually progresses from androgen-dependent to -independent tumor, making androgen ablation therapy ineffective; however, mechanisms for development of androgen-independent disease are unclear. More than 80% clinically androgen-independent prostate tumors show high AR expression. In some prostate cancers, AR levels are increased because of gene amplification/overexpression whereas, in others, it is mutated; nonetheless, its involvement in disease transition to androgen-independent growth and subsequent failure of endocrine therapy are misunderstood. Zhao et al. showed that in prostate cancer cells from patient who failed androgen ablation therapy, doubly mutated AR functioned as high-affinity cortisol/cortisone receptor (AR^{ccr}).[57] Cortisol, main circulating glucocorticoid, and its metabolite cortisone stimulate growth of prostate cancer cells and increase PSA secretion in androgen absence. Physiological concentrations of free cortisol and total cortisone, in men, exceeded AR^{ccr} binding affinity and would activate receptor, promoting prostate cancer cell proliferation. They showed mechanism for androgen-independent growth of advanced prostate cancer. Understanding mechanism, and recognizing presence of glucocorticoid-responsive AR mutants, are important for therapy form development for treatment of prostate cancer subset.

6.4.8 THE DEVELOPMENT OF ANDROGEN-INDEPENDENT PROSTATE CANCER

Feldman and Feldman reviewed development of androgen-independent prostate cancer.[58] Normal prostate and early-state gland cancers depend on androgens for growth/survival, and androgen ablation therapy causes them to regress. Cancers that are not cured by surgery eventually become androgen independent, rendering antiandrogen therapy ineffective; but, how does androgen independence arise? They predicted that understanding pathways, which lead to development of androgen-independent prostate cancer, will pave way to effective therapies for untreatable diseases.

6.4.9 REDUCED 1α-HYDROXYLASE ACTIVITY IN PROSTATE CANCER CELLS CORRELATION

Evidence from epidemiological, molecular and genetic studies suggests Vit-D role in prostate cancer development/progression. In experimental models and clinical trials, $1,25(OH)_2D_3$ exerted antiproliferative, prodifferentiating and antimetastatic/invasive effects on prostatic epithelial cells. Because direct clinical application of $1,25(OH)_2D_3$ is limited by hypercalcemia adverse effect, Hsu et al. investigated potential therapeutic utility of its lesser calcemic precursor $25(OH)D_3$, which is converted locally within prostate to the former by 1α-hydroxylase (CYP27B1), which activity quantification in human prostatic epithelial cells, by enzyme/substrate reaction analyses, revealed decreased activity in cells derived from adenocarcinomas, compared with normal/BPH tissues.[58] In growth assays, they found that $25(OH)D_3$ inhibited normal/BPH cells growth similarly to $1,25(OH)_2D_3$. In contrast, in primary cultures of cancer cells and established cell lines, $25(OH)D_3$ antiproliferative action resulted lesser pronounced than $1,25(OH)_2D_3$. Their results indicated that growth inhibition by $25(OH)D_3$ depends on endogenous CYP27B1 activity, which is deficient in prostate cancer cells; finding presents ramifications for disease prevention/therapy with Vit-D compounds.

6.4.10 IGFBP-3 MEDIATES $1,25(OH)_2D_3$ GROWTH INHIBITION IN LNCAP PROSTATE CANCER

Boyle et al. determined that insulin-like growth factor-binding protein-3 (IGFBP-3) induction, by $1,25(OH)_2D_3$, is necessary component of

Vit-mediated growth inhibition of LNCaP human prostate cancer cells; induction of cyclin-dependent kinase inhibitory protein p21/WAF1/CIP1 by 1,25$(OH)_2D_3$ is mediated by IGFBP-3.[60] They determined IGFBP-3 induction by 1,25$(OH)_2D_3$, by enzyme-linked immunosorbent assay for IGFBP-3 protein and Northern blot analysis for IGFBP-3 mRNA; they controlled LNCaP growth assays measuring DNA content; they found out IGFBP-3 contribution towards 1,25$(OH)_2D_3$-mediated growth inhibition, adding either antisense oligonucleotides or immunoneutralizing antibodies to culture media of growth assays; they verified WAF1/CIP1 regulation by Western blot analysis. Adding 1,25$(OH)_2D_3$ to LNCaP showed that Vit upregulated IGFBP-3 at cell mRNA/protein levels threefold over control; adding IGFBP-3 protein to LNCaP growth medium inhibited growth; adding IGFBP-3 antisense oligonucleotides or antibodies, directed towards it, abolished 1,25$(OH)_2D_3$ growth inhibition indicating that effect is IGFBP-3 dependent; in order to connect mechanisms of IGFBP-3/1,25$(OH)_2D_3$-mediated growth inhibition, they showed that IGFBP-3 upregulates WAF1 protein expression to twofold over control; adding IGFBP-3 immunoneutralizing antibody prevented WAF1 1,25$(OH)_2D_3$-induced upregulation. The 1,25$(OH)_2D_3$ upregulates IGFBP-3 in LNCaP at mRNA/protein levels; 1,25$(OH)_2D_3$ growth inhibition on LNCaP depends on active IGFBP-3, as evidenced by loss of growth inhibition induced by IGFBP-3 antisense oligonucleotide and immunoneutralization experiments; IGFBP-3/1,25$(OH)_2D_3$ connection lies in cyclin-dependent kinase inhibitory protein WAF1, since IGFBP-3/1,25$(OH)_2D_3$ each upregulate protein and both inhibit LNCaP growth; they hypothesized that action mechanism by which IGFBP-3/1,25$(OH)_2D_3$ induce growth inhibition is WAF1 induction, because IGFBP-3 immunoneutralizing antibodies abrogate 1,25$(OH)_2D_3$-mediated WAF1 upregulation and growth inhibition.

6.4.11 INHIBITION OF PROSTATE CANCER GROWTH BY VITAMIN D: EXPRESSION REGULATION

Prostate cancer cells express VDRs, and 1,25$(OH)_2D_3$ inhibits growth of epithelial cells derived from normal, BPH and gland neoplasia, and established disease cell lines.[61] Growth inhibition of 1,25$(OH)_2D_3$ in cell cultures are modulated tissue by presence/activities of enzymes 25$(OH)D_3$ CYP24, which indicates 1,25$(OH)_2D_3$/25$(OH)D_3$ CYP27B1 inactivation, which catalyzes synthesis. In LNCaP human prostate cancer cells, 1,25$(OH)_2D_3$ exerts antiproliferative activity by cell-cycle arrest via IGFBP-3 expression induction which, in turn, increases levels of cell-cycle inhibitor p21 leading to growth arrest. The cDNA microarray analyses of primary prostatic epithelial and gland cancer cells reveal that 1,25$(OH)_2D_3$ regulates target genes, expanding anticancer activity mechanisms and raising potential therapeutic targets, which genes are involved in growth regulation, protection from oxidative stress (OS) and cell–cell/matrix interactions. A clinical trial showed that 1,25$(OH)_2D_3$ can slow PSA rise rate in prostate cancer patients, showing concept proof that it exhibits therapeutic activity in men. Research on 1,25$(OH)_2D_3$/analogues role for prostate cancer therapy/chemoprevention is pursued.

6.4.12 NUCLEAR RECEPTOR LIGANDS–VITAMIN-D SIGNALING PATHWAY INTERACTION

Hormonal ligands and/or nuclear receptors that mediate their actions were targeted for prostate cancer therapy.[62] Androgens, ligands for AR, are critical for prostate cancer growth. Its production inhibition was treatment mainstay for advanced prostate cancer for decades. Other tested targets include retinoid receptors (RAR/RXR), glucocorticoid receptors (GRs), oestrogen receptors (ORs) and peroxisome proliferator-activated receptors (PPARs). The 1,25$(OH)_2D_3$, acting via VDR, presents tumor suppressive activities in prostate, for example, proliferation inhibition, apoptosis/differentiation induction and cellular invasion reduction. Because of properties, 1,25$(OH)_2D_3$ and lesser hypercalcemic analogues are evaluated as agents to prevent/treat prostate cancer. Androgens, retinoids, glucocrticoids, oestrogens and PPAR agonists in/directly impact Vit-D signaling pathways and vice versa. In order to design effective strategies to use 1,25$(OH)_2D_3$ to prevent/treat prostate cancer, interactions of other nuclear receptors/ligands with Vit-D signaling pathway need to be considered.

6.4.13 FULVESTRANT DOWNREGULATES ANDROGEN RECEPTOR EXPRESSION

The AR plays role in prostate cancer development/progression. Targeting AR for downregulation results strategy for treating prostate cancer, especially hormone-refractory/androgen-independent one. Bhattacharyya et al. showed that antioestrogen fulvestrant (ICI-182780, ICI) suppressed AR expression in human prostate cancer cells, for example, androgen-independent ones.[63]

In LNCaP cells, ICI (10 μM) treatment decreased AR mRNA expression by 43% after 24 h and protein one by 50% after 48 h. They examined AR downregulation mechanism by ICI in LNCaP cells. The ICI did not bind to T877A-mutant AR present in LNCaP cells nor did it promote AR proteasomal degradation; it did not affect AR mRNA/protein half-life; however, it decreased activity of AR promoter-luciferase reporter plasmid transfected into LNCaP cells, suggesting direct repression of AR gene transcription. As result of AR downregulation by ICI, androgen induction of PSA mRNA and protein expression were attenuated. The LNCaP cell proliferation was inhibited by ICI treatment: after 6 days, 70% growth inhibition resulted in androgen-stimulated LNCaP cells. They demonstrated that antioestrogen ICI is potent AR downregulator, which causes inhibition of prostate cancer cell growth; they showed that AR downregulation by ICI results effective for treatment of all prostate cancers, especially AR-dependent androgen-independent one.

6.4.14 A PHASE-II TRIAL OF 1,25$(OH)_2D_3$ AND NAPROXEN IN RECURRENT PROSTATE CANCER

Androgen-deprivation therapy is used in patients with progressive prostate cancer, but can be associated with adverse effects. Srinivas and Feldman determined whether 1,25$(OH)_2D_3$/naproxen treatment is effective, in safely delaying prostate cancer growth/progression in men with early recurrent disease.[64] Patients with biochemical relapse, after local therapy for prostate cancer, were treated with high dose 1,25$(OH)_2D_3$ (DN101, Novacea, 45 μg once per week) and naproxen (375 mg twice daily) for 1 year, and followed with serum PSA levels and imaging studies. 21 patients were enrolled in trial. Four patients met criteria for progression with PSA doubling time (PSADT), which decreased while on therapy. 14 patients presented PSADT prolongation compared to baseline. Combination therapy with weekly 1,25$(OH)_2D_3$ and daily naproxen resulted tolerated by most patients, and PSADT prolongation was achieved in 75% patients.

6.4.15 PROSTAGLANDIN SYNTHESIS/ACTIONS INHIBITION BY GENISTEIN AND SOY

Soy constitutes a food recommended in Alzheimer-patient nutrition. On one hand, it shows a high content in lecithin that can help to strengthen acetylcholine levels. On the other, its isoflavones, for example, phytoestrogen genistein and dadzein, present mainly phytooestrogenic (it contains natural oestrogens) and antioxidant properties, so that its utility is high. Genistein is inhibitor of enzyme glycogen synthase kinase 3 (GSK-3), which is overexpressed in Alzheimer-patient brains and is related to all pathological processes, for example, amyloid plates, neurofibrillar coils and neuronal death, so that isoflavones can result interesting for Alzheimer disease prevention. Abdallah et al. reported chemical reaction of soybean flavonoids with DNA.[65] Soy and genistein inhibit prostate cancer development/progression. In cultured cells and prostate cancer patients, Swami et al. revealed pathway for genistein actions: synthesis/bioaction inhibition of prostaglandins (PGs), disease growth stimulators.[66] In cell-culture experiments, genistein decreased cyclooxygenase-2 (COX-2) mRNA/protein expression in LNCaP/PC-3 prostate cancer cell lines, and primary gland epithelial cells, and increased 15-hydroxyprostaglandin dehydrogenase (15-PGDH) mRNA levels in primary gland cells; it reduced PG E_2 (PGE_2) secretion by cells; it decreased EP4/FP PG receptor mRNAs providing additional mechanism for PG bioeffects suppression; it attenuated growth stimulatory effects of exogenous PGs and endogenous ones, derived from precursor arachidonic acid. They performed pilot randomized double blind clinical study in which placebo, or soy isoflavone supplements, were given to prostate cancer patients in neo-adjuvant setting for 2 weeks before prostatectomy. Gene expression changes were measured in prostatectomy specimens. In prostate cancer patients ingesting isoflavones, they observed decreases in prostate COX-2 mRNA and increases in p21 mRNA. There were correlations between COX-2 mRNA suppression, p21 mRNA stimulation and serum isoflavone levels. They proposed that PG pathway inhibition contributes to beneficial effect of soy isoflavones, in prostate cancer chemoprevention/treatment.

6.4.16 VITAMIN-D METABOLISM AND ACTION IN THE PROSTATE: IMPLICATIONS

Swami et al. reviewed Vit-D metabolism/action in prostate with implications for health/disease.[67] Prostate neoplasia is second most common cancer in men worldwide. Epidemiological, molecular and cellular studies implicated Vit-D deficiency as risk factor for prostate cancer development/progression. Studies using cell-culture systems and animal models suggest that Vit-D acts to reduce prostate cancer growth, via cellular progression/differentiation regulation; however, although preclinical studies indicate antineoplasic activity, therapeutic benefit proof in men is lacking. Antiproliferative/prodifferentiating Vit-D properties were attributed to 1,25$(OH)_2D_3$, its hormonally active form, acting via VDR. Two key enzymes mediate Vit-D metabolism in target tissues: CYP27B1, which catalyzes 1,25$(OH)_2D_3$ synthesis from 25(OH)D, and CYP24, which catalizes initial step in 1,25$(OH)_2D_3$ conversion to lesser active metabolites. Factors affect 1,25$(OH)_2D_3$ synthesis/catabolism balance and manoeuvres, for example, its combination therapy with other drugs, were explored to treat prostate cancer and reduce risk.

6.4.17 *VITAMIN D: HEMATOPOIESIS/IMMUNE SYSTEM AND CLINICAL APPLICATIONS*

It was reviewed Vit-D, its effect on hematopoiesis and immune system, and clinical applications.[68] After maximum levels of Vit-D are achieved, UV converts pre-D to lumisterol and tachysterol, preventing higher levels to be reached (cf. Fig. 6.8).

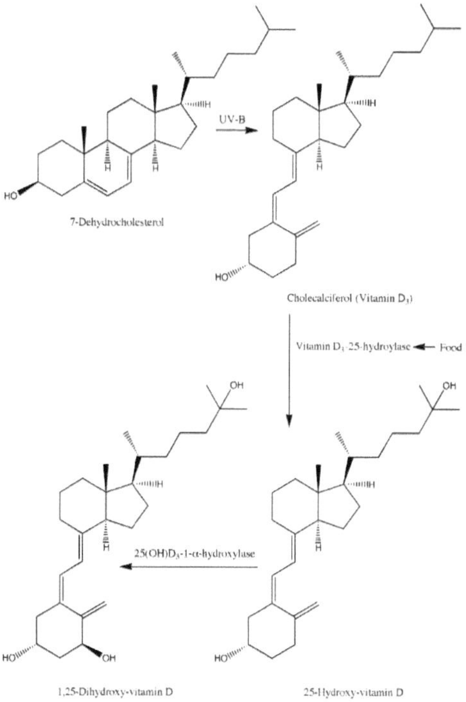

FIGURE 6.8 Vit-D synthesis and metabolic pathway: Vit-D metabolites are transported in blood bound primarily to Vit-D binding protein (DBP) and albumin.

6.5 FINAL REMARKS

From the present results and discussion, the following final remarks can be drawn:

1. Improving vitamin-D undergoes cancer protection; recommendations increased for serum levels, and sun exposure intensities 1000–4000 IU intakes lead to more healthy level of serum vitamin at 75 nM, which defends versus breast cancer, etc. Inverse relation exists between serum vitamin levels and CRC; no breast/prostate disease association appeared. Gene *CST5* acts as tumor suppressor with unpredicted effects, which contributes to vitamin-D_3 antitumor action. Ectopic cystatin-D modifies gene expression profile of HCT116; changes explain its antitumor action independently of cathepsin inhibition. Vitamin-controlled genes are related to protein degradation regulating integrity/stability.
2. Results support a protective effect of vitamin-D versus colon cancer.
3. Human prostate carcinoma cells LNCaP/DU-145/PC-3 possess specific vitamin-D receptor; vitamin treatment elicits differentiation exerting benefit.
4. Vitamin-D_3 analogues that exhibit reduced calcemic activity elicit antiproliferation in LNCaP/PC-3; its-receptor binding is critical. Antiproliferation of analogues with lesser calcemic activity shows therapy.
5. There are in vitro/vivo differentiating effects of vitamin D_3/ Ro25-6760 on prostate cancer; drugs undergo benefit.
6. Liarozole augments vitamin-D_3 inhibition of DU-145 growth; mechanism results because of CYP24 inhibition, increasing vitamin half-life and receptor homologous upregulation. Vitamin/liarozole combination therapy serves as prostate cancer treatment.
7. Understanding androgen-independent growth of advanced prostate cancer, and recognizing presence of glucocorticoid responsive androgen receptor mutants, result important for therapy.
8. Understanding pathways, which lead to development of androgen-independent prostate cancer, will pave way to untreatable-disease therapies.
9. Growth inhibition by vitamin D_3 depends on CYP27B1 activity, which is deficient in prostate cancer cells; there are disease prevention/therapy applications with vitamin compounds.
10. Vitamin D_3 upregulates IGFBP-3 in LNCaP at mRNA/protein levels; growth inhibition on LNCaP depends on active IGFBP-3, as

evidenced by its loss induced by IGFBP-3 antisense oligonucleotide and immunoneutralization; possible connection lies in cyclin-dependent kinase inhibitory protein WAF1, since each upregulates protein and both inhibit LNCaP growth. The IGFBP-3/vitamin induce growth inhibition via WAF1 induction, because the former immunoneutralizing antibodies abrogate the latter-mediated WAF1 upregulation and growth inhibition.

11. Vitamin D_3 slows rate of prostate-specific antigen rise in gland cancer patients proving therapy; there is vitamin/analogues-role research for disease treatment/chemoprevention.
12. Antioestrogen fulvestrant results potent androgen receptor down-regulator, which inhibits prostate cancer cell growth; downregulation undergoes effective for treatment of disease, especially androgen receptor-dependent androgen-independent one.
13. Weekly vitamin D_3/daily naproxen combination therapy resulted in prostate-specific antigen prolongation, doubling time in 75% men.
14. Prostaglandin pathway inhibition contributes to genistein benefit in prostate cancer chemoprevention/treatment.

Further work will deal with stem cells and their connection with cancer, prostate disease, MEK/ERK pathway overactivation in liver tumors and resistance to transforming growth factor-β-induced death.

ACKNOWLEDGMENTS

The authors thank support from Generalitat Valenciana (Project No. PROMETEO/2016/094) and Universidad Católica de Valencia San Vicente Mártir (Project No. 2019-217-001).

KEYWORDS

- **cancer prognosis**
- **apoptosis**
- **prostate cancer patient**
- **vitamin-D receptor (VDR)**
- **vitamin-D metabolism**
- **calcium homeostasis**

REFERENCES

1. Bonnet, R. *Chemical Aspects of Photodynamic Therapy*; Gordon and Breach: Amsterdam, 2000.
2. Dusso, A. S.; Brown, A. J.; Slatopolsky, E. Vitamin D. *Am. J. Physiol. Renal Physiol.* **2005,** *289*, F8–F28.
3. Tjellesen, L.; Christiansen, C. Vitamin D Metabolites in Normal Subjects During One Year. A Longitudinal Study. *Scand. J. Clin. Lab. Invest.* **1983,** *43*, 85–89.
4. Hine, T.J.; Roberts, N. B. Seasonal Variation in Serum 25-hydroxy Vitamin D_3 Does Not Affect 1,25-dihydroxy Vitamin D. *Ann. Clin. Biochem.* **1994,** *31* (Pt. 1), 31–34.
5. Garland, C. F.; Garland, F. C.; Shaw, E. K.; Comstock, G.W.; Helsing, K.J.; Gorham, E. D. Serum 25-hydroxyvitamin D and Colon Cancer: Eight-year Prospective Study. *Lancet* **1989,** *334*, 1176–1178.
6. Cândido, F.G.; Bressan, J. Vitamin D: Link Between Osteoporosis, Obesity, and Diabetes? *Int. J. Mol. Sci.* **2014,** *15*, 6569–6591.
7. Józwicki, W.; Brozyna, A. A.; Siekiera, J.; Slominski, A. T. "Expression of Vitamin D Receptor (VDR) Positively Correlates with Survival of Urothelial Bladder Cancer Patients. *Int. J. Mol. Sci.* **2015,** *16*, 24369–24386.
8. Umar, M.; Sastry, K. S.; Chouchane, A. I. Role of Vitamin D Beyond the Skeletal Function: A Review of the Molecular and Clinical Studies. *Int. J. Mol. Sci.* **2018,** *19*, 1618–1-28.
9. Kutner, A.; Brown, G. Vitamins D: Relationship Between Structure and Biological Activity. *Int. J. Mol. Sci.* **2018,** *19*, 2119–1-11.
10. Zmijewski, M.A. Vitamin D and Human Health. *Int. J. Mol. Sci.* **2019,** *20*, 145–1-6.
11. Torrens, F; Castellano, G. Modelling of Complex Multicellular Systems: Tumour–immune Cells Competition. *Chem. Central J.* **2009,** *3* (Suppl. I), 75–1-1.
12. Torrens, F.; Castellano, G. Information Theoretic Entropy for Molecular Classification: Oxadiazolamines as Potential Therapeutic Agents. *Curr. Comput.-Aided Drug Des.* **2013,** *9*, 241–253.
13. Torrens, F.; Castellano, G. Molecular Classification of 5-amino-2-aroylquinolines and 4-aroyl-6,7,8-trimethoxyquinolines as Highly Potent Tubulin Polymerization Inhibitors. *Int. J. Chemoinf. Chem. Eng.* **2013,** *3* (2), 1–26.
14. Estrela, J.M.; Mena, S.; Obrador, E.; Benlloch, M.; Castellano, G.; Salvador, R.; Dellinger, R.W. Polyphenolic Phytochemicals in Cancer Prevention and Therapy: Bioavailability *versus* Bioefficacy. *J. Med. Chem.* **2017,** *60*, 9413–9436.
15. Torrens, F.; Castellano, G. Molecular Classification of Antitubulin Agents with Indole Ring Binding at Colchicine-Binding Site. In *Molecular Insight of Drug Design*; Parikesit, A. A., Ed.; InTechOpen: Vienna, 2018; pp 47–67.
16. Torrens, F.; Castellano, G. Molecular Classification of 2-Phenylindole-3-carbaldehydes as Potential Antimitotic Agents in Human Breast Cancer Cells. In *Theoretical Models and Experimental Approaches in Physical Chemistry: Research Methodology and Practical Methods*; Haghi, A. K., Thomas, S., Praveen, K. M., Pai, A. R., Eds.; Apple Academic–CRC: Waretown, NJ, in Press.
17. Torrens, F.; Castellano, G. Cancer and Hypotheses on Cancer. In Molecular Chemistry and Biomolecular Engineering: Integrating Theory and Research with Practice; Pogliani, L., Torrens, F., Haghi, A. K., Eds.; Apple Academic–CRC: Waretown, NJ, in Press.

18. Torrens, F.; Castellano, G. Precision Personalized Medicine from Theory to Practice: Cancer. In Green Chemistry and Biodiversity: Principles, Techniques, and Correlations; Aguilar, C. N., Ameta, S. C., Haghi, A. K., Eds.; Apple Academic–CRC: Waretown, NJ, in Press.
19. Torrens, F.; Castellano, G. AIDS Destroys Immune Defences: Hypothesis. *New Front. Chem.* **2014,** *23*, 11–20.
20. Torrens-Zaragozá, F.; Castellano-Estornell, G. Emergence, Spread and Uncontrolled Ebola Outbreak. *Basic Clin. Pharmacol. Toxicol.* **2015,** *117* (Suppl. 2), 38–38.
21. Torrens, F.; Castellano, G. 2014 Spread/Uncontrolled Ebola Outbreak. **New Front. Chem. 2015,** 24, 81–91.
22. Torrens, F.; Castellano, G. Ebola Virus Disease: Questions, Ideas, Hypotheses and Models. *Pharmaceuticals* **2016,** *9*, 14–6-6.
23. Torrens, F.; Castellano, G. Clinical Translational Research: Cancer, Diabetes, and Cardiovascular Disease. In *Physical Biochemistry, Biophysics, and Molecular Chemistry: Applied Research and Interactions*; Torrens, F., Mahapatra, D. K., Haghi, A. K., Eds.; Apple Academic–CRC: Waretown, NJ, in Press.
24. Torrens, F.; Castellano, G. Primary Health, Smoking Habits and Immunotherapy in Cancer. In *Environmental Technology and Engineering Techniques: Basic Concepts and Health Interventions*; Khalaf, M. N., Smirnov, M. O., Kannan, P., Haghi, A. K., Eds.; Apple Academic–CRC: Waretown, NJ, in Press.
25. Torrens, F.; Castellano, G. Epigenomics, Epithelial Plasticity, Clinical Genetics and Rare Diseases. In *Environmental Technology and Engineering Techniques: Basic Concepts and Health Interventions*; Khalaf, M. N., Smirnov, M. O., Kannan, P., Haghi, A. K., Eds.; Apple Academic–CRC: Waretown, NJ, in Press.
26. Torrens, F.; Castellano, G. Predictive in Silico Global Metabolism Analysis: Disease Mechanism. In *Processing and Characterization of Green Engineered and Sustainable Materials*; Haghi, A. K., Ed.; Apple Academic–CRC: Waretown, NJ, in Press.
27. Torrens, F.; Castellano, G. Intracellular Proteolysis and Neoplasia: Ubiquitin-Proteasome System, Cell Cycle and Cancer. In *Natural Pharmaceuticals and Green Microbial Technology: Health, Environment and Resource Recovery*; Mahapatra, D. K.; Shinde, R. S., Haghi, A. K., Eds.; Apple Academic–CRC: Waretown, NJ, in Press.
28. Whitney, C. K. On the Several Molecules and Nanostructures of Water. *Int. J. Mol. Sci.* **2012,** *13*, 1066–1094.
29. Gandzha, I.; Kadeisvily, J. *New Sciences for a New Era: Mathematical, Physical and Chemical Discoveries of Ruggero Maria Santilli*; Sankata, Tebahal (Kathmandu, Nepal), 2010.
30. de Petris, L.; Bergfeldt, K.; Hising, C.; Lundqvist, A.; Tholander, B.; Pisa, P.; van der Zanden, H. G. M.; Masucci, G. Correlation Between HLA-A2 Gene Frequency, Latitude, Ovarian and Prostate Cancer Mortality Rates. *Med Oncol.* **2004,** *21*, 49–52.
31. Giovannucci, E. The Epidemiology of Vitamin D and Cancer Incidence and Mortality: A Review (United States). *Cancer Causes Control*, **2005,** *16*, 83–95.
32. Ingraham, B. A.; Bragdon, B.; Nohe, A. Molecular Basis of the Potential of Vitamin D to Prevent Cancer. *Curr. Med. Res. Opin.* **2008,** *24*, 139–149.
33. Gandini, S.; Boniol, M.; Haukka, J.; Byrnes, G.; Cox, B.; Sneyd, M. J.; Mullie, P.; Autier, P. Meta-analysis of Observational Studies of Serum 25-hydroxyvitamin D Levels and Colorectal, Breast and Prostate Cancer and Colorectal Adenoma. *Int. J. Cancer* **2011,** *128*, 1414–1424.

34. Álvarez-Díaz, S. Personal Communication.
35. Weinberg, R. A. *The Biology of Cancer*. Garland: Abingdon, UK, 2007.
36. van de Wetering, M.; Sancho, E.; Verweij, C.; de Lau, W.; Oving, I.; Hurlstone, A.; van der Horn, K.; Batlle, E.; Coudreuse, D.; Haramis, A.-P.; Tjon-Pon-Fong, M.; Moerer, P.; van den Born, M.; Soete, G.; Pals, S.; Eilers, M.; Medema, R.; Clevers, H. The β-catenin/TCF-4 Complex Imposes a Crypt Progenitor Phenotype on Colorectal Cancer Cells. *Cell* **2002,** *111*, 241–250.
37. Norman, A.W. From Vitamin D to Hormone D: Fundamentals of the Vitamin D Endocrine System Essential for Good Health. *Endocr. Rev*. **2012,** *33*, 456–492.
38. Rosen, C. J.; Adams, J. S.; Bikle, D. D.; Black, D. M.; Demay, M. B.; Manson, J.-A. E.; Murad, M. H.; Kovacs, C. S. The Nonskeletal Effects of Vitamin D: An Endocrine Society Scientific Statement. *Endocr. Rev*. **2012,** *33*, 456–492.
39. Pálmer, H. G.; González-Sancho, J. M.; Espada, J.; Berciano, M. T.; Puig, I.; Baulida, J.; Quintanilla, M.; Cano, A.; de Herreros, A. G.; Lafarga, M.; Muñoz, A. Vitamin D_3 Promotes the Differentiation of Colon Carcinoma Cells by the Induction of E-cadherin and the Inhibition of β-catenin Signaling. *J. Cell Biol*. **2001,** *154*, 369–387.
40. Ordóñez-Morán, P.; Larriba, M. J.; Pálmer, H. G.; Valero, R. A.; Barbáchano, A.; Duñach, M.; García de Herreros, A.; Villalobos, C.; Berciano, M. T.; Lafarga, M.; Muñoz, A. RhoA–ROCK and p38MAPK-MSK1 Mediate Vitamin D Effects on Gene Expression, Phenotype, and Wnt Pathway in Colon Cancer Cells. *J. Cell Biol*. **2008,** *183*, 697–710.
41. Cano, A.; Pérez-Moreno, M. A.; Rodrigo, I.; Locascio, A.; Blanco, M. J.; del Barrio, M. G.; Portillo, F.; Nieto, M. A. The Transcription Factor Snail Controls Epithelial–mesenchymal Transitions by Repressing E-cadherin Expression. *Nat. Cell Biol*. **2000,** *2*, 76–83.
42. Nieto, M. A. The Snail Superfamily of Zinc-finger Transcription Factors. *Nat. Rev. Mol. Cell Biol*. **2002,** *3*, 155–166.
43. Pálmer, H. G.; Larriba, M. J.; García, J. M.; Ordóñez-Morán, P.; Peña, C.; Peiró, S.; Puig, I.; Rodríguez, R.; de la Fuente, R.; Bernad, A.; Pollán, A.; Bonilla, F.; Gamallo, C.; García de Herreros, A.; Muñoz, A. The Transcription Factor SNAIL Represses Vitamin D Receptor Expression and Responsiveness in Human Colon Cancer. *Nat. Med*. **2004,** *10*, 917–919.
44. Peña, C.; García, J. M.; Silva, J.; García, V.; Rodríguez, R.; Alonso, I.; Millán, I.; Salas, C.; García de Herreros, A.; Muñoz, A.; Bonilla, F. E-cadherin and Vitamin D Receptor Regulation by SNAIL and ZEB1 in Colon Cancer: Clinicopathological Correlations. *Hum. Mol. Genet*. **2005,** *14*, 3361–3370.
45. Peinado, H.; Olmeda, D.; Cano, A. Snail, ZEB and bHLH Factors in Tumour Progression: An Alliance Against the Epithelial Phenotype? *Nat. Rev. Cancer* **2007,** *7*, 415–428.
46. Álvarez-Díaz, S.; Valle, N.; García, J. M.; Peña, C.; Freije, J. M. P.; Quesada, V.; Astudillo, A.; Bonilla, F.; López-Otín, C.; Muñoz, A. Cystatin D Is a Candidate Tumor Suppressor Gene Induced by Vitamin D in Human Colon Cancer Cells. *J. Clin. Invest*. **2009,** *119*, 2343–2358.
47. Cristobo, I.; Larriba, M. J.; de los Ríos, V.; García, F.; Muñoz, A.; Casal, J. I. Proteomic Analysis of 1α,25-dihydroxyvitamin D_3 Action on Human Colon Cancer Cells Reveals a Link to Splicing Regulation. *J. Proteomics* **2011,** *75*, 384–397.
48. Dixon, K. M.; Tongkao-On, W.; Sequeira, V. B.; Carter, S. E.; Song, E. J.; Rybchyn, M. S.; Gordon-Thomson, C.; Mason, R. S. Vitamin D and Death by Sunshine. *Int. J. Mol. Sci*. **2013,** *14*, 1964–1977.

49. Skowronski, R. J.; Peehl, D. M.; Feldman, D. Vitamin D and Prostate Cancer: 1,25-dihydroxyvitamin D_3 Receptors and Actions in Human Protease Cancer Cell Lines. *Endocrinology*. **1993,** *132*, 1952–1960.
50. Kumagai, J.; Fujimura, T.; Takahashi, S.; Urano, T.; Ogushi, T.; Horie-Inoue, K.; Ouchi, Y.; Kitamura, T.; Muramatsu, M.; Blumberg, B.; Inoue, S. Cytochrome P450 2B6 Is a Growth-inhibitory and Prognostic Factor for Prostate Cancer. *Prostate* **2007,** *67*, 1029–1037.
51. Skowronski, R. J.; Peehl, D. M.; Feldman, D. Actions of Vitamin D_3, Analogs on Human Prostate Cancer Cell Lines: Comparison with 1,25-dihydroxyvitamin D_3. *Endocrinology* **1995,** *136*, 20–26.
52. Feldman, D.; Skowronski, R. J.; Peehl, D. M. Vitamin D and Prostate Cancer. *Adv. Exp. Med. Biol.* **1995,** *375*, 53–63.
53. Getzenberg, R. H.; Light, B. W.; Lapco, P. E.; Konety, B. R.; Nangia, A. K.; Acierno, J. S.; Dhir, R.; Shurin, Z.; Day, R. S.; Trump, D. L.; Johnson, C. S. Vitamin D Inhibition of Prostate Adenocarcinoma Growth and Metastasis in the Dunning Rat Prostate Model System. *Urology* **1997,** *50*, 999–1006.
54. Gross, C.; Stamey, T.; Hancock, S.; Feldman, D. Treatment of Early Recurrent Prostate Cancer with 1,25-dihydroxyvitamin D_3 (Calcitriol). *J. Urol.* **1998,** *159*, 2035–2040.
55. Ly, L. H.; Zhao, X.-Y.; Holloway, L.; Feldman, D. Liarozole Acts Synergistically with 1α,25-dihydroxyvitamin D_3 to Inhibit Growth of DU 145 Human Prostate Cancer Cells by Blocking 24-hydroxylase Activity. *Endocrinology* **1999,** *140*, 2071–2076.
56. Feldman, D.; Zhao, X.-Y.; Krishnan, A. V. Edotorial/Mini-review: Vitamin D and Prostate Cancer. *Endocrinology* **2000,** *141*, 5–9.
57. Zhao, X.-Y.; Malloy, P. J.; Krishnan, A. V.; Swami, S.; Navone, N. M.; Peehl, D. M.; Feldman, D. Glucocorticoids Can Promote Androgen-independent Growth of Prostate Cancer Cells through a Mutated Androgen Receptor. *Nat. Med.* **2000,** *6*, 703–706.
58. Feldman, B. J.; Feldman, D. The Development of Androgen-independent Prostate Cancer. *Nat. Rev.* **2001,** *1*, 34–45.
59. Hsu, J.-Y.; Feldman, D.; McNeal, J. E.; Peehl, D. M. Reduced 1α-hydroxylase Activity in Human Prostate Cancer Cells Correlates with Decreased Susceptibility to 25-hydroxyvitamin D_3-induced Growth Inhibition. *Cancer Res.* **2001,** *61*, 2852–2856.
60. Boyle, B. J.; Zhao, X.-Y.; Cohen, P.; Feldman, D. Insulin-like Growth Factor Binding Protein-3 Mediates 1α,25-dihydroxyvitamin D_3 Growth Inhibition in the LNCaP Prostate Cancer Cell Line through p21/WAF1. *J. Urol.* **2001,** *165*, 1319–1324.
61. Krishnan, A. V.; Peehl, D. M.; Feldman, D. Inhibition of Prostate Cancer Growth by Vitamin D: Regulation of Target Gene Expression. *J. Cell. Biochem.* **2003,** *88*, 363–371.
62. Peehl, D. M.; Feldman, D. Interaction of Nuclear Receptor Ligands with the Vitamin D Signaling Pathway in Prostate Cancer. *J. Steroid Biochem. Mol. Biol.* **2004,** *92*, 307–315.
63. Bhattacharyya, R. S.; Krishnan, A. V.; Swami, S.; Feldman, D. Fulvestrant (ICI 182,780) Down-regulates Androgen Receptor Expression and Diminishes Androgenic Responses in LNCaP Human Prostate Cancer Cells. *Mol. Cancer Ther.* **2006,** *5*, 1–11.
64. Srinivas, S.; Feldman, D. A Phase II Trial of Calcitriol and Naproxen in Recurrent Prostate Cancer. *Anticancer Res.* **2009,** *29*, 3605–3610.
65. Abdallah, H. H.; Mavri, J.; Repiç, M.; Lee, V. S.; Wahab, H. A. Chemical Reaction of Soybean Flavonoids with DNA: A Computational Study Using the Implicit Solvent Model. *Int. J. Mol. Sci.* **2012,** *13*, 1269–1283.

66. Swami, S.; Krishnan, A. V.; Moreno, J.; Bhattacharyya, R. S.; Gardner, C.; Brooks, J. D.; Peehl, D. M.; Feldman, D. Inhibition of Prostaglandin Synthesis and Actions by Genistein in Human Prostate Cancer Cells and by Soy Isoflavones in Prostate Cancer Patients. *Int. J. Cancer* **2009,** *124*, 2050–2059.
67. Swami, S.; Krishnan, A. V.; Feldman, D. Vitamin D Metabolism and Action in the Prostate: Implications for Health and Disease. *Mol. Cell. Endocrinol.* **2011** in press.
68. Medrano, M.; Carrillo-Cruz, E.; Montero, I.; Perez-Simon, J. A.; Vitamin D: Effect on Haematopoiesis and Immune System and Clinical Applications. *Int. J. Mol. Sci.* **2018,** *19*, 2663–1-25.

CHAPTER 7

Stem Cells and Their Connection with Cancer: Prostate Disease

FRANCISCO TORRENS[1*] and GLORIA CASTELLANO[2]

[1]*Institut Universitari de Ciència Molecular, Universitat de València, Edifici d'Instituts de Paterna, PO Box 22085, E-46071 València, Spain*

[2]*Departamento de Ciencias Experimentales y Matemáticas, Facultad de Veterinaria y Ciencias Experimentales, Universidad Católica de Valencia, San Vicente Mártir, Guillem de Castro-94, E-46001 València, Spain*

[*]*Corresponding author. E-mail: torrens@uv.es*

ABSTRACT

Acute myeloid leukemia enables research on different-oncogene biofunction, in leukemic transformation/progression and allows devising healing strategies, for example, gene/antisense therapies, and so on, aimed at leukemic stem cell. Chain net, in subventricular zone, carries neuronal precursors from caudal subventricular zone to rostral parts of adult brain. It should be determined what guides migration via subventricular zone net, which other vertebrate brains contain pathway array and biosignificance of such an extensive neuronal-precursor net, many destined for single brain region: olfactory bulb. Despite robust germinal capacity, no evidence of cells migrating in chains exists along subventricular zone or olfactory peduncle to bulb, which is explained by long distance that separates olfactory bulbs from cerebrum in man or micro-osmatic capabilities. Findings raise fate question of cells born in human subventricular zone. Identification of large, anatomically discrete population of proliferating, multipotent human astrocytes leads to understanding role that human neural stem cells act in tumorigenic demyelination/degeneration. Planar polarity of ciliated ependymal cells is essential for formation of chemorepulsive-factor gradients, which guide neuroblast migration in adult brain. Polarized ciliated cells contribute important vectorial information for body plan development. Polarized epithelia and motile cilia in brain serve as directional-information conveyors for neuronal migration.

7.1 INTRODUCTION

Understanding of the stability of cell phenotype was transformed in a totally unexpected manner, for example, cellular reprogramming tells that cell fate is plainly far less rigidly determined than was previously believed; however, the limitations to our opportunities to change cell fate are still not clear. Is it possible, for example, to produce any cell type from any donor cell type, given the introduction of the necessary regulatory molecules and an appropriate culture environment? Cell reprogramming created extraordinary opportunities for research. In time, reprogramming studies may offer new drug treatments, which prevent degenerative changes or induce stem/progenitor cells to differentiate into the affected lineage in a specific condition. Cell transplantation might also become available for some degenerative diseases. The pace/scope of these advances is clearly unpredictable. In all likelihood, cell therapy will not be possible for all degenerative/genetic diseases for reasons that one cannot yet foresee. Furthermore, the development of treatments will almost definitely happen over a period of decades and not merely years. Despite these notes of caution, one can be optimistic that the place of advance will not slacken and that scientists can hope to deliver the therapeutic gold that patients so urgently seek. The *Philosophical Transactions of the Royal Society* published an special issue on the evolving biology of cell reprogramming.[1]

A question (Q) follows:

Q1. How research guides into stem cell (SCs) biological knowledge and about its use in animal models?

Bioethical considerations are discussed.

In earlier publications, it was informed the modeling of complex multicellular systems and tumor-immune cells competition,[2] information theoretic entropy for molecular classification of oxadiazolamines as potential therapeutic agents,[3] molecular classification of 5-amino-2-aroylquinolines and 4-aroyl-6,7,8-trimethoxyquinolines as highly potent tubulin polymerization inhibitors,[4] polyphenolic phytochemicals in cancer prevention, therapy, bioavailability versus bioefficacy,[5] molecular classification of antitubulin agents with indole ring binding at colchicine-binding site (BS),[6] molecular classification of 2-phenylindole-3-carbaldehydes as potential antimitotic agents in human breast cancer cells,[7] cancer, its hypotheses,[8] precision personalized medicine from theory to practice and cancer.[9] It was reported how human immunodeficiency virus/acquired immunodeficiency syndrome destroy immune defenses, hypothesis,[10] 2014 emergence, spread, uncontrolled Ebola outbreak,[11,12] Ebola virus disease, questions, ideas, hypotheses, models,[13] clinical translational research, cancer, diabetes, cardiovascular disease,[14] primary health, smoking habits, immunotherapy in cancer,[15] epigenomics, epithelial plasticity, clinical genetics, rare diseases,[16] predictive *in silico* global metabolism analysis, disease mechanism,[17] intracellular proteolysis, neoplasia, ubiquitin-proteasome system, cell cycle and cancer,[18] vitamin D, analogs, drugs and their relevance to cancer inhibition.[19] Present report reviews state of the art of SCs and their connection with cancer, especially prostate disease. The aim of this work is to initiate a debate by suggesting a number of questions, which can arise when addressing subjects of SCs, pattern formation in nervous system, the possible versus the actual, neural-system aging, neurons regeneration, the mother of all cells, the pluripotent stem cells (PSCs), mother/daughter-cell stories and how to organize our brain. It was provided, when possible, answers (A), hypotheses (H), and facts (F) on these materials.

7.2 METHOD

Carcinogenesis shows identifying-biomarker opportunities (cf. Fig. 7.1).[20]

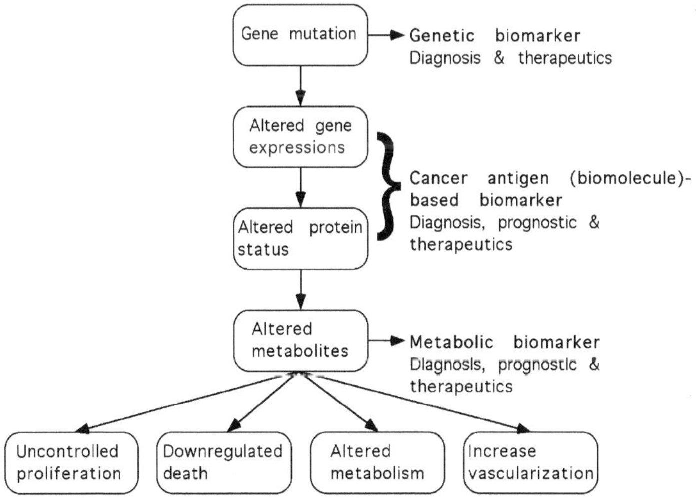

FIGURE 7.1 The process of carcinogenesis, showing opportunities of identifying biomarker.

7.3 RESULTS: STEM CELLS AND THEIR RELEVANCE TO THE INHIBITION OF CANCER

7.3.1 ACUTE MYELOID LEUKEMIA AS HIERARCHY FROM PRIMITIVE HEMATOPOIETIC CELL

On acute myeloid leukemia (AML), little consensus exists about target cell within hematopoietic SC hierarchy, which is susceptible to leukemic transformation, or mechanism that underlies phenotypic, genotypic, and clinical heterogeneity. Bonnet and Dick showed that cell capable of initiating human AML in nonobese diabetic mice, with severe combined immunodeficiency disease (NOD/SCID mice, termed SCID leukemia-initiating cell, SL-IC), possesses differentiative/proliferative capacities and potential for self-renewal expected of leukemic SC.[21] The SL-ICs from AML subtypes, regardless of heterogeneity in maturation characteristics of leukemic blasts, were cluster of differentiation $(CD)34^{++}CD38^{-}$ similar to cell-surface phenotype of normal SCID-repopulating cells, showing that normal primitive rather than committed progenitor ones are target for leukemic transformation; they were able to differentiate *in vivo* into leukemic blasts indicating that leukemic clone is organized as hierarchy.

7.3.2 THE MESELSON–STAHL EXPERIMENT LIVES ON

Watson and Crick's article closed with clue *that the specific pairing we have postulated immediately suggests a possible copying mechanism for the genetic material.*[22] Their follow-up report explained probable semiconservative route: double helix parts and each half assembles new nucleotides.[23] Bacterial transformation, and bacteriophage infection, indicates that DNA transmits hereditary information and directs its own replication, which mechanism hypotheses differ in predictions they make, concerning distribution among progeny molecules of atoms derived from parental molecules.[24] Survival of rapidly renewing tissues of long-lived animals, for example, man, requires that they be protected versus natural selection of fitter variant cells (cancer).[25] In SC age, immortal strand hypothesis continues (cf. Fig. 7.2).[26]

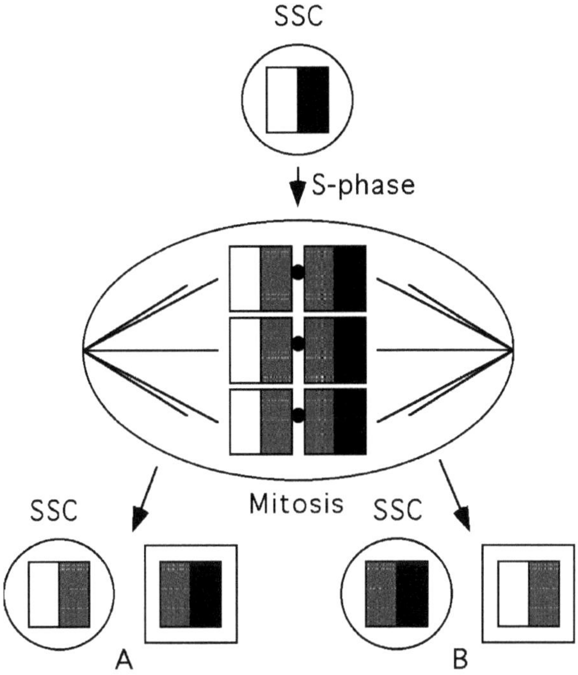

FIGURE 7.2 Immortal strand hypothesis: (a) older DNA strands are preferentially sorted into somatic stem cells (SSCs) when they self-renew; if adult SSCs undergo random chromosome segregation, for example, (B), chromosomes assort with equal frequency to SSCs/differentiated sisters (squares).

7.3.3 NET OF TANGENTIAL PATHWAYS FOR NEURONAL MIGRATION IN ADULT BRAIN

Cells in adult-mammal brains continue to proliferate in subventricular zone (SVZ) throughout lateral ventricle wall. Alvarez-Buylla group showed, via whole-mount dissections of adult-mice wall, that SVZ is organized as net of neural-precursor chains, which are immunopositive to polysialylated form of neural cell adhesion molecule, present at plasticity sites, and neuron-specific class III β-tubulin TuJ1, early neuronal marker.[27] Most chains orient along rostrocaudal axis and many join rostral migratory stream (RMS), which terminates in olfactory bulb. Via focal 1,1′-dioctadecyl-3,3,3′3′-tetramethylindocarbocyanine perchlorate microinjections, and transplantation of SVZ cells carrying neuron-specific reporter gene, they showed that cells originating at different SVZ rostrocaudal levels migrate rostrally and reach olfactory bulb, where they differentiate into neurons; they revealed pathway extensive net for tangential chain migration of neuronal precursors, throughout lateral ventricle wall in adult mammalian brain.

7.3.4 BASAL RADIAL GLIA IN SUBVENTRICULAR ZONE AND MECHANISMS OF CEREBRAL CORTEX EXPANSION AND FOLDING

Cerebral cortex of large mammals undergoes massive surface area expansion/folding during development. Specific mechanisms to orchestrate cortex growth in surface area rather than thickness are likely to exist, but they were not identified. Analyzing multiple species, Reillo *et al.* identified specialized type of progenitor cell, which is exclusive to mammals with folded cerebral cortex, which they named intermediate radial glia cell (IRGC).[28] The IRGCs express Pax6 but not Tbr2 have radial fiber contacting pial surface but not ventricular surface and are found in both inner SVZ and outer-subventricular zone (OSVZ). They found that IRGCs are massively generated in OSVZ, augmenting radial fiber numbers. Fanning out of expanding radial fiber scaffold promotes tangential dispersion of radially migrating neurons, allowing for growth in surface area of cortical sheet. Tangential expansion of particular cortical regions was preceded by high proliferation in underlying OSVZ, whereas experimental IRGC reduction impaired tangential neuron dispersion and resulted in smaller cortical surface. Generation of IRGCs plays role in tangential expansion of mammalian cerebral cortex. The SVZ progenitors are developing-neocortex hallmark. Studies described SVZ-progenitor type that retains basal process at mitosis, sustains expression of radial glial markers, and is capable of self-renewal. Progenitors, referred to as basal radial glia (bRG), occur at high relative abundance in gyrencephalic primates (human)/nonprimates (ferret) SVZ but not lissencephalic rodents (mouse). Kelava *et al.* analyzed bRG-cell occurrence in embryonic neocortex of common marmoset *Callithrix jacchus*, near-lissencephalic primate.[29] Cells bRG, expressing Pax6, Sox2 (but not Tbr2), glutamate aspartate transporter and glial fibrillary acidic protein, and retaining basal process at mitosis, occur at similar relative abundance in marmoset SVZ as in human/ferret. Progenitor proportion in M-phase was lower in embryonic marmoset than developing ferret neocortex, raising possibility of longer cell cycle. Fitting gyrification indices of 26 anthropoid species to evolutionary model suggested that marmoset evolved from gyrencephalic ancestor. Their results suggested that high relative abundance of bRG cells is necessary, but is not sufficient, for gyrencephaly and that marmoset's lissencephaly evolved secondarily by changing progenitor parameters other than progenitor type.

7.3.5 NEUROGENESIS IN THE ADULT HUMAN HIPPOCAMPUS: ARE NEURONS PRODUCED?

Genesis of new cells, for example, neurons, in adult human brain was not shown. Eriksson *et al.* investigated whether neurogenesis occurs in adult human brain, in regions previously identified as neurogenic in adult rodents/monkeys.[30] Human brain tissue was obtained postmortem from patients who were treated with thymidine analog bromodeoxyuridine (BrdU), which labels DNA during S-phase. Via immunofluorescent labeling for BrdU and one neuronal marker, NeuN, calbindin, or neuron-specific enolase, they showed that new neurons, defined by markers, are generated from dividing progenitor cells in dentate gyrus of adult humans. They indicated that human hippocampus retains ability to generate neurons throughout life. García-Verdugo answered the question: Are new neurons produced in our brain?[31]

7.3.6 ASTROCYTE RIBBON IN HUMAN BRAIN CONTAINS NEURAL STEM CELLS BUT LACKS CHAIN MIGRATION

The SVZ is a major source of adult neural stem cells (NSCs) in rodent brain, generating thousands of olfactory bulb neurons every day. If adult human brain contains comparable germinal region, this could have implications for neuroregenerative therapy. SCs were isolated from human brain but adult NSC identity, organization, and function in SVZ are unknown. Alvarez-Buylla group described SVZ-astrocyte ribbon lining lateral ventricles of adult human brain, which proliferate *in vivo* and behave as multipotent progenitor cells *in vitro*; it was not observed in other vertebrates.[32] They found no migrating-neuroblast chain evidence in SVZ/pathway to olfactory bulb; they identified SVZ astrocytes as NSCs in unique-organization niche in adult human brain.

7.3.7 NEW NEURONS FOLLOW THE FLOW OF CEREBROSPINAL FLUID IN THE ADULT BRAIN

In adult brain, SVZ-born neuroblasts migrate from lateral ventricle walls to olfactory bulb. How do cells orient over such a long distance via complex territories? Alvarez-Buylla group showed that neuroblast migration parallels cerebrospinal fluid (CSF) flow.[33] Ependymal cilium beating is required for normal CSF flow, concentration gradient formation of CSF guidance molecules, and neuroblast directional migration. They suggested that polarized epithelial cells contribute vectorial information for guidance of young migrating neurons.

7.3.8 CORRIDORS OF MIGRATING NEURONS IN THE BRAIN AND DECLINE DURING INFANCY

The SVZ of many adult nonhuman mammals generates large numbers of neurons destined for olfactory bulb. Along lateral ventricle walls, immature neuronal progeny migrate in tangentially oriented chains, which coalesce into RMS connecting SVZ to olfactory bulb. Adult human SVZ, in contrast, contains hypocellular gap layer separating ependymal lining from periventricular astrocytes ribbon. Some SVZ astrocytes function as neural SCs *in vitro* but their function *in vivo* remains controversial. Initial report found few SVZ proliferating cells and rare migrating immature neurons in adult-human RMS; in contrast, study indicated robust proliferation/migration in human SVZ/RMS. Alvarez-Buylla group found that infant-human SVZ/RMS contain extensive corridor of migrating immature neurons, before age 18 months but, contrary to previous reports, germinal activity subsides in older children and is nearly extinct by adulthood.[34] During limiting neurogenesis window, not all new neurons in human SVZ are destined for olfactory bulb. Findings reveal robust streams of tangentially migrating immature neurons in human early postnatal SVZ/cortex. Pathways represent potential targets of neurological injuries affecting neonates.

7.3.9 EXTENSIVE MIGRATION OF YOUNG NEURONS INTO INFANT HUMAN FRONTAL LOBE

The first few months after birth, when a child begins to interact with the environment, are critical to human brain development. The human frontal lobe is important for social behavior and executive function; it rose in size and complexity relative to other species but the processes that contributed to the expansion are unknown. Alvarez-Buylla group studies of postmortem infant-human brains revealed a collection of neurons that migrate and integrate widely into the frontal lobe during infancy.[35] Chains of young neurons move tangentially close to the walls of the lateral ventricles and along blood vessels. The cells then individually disperse long distances to reach cortical tissue, where they differentiate and contribute to inhibitory circuits. Late-arriving interneurons could contribute to developmental plasticity, and the disruption of their postnatal migration or differentiation may underlie neurodevelopmental disorders.

7.3.10 DOES ADULT NEUROGENESIS PERSIST IN THE HUMAN HIPPOCAMPUS?

New neurons continue generated in the subgranular zone (SGZ) of the dentate gyrus of the adult mammalian hippocampus, which process was linked to learning and memory, stress, and exercise, and are thought to be altered in neurological disease. In humans, studies suggested that hundreds of new neurons are added to the adult dentate gyrus every day, whereas others found many fewer putative new neurons. Despite the discrepancies, it is generally believed that the adult human hippocampus continues to generate neurons. Alvarez-Buylla group showed that a defined population of progenitor cells does not coalesce in SGZ during human fetal or postnatal development.[36] They found that the number of proliferating progenitors and young neurons in the dentate gyrus declines sharply during the first year of life, and only a few isolated young neurons are observed by seven and 13 years of age. In adult patients with epilepsy and healthy adults (18–77 years; n = 17 postmortem samples from controls; n = 12 surgical resection samples from patients with epilepsy), young neurons were not detected in the dentate gyrus. In the monkey (Macaca mulatta) hippocampus, proliferation of neurons in SGZ was found in early postnatal life, but this diminished during juvenile development as neurogenesis decreased. They concluded that recruitment of young neurons to the primate hippocampus decreases rapidly during the first years of life and that neurogenesis in the dentate gyrus does not continue, or is rare, in adult humans. The early decline in hippocampal neurogenesis raises questions about how the function of the dentate gyrus differs between humans and other species in which adult hippocampal neurogenesis is preserved. Does adult neurogenesis persist in the human hippocampus?[37]

7.3.11 STEM CELLS AND PATTERN FORMATION IN NERVOUS SYSTEM: POSSIBLE VERSUS ACTUAL

The NSCs are broadly defined as multipotent, self-renewing progenitor cells: in central nervous system (CNS), they generate neurons, astrocytes, and oligodendrocytes; in peripheral nervous system, neurons, Schwann cells, and other neural crest derivatives, for example, smooth muscle cells. Progress was made in identifying/understanding their properties; there was quantal increase in attention afforded SCs. Anderson reviewed SCs and pattern formation in nervous system and proposed Q/A/H/F.[38]

Q1. What accounts for the heightened level of interest in SCs?

Q2. What do people really know about these cells?

Q3. What remains to be learned about these cells?

Q4. How do NSCs choose their fates?

Q5. To what extent does cell-autonomous programs in NCSs, versus sequential presentation of instructive signals, underlie neurons/glia orderly generation in developing brain?

Q6. What determines the self-renewal capacity of NSCs?

Q7. Which cells in the brain are SCs?

Q8. Are their properties immutable or do they change over time?

Q9. In addition, just how broad is the developmental potential of NSCs?

Q10. Is there a unitary NSC for all the different types of neurons in the brain?

Q11. How can people's NSC biology understanding be better integrated with their growing knowledge of neural pattern formation?

Q12. What do progenitor cells do?

Q13. How do they do it?

Q14. What *can* SCs and progenitor cells do, given the opportunity?

Q15. Is NSC biology becoming an applied-neuroscience field, one in which it does not matter whether isolated-SC behavior is an accurate reflection of their normal properties *in vivo*, so long as they are therapeutically useful?

Q16. Have cell-lineage/commitment issues become conceptually irrelevant to understanding pattern formation in developing nervous system?

H1. Individual SCs necessarily self-renew via asymmetric divisions.

H2. (Qian, Shen, Goderie, He, Capela, Davis and Temple, 2000). Self-renewal indeed occurred.

Q17. Are asymmetric cell divisions self-renewing or differentiative?

Q18. Are there any self-renewal factors for NSCs?

H3. Autocrine factors cooperate with mitogens, for example, fibroblast growth factor 2 (FGF2), to promote self-renewal.

Q19. Can activity of factors that maintain NSCs multipotency be uncoupled from SC mitogens?

H4. (Wang and Barres, 2000). Notch signaling in some kinds of neural progenitor cells promotes glial differentiation rather than self-renewal.

Q20. Is this generally true of Notch signaling in all NSCs?

Q21. Does the fact that, *in vivo*, neurons are usually generated before astrocytes/oligodendrocytes reflect an intrinsic *timer* operating in NSCs or sequential presentation of first neurogenic and then gliogenic instructive signals?

H5. Cell-autonomous mechanisms of fate determination are at work.

H6. Alternative H (Sternberg and Horvitz, 1989): Such stereotypy reflects reproducible local cell–cell interactions.

H7. Neurogenesis and gliogenesis proceed via committed precursors.

Q22. Do multipotent NSCs *throw off* postmitotic progeny that directly differentiate into neurons or generate neurons via a proliferating, committed neuronal progenitor?

H8. (Metcalf, 1989, 1991). Many cell-extrinsic factors act instructively, promoting choice of one fate at expense of others rather than selectively supporting survival/proliferation of lineage-committed progenitors.

Q23. Does Notch promote gliogenesis directly or indirectly promoting an irreversible loss of neurogenic capacity?

H9. (Johe, 1996; Poncet, 1996; Orentas, 1999; Lu, 2000). Multiple extracellular signals promote oligodendrocyte differentiation [e.g., thyroid hormone, Sonic Hedgehog (Shh)].

Q24. Is Shh a true instructive signal for oligodendrocyte fate or does it act as a mitogen/survival factor for oligodendrocyte precursors?

H10. Bone morphogenetic proteins (BMPs) are necessary/sufficient for autonomic neurogenesis.

Q25. Does fact that embryonic cortical NSCs spontaneously generate neuronal precursors in a defined medium reflect an autonomous, *default* program or a cell-heritable memory of an earlier exposure to an instructive signal?

Q26. How do effects of instructive differentiation factors on NSCs relate to production of lineage-restricted precursors?

H11. Simplifying H: Such factors promote the generation of such restricted precursors.

Q27. What does it happen when NSCs are exposed concurrently to multiple, competing instructive signals?

H12. The NSCs behave like dynamic microprocessors, whose output is a function not only of absolute/relative concentrations of ambient instructive signals but also of an internal integration function that computes cells response to such signals.

H13. This internal integration function changes in NSCs with time.

Q28. Are new neurons formed in the brains of adult mammals?

H14. (Altman, 1962, 1969; Altman and Das, 1966). Neurogenesis occurs in regions of adult rodent brain.

Q29. Does neurogenesis occur in the adult primate cerebral cortex?

H15. (Reynolds and Weiss, 1992; Richards, 1992; Lois and Alvarez-Buylla, 1993; Morshead, 1994; Gage, 1995; Gritti, 1996; Palmer, 1997). Multipotent, neurons, and glia self-renewing progenitors are cultured from adult brain regions.

H16. Such adult neuronal precursors are SCs.

Q30. What is the location and identity of such SCs in the brain?

Q31. What cells are SCs in the adult brain?

Q32. What is the identity and location of CNS NSCs in the adult brain?

H17. 5-Bromo-2′-deoxyuridine (BrdU)-labeled cells divide in a specific *niche* that is tightly associated with blood vessels.

Q33. Do brain endothelial cells also provide growth factors for NSCs?

F1. Dividing endothelial cells are closely associated with dividing neural progenitors.

Q34. Is there a bidirectional communication between the two cells types?

Q35. Are there SCs in nonneurogenic regions of the adult brain?

Q36. Do SCs exist only in adult-brain regions that actively undergo neurogenesis or are there also latent SC populations in other brain regions?

Q37. Could endogenous SCs reactivation be used to treat brain injury/disease in regions that do not undergo natural turnover, for example, striatum for Huntington's disease?

Q38. Do SCs from nonneurogenic regions exhibit stem-like properties *in vitro* but not *in vivo* as a reprogramming/dedifferentiation consequence?

Q39. Does adult neurogenesis occur at a fixed, constitutive rate or can it be influenced by organism's environment/behavior?

H18. Genes control the rate of adult neurogenesis as well.

Q40. Is the magnitude of adult neurogenesis high enough to be physiologically significant?

Q41. Can endogenous SCs be mobilized to replace degenerating neurons?

H19. Inducible neurogensis can be augmented/directed to particular regions for therapy.

Q42. How broad is the developmental potential of CNS SCs?

Q43. Is there a generic SC for the entire CNS?

Q44. Is there an analogous, generic SC for the entire mature nervous system?

H20. Ex-vivo expansion of neural progenitor cells causes developmental reprogramming.

Q45. Do broad developmental potentials revealed by such transplantation experiments reflect CNS-SCs properties *in situ*?

Q46. Are SCs capable of generating such novel derivatives in response to endogenous cues that normally elicit derivatives differentiation directly transplanting them to relevant sites *in vivo* without any intervening culture manipulations?

H21. (Kintner and Brockes, 1984). Reprogramming/dedifferentiation is detectable by unusual combinations of antigenic marker expression.

H22. (Kondo, 2000). Fact that oligodendrocyte precursors grown under certain culture conditions can generate neurons reflects reprogramming.

Q47. How many classes of neurons can CNS NSCs generate?

Q48. Do neurogenic SCs exist in the brain nonneurogenic regions from which they can be cultured?

Q49. Have NSCs nonneural potential?

Q50. The SC potential: Can anything make anything?

H23. The SCs from one tissue/lineage give rise to differentiated cell types of other organs, to which they do not normally contribute.

H24. (Clarke, 2000). The NSCs contribute to multiple tissues from all three germ layers, if transplanted into sufficiently early stages host embryos.

Q51. However, what do such observations really mean?

Q52. Are *NSCs* actually PSCs *in situ*, but ones that generate neural derivatives only because they are located in brain?

Q53. Does the in-vitro cultivation necessary to isolate CNS NSCs result in their reprogramming?

H25. Uncultured neural progenitors are more restricted in neurons subtypes they can generate.

H26. NSCs exist that are multipotent/self-renewing but that are nevertheless restricted with respect to neuronal subtypes they are able to generate.

H27. (Gray and Sanes, 1992; Noctor, 2001). Embryonic radial glial cells E14.5 are NSCs by in-vivo lineage analyses.

Q54. Were neurogenic radial glia committed to a neuronal fate or multipotent SCs that could not make transition from neurogenesis to gliogenesis under ambient culture conditions?

H28. Multipotent NSCs are distinct from neurogenic radial glia.

Q55. How to integrate the disembodied/context divergent perspectives on neural development?

Q56. How do inductive signals act, at level of progenitor-cell fields/groups, to control differentiation of distinct neuronal subtypes?

Q57. How does differentiation control at level of individual progenitor cells relate to lineage/developmental-restriction issues?

Q58. Why has neuroembryology/NSC biology come to have such different intellectual/experimental orientations?

Q59. How has neuroembryology/NSC biology come to have such different intellectual/experimental orientations?

Q60. How do some discoveries/concepts provided by neuroembryology bear on some assumptions that underlie SC bioapproach to neural development?

Q61. Is molecular regionalization of CNS ventricular zone compatible with a generic NSC concept?

Q62. Are all mature cell types of the nervous system derived from a unitary SC?

H29. The SCs are not a minor subpopulation that does not express such factors.

H30. The SCs in different regions of developing nervous system are likely distinct from one another in their program of transcription factor gene expression.

Q63. Are progenitor cells expressing particular homeodomain factors committed to their fate?

H31. The transplanted population is homogeneous.

Q64. How will further progress in NSC biology require integration of hematopoietic-style approach with concepts of regional progenitor identity provided by neuroembryology?

Q65. Are region-specific patterns of homeodomain transcription factor expression maintained/deregulated in CNS SCs following their dissociation/isolation?

H32. It is possible to isolate prospectively CNS SCs that express a particular homeodomain gene via its enhancer sequences to drive green fluorescent protein (GFP) expression in transgenic animals.

Q66. Are they restricted in the subtypes of neurons they can generate?

H33. Such cell transplantation provides most direct test of whether they are restricted in neurons subtypes they can generate.

H34. Differentiation of a desired cell type is achieved in dissociated cell culture if appropriate cocktail of soluble/substrate-associated factors is provided.

H35. Complex three-dimensional (3D) patterning mechanisms of developing neural tube are crucial for proper neuronal subtype specification and difficult to reconstitute in dissociated NSC cultures.

H36. Value of *SC concept* is not negated for nervous system but rather adapted to different anatomical/developmental constraints on that tissue.

Q67. Why does it matter whether or not a generic CNS SC exists?

A67. It has important implications for both understanding development and treating disease.

Q68. If isolated SCs can generate virtually any neuron/glial-cell kind when transplanted into brain, does it matter if pluripotency is an isolation/growth-methodology artifact?

Q69. How does neural development actually work?

Q70. If isolated-NSCs pluripotency were indeed caused by tissue dissociation/ex-vivo expansion, how does this occur at molecular level?

Q71. If isolated-NSCs pluripotency were indeed caused by tissue dissociation/ex-vivo expansion, why does this occur at molecular level?

Q72. Neuroembryology and NSC biology: Shall ever the twain meet?

Q73. What extent does expression of particular transcription factors signify commitment to, or restriction from, particular fates to?

Q74. Is there a unitary cascade/net of growth/transcription factors that SCs must be subject to generate every neuronal subtype or can a given neuron type be generated via multiple routes?

H37 [Challenge (C)1]. Mere neuroepithelial-tissue dissociation that is necessary to isolate CNS NSCs deregulates spatial patterning mechanisms that impose regional identity on progenitor cells in different brain parts.

H38 (C2). Differentiation of particular neuronal/glial subtypes *in vivo* occurs within a highly organized 3D matrix of graded positional signals, which are lost/difficult to reconstitute in dissociated NSC cultures.

Q75. Does NSC biology become an applied science, with much to gain from but little to contribute to people's understanding of patterning mechanisms underlying neural development?

Q76. Which developmental-program features are intrinsic to individual cells/their progeny and independent of surrounding patterning information?

Q77. Which extent do differentiation/patterning signals act exclusively by instructing specific cell fates as opposed to promoting prespecified-progenitors survival/proliferation to?

Q78. How do developmental changes in cell-intrinsic programs restrict progenitors responses to cell-extrinsic signals?

Q79. How do transcription factors affect individual-cells behavior within *fields*/groups of presumably equipotent cells?

Q80. Do transcription factors invariably act cell autonomously to control cell identities or can they also act noncell autonomously influencing interactions with other cells?

Q81. Do morphogens/transcription factors change genealogy by which SCs generate their differentiated derivatives or simply derivative type but not lineage?

Q82. How does expression of various transcription factors map onto various types of SCs/progenitor cells that were identified by cell biostudies?

Q83. How do all transcription factors affect cellular processes (e.g., migration, delamination from neuroepithelium, adhesion, and cell division/morphology)?

Q84. What extent are such processes key mediators of the determination of cell fate to?

Q85. How do SCs behave in absence of kind of dynamic patterning mechanisms that control embryonic neural development?

7.3.12 TELOMERE SHORTENING/CHROMOSOMAL INSTABILITY ABROGATE NEURAL STEM CELLS PROLIFERATION

Chromosome integrity is essential for cell viability, and highly proliferative cell types require active telomere elongation mechanisms to grow indefinitely; telomerase activity deletion, in genetically modified murine strain, results in growth impairments in highly proliferative cell populations. Ferrón *et al.* showed that telomere attrition impairs in-vitro proliferation of adult NSCs, isolated from SVZ of telomerase-deficient adult mice.[39] Reduced proliferation of postnatal neurogenic progenitors was observed *in vivo*, in absence of exogenous mitogenic stimulation. Severe telomere erosion, resulting in chromosomal abnormalities and tumor protein 53 (p53) nuclear accumulation, did not affect in-vitro proliferative potential of embryonic

NSCs. They showed that intrinsic differences exist between embryonic/adult neural progenitor cells, in their response to telomere shortening, and populations of tissue-specific SCs bypass DNA damage checkpoints.

7.3.13 PIGMENT EPITHELIUM-DERIVED FACTOR IS A NICHE SIGNAL FOR NEURAL STEM CELL RENEWAL

Adult SCs are characterized by self-renewal/multilineage differentiation, and properties are regulated by signals from adjacent differentiated cell types and extracellular matrix molecules, which collectively define SC *niche*. Self-renewal is essential for lifelong persistence of SCs but its regulation is misunderstood. In mammalian brain, neurogenesis persists in two germinal areas, SVZ and hippocampus, where continuous postnatal neuronal production is supported by NSCs. Ramírez-Castillejo *et al.* showed that pigment epithelium-derived factor (PEDF) is secreted by murine SVZ components and promotes adult-NSC self-renewal, *in vitro*.[40] Intraventricular PEDF infusion activated slowly dividing SCs, whereas endogenous-PEDF blockade decreased their cycling. They showed that PEDF is niche-derived regulator of adult NSCs and evidenced role for PEDF protein in NSC maintenance.

7.3.14 VASCULAR NICHE FACTOR PIGMENT EPITHELIUM-DERIVED FACTOR MODULATES NOTCH-DEPENDENT STEMNESS IN SUBEPENDYMAL ZONE

Andreu-Agulló *et al.* addressed question of how SC microenvironments regulate self-renewal.[41] They found that Notch was active in astoglia-like NSCs but not in transit-amplifying progenitors of murine subependymal zone (SEZ), and that level of Notch transcriptional activity correlated with self-renewal/multipotency. Dividing NSCs balanced renewal with commitment via controlled segregation of Notch activity, leading to biased expression of known (*Hes1*)/unknown (*Egfr*) Notch target genes in daughter cells. The PEDF enhanced Notch-dependent transcription in cells with low Notch signaling, subverting asymmetrical-division output to production of two highly self-renewing cells. Mechanistically, PEDF induced noncanonical activation of nuclear factor (NF)-κB pathway, dealing to dismissal of transcriptional nuclear receptor corepressor (N-CoR) from specific Notch-responsive promoters. They provided basis for stemness regulation in vascular niches and indicated that Notch/PEDF cooperate to regulate self-renewal.

7.3.15 GLIAL PRECURSORS CLEAR SENSORY NEURON CORPSES DURING DEVELOPMENT

During peripheral ganglia development, 50% neurons that are generated undergo apoptosis; how massive corpse numbers are removed is unknown. Wu *et al.* found that satellite glial cell precursors are primary phagocytic cells for apoptotic corpse removal, in developing murine dorsal root ganglia (DRGs).[42] Confocal/electron microscopic analyses revealed that glial precursors, rather than macrophages, were responsible for clearing most dead DRG neurons. They identified engulfment receptor Jedi-1, and multiple epidermal growth factor (EGF)-like-domains 10 (MEGF10) purported engulfment one, as homologs of invertebrate engulfment ones Drape/CED-1 expressed in glial precursor cells. Expression of Jedi-1/MEGF10 in fibroblasts facilitated binding to dead neurons, and knocking down either protein in glial cells or overexpressing truncated forms, lacking intracellular domain, inhibited apoptotic neuron engulfment. Results suggest cellular/molecular mechanism by which neuronal corpses are culled during DRG development.

7.3.16 THE MYTH OF PROMETHEUS AND REGENERATIVE MEDICINE

SCs present in various organs of adult mammals, for example, man, contribute to continuous cell renewing of tissues.[43] Isolating cells from various tissues, their plasticity, and expansion capacity show that SCs are used to obtain differentiated cells of desired type and, in sufficient number in culture plate, for their later transplantation in patients with pathologies that follow with cell losses, for example, degenerative diseases. It is tried to obtain cells specific of nervous system [dopaminergic neurons for Parkinson's disease, γ-aminobutyric acid (GABA)ergic ones for Huntington's chorea, oligodendroglias for demyelinating diseases, for example, multiple sclerosis, etc.], trying to induce differentiation of NSCs previously expanded *ex vivo*. Cultures seem not to reproduce conditions found in specific micro-environments/niches, in which SCs are placed in cerebrum and that regulate their plasticity/persistence. Many signals that regulate SC behavior in niches are unknown, and it lacks much to understand basic biology of adult SCs. Knowledge derived from study of cells in adult tissues leads not only to improve cultures but also to attempt controlled reactivating of endogenous SCs and eliminate transplantation.

7.3.17 KINASE-DYRK1A CRITICAL REGULATED SEGREGATION DURING ASYMMETRIC NEURAL STEM CELL DIVISION

SC division can result in two sibling cells exhibiting differential mitogenic and self-renewing potential. Ferrón *et al.* evidenced that dual-specificity kinase Dyrk1A is part of molecular pathway, involved in regulation of biased epidermal growth factor receptor (EGFR) signaling, in progeny of dividing NSCs of adult SEZ.[44] They showed that EGFR asymmetry requires regulated sorting, and normal *Dyrk1a* dosage is necessary to sustain EGFR in both daughters of symmetrically dividing progenitor. The Dyrk1A is asymmetrically/symmetrically distributed during mitosis, and biochemical analyses indicate that it prevents endocytosis-mediated EGFR degradation by mechanism that requires phosphorylation of EGFR signaling modulator Sprouty2. The *Dyrk1a* heterozygous NSCs exhibit defects in self-renewal, EGF-dependent cell-fate decisions and long-term persistence *in vivo*, showing that symmetrical divisions play role in SEZ-reservoir maintenance.

7.3.18 DLK1-IMPRINTING POSTNATAL LOSS IN STEM CELLS/NICHE ASTROCYTES REGULATES NEUROGENESIS

Gene is considered to be imprinted if only copy inherited from mother/father is expressed throughout life; however, imprinted gene *Dlk1* disobeys rule during postnatal neurodevelopment.[45] Gene for atypical NOTCH ligand δ-like homolog-1 (*Dlk1*) encodes membrane-bound/secreted isoforms, which function in developmental processes *in vitro/vivo*. Member *Dlk1* of imprinted gene cluster is expressed from paternally inherited chromosome. Ferrón *et al.* showed that mice, which are deficient in *Dlk1*, present defects in postnatal neurogenesis in SVZ: developmental continuum that results in mature neuron depletion in olfactory bulb.[46] They showed that DLK1 is secreted by niche astrocytes, whereas its membrane-bound isoform is present in NSCs and required for inductive effect of secreted DLK1 on self-renewal; they found that requirement exists for *Dlk1* to be expressed from maternal/paternally inherited chromosomes. Selective absence of *Dlk1* imprinting, in NSCs/niche astrocytes, is associated with postnatal acquisition of DNA methylation at germ-line-derived imprinting control region. Results emphasize molecular relations between NSCs and microenvironment niche astrocytes, identifying signaling system encoded by single gene that functions coordinately in both. Genomic imprinting modulation, in SC environment, adds epigenetic regulation level to niche establishment/maintenance, raising questions about imprinting adaptability, function, and evolution in specific developmental contexts.

7.3.19 STEM CELLS: BIOLOGY AND THERAPY

SCs present multilineage differentiation and self-perpetuation capacities.[47] Their plasticity/expansion capacity, *ex vivo*, possibilities obtaining pure cell populations with chosen phenotype and in number sufficient for directed cell therapy in context, for example, degenerative diseases. Differentiation is tried in SC populations previously expanded *ex vivo* with mitogens. Study of signals/mechanisms involved in terminal differentiation of SCs must be accompanied of expansion phase analysis, which precedes their transplantation. Although the greatest differentiating potential takes place in embryonic stem cells (ESCs), their existence in various tissues, for example, neural ones obtained from adult cerebrum, possibilities obtaining information relative to proper signals of niche/microenvironment that cells occupy. Signals that regulate SC behavior, at niches, are unknown but it is clear that their actions determine the presence of controlled set of SCs, during all individual life, and progeny production in homeostasis conditions, and modulate/induce regenerative response in cell depletion conditions. Knowledge derived from SSCs could cause reactivating endogenous ones eliminating transplantation.

7.3.20 *Bmp2* TRANSCRIPTIONAL REPRESSION BY P21WAF1/CIP1 LINKS QUIESCENCE TO NEURAL STEM CELL MAINTENANCE

Relative quiescence and self-renewal are defining features of adult SCs but their potential coordination remains unclear. Subependymal NSCs lacking cyclin-dependent kinase (CDK) inhibitor (CKI) 1a (p21) exhibit rapid expansion, which is followed by their permanent loss later in life. Porlan *et al.* showed that transcription of the gene encoding *Bmp2* in NSCs is under the direct negative control of p21 via actions, which are CDK independent.[48] Loss of p21 in NSCs results in increased levels of secreted *Bmp2*, which induce premature terminal differentiation of multipotent NSCs into mature non-neurogenic astrocytes in an autocrine and/or paracrine manner. They showed that the cell-nonautonomous *p21*-null phenotype is modulated by the Noggin-rich environment of the subependymal niche. They described a dual function providing a physiological example of combined cell-autonomous and -nonautonomous functions of p21 with implications in self-renewal, linking the relative quiescence of adult SCs to their longevity and potentiality.

7.3.21 p27Kip1 REGULATES THE POPULATION OF SUBEPENDYMAL ZONE BIOPOTENTIAL PROGENITORS

In mammalian adult brain, two zones exist that maintain their neurogenic activity thanks to the existence of NSCs with a capacity of self-renewal and producing differentiated progeny. Fariñas group focused on the study of NSCs from SEZ, which generate neuroblasts and oligodendroblasts that migrate to the olfactory bulb and *corpus callosum*, respectively.[49] The NSCs generate their progeny via neural progenitors of quick amplification. The cells kinetics and self-renewing capacity are regulated by extrinsic factors produced by the microenvironment and intrinsic regulators, for example, CKIs. Their general objective is the study of the function of CKI *p27Kip1* (p27, encoded by *Cdkn1b*) in the dynamics of neural progenitors and its possible role as transcriptional regulator of different multipotency-related factors. The SEZ-isolated NSCs are capable of dividing *in vitro* generating clone aggregates called *neurospheres*, which could be submitted to a differentiation protocol, so that it is possible to study the different present populations via immunocytochemistry. Their in-vivo studies departed from permolten and histologically processed brains to study different-markers presence via immunohistochemistry. They used mice wild type and mutant for *Cdkn1b*. *In vitro*, they detected a biopotential population of neural progenitors characterized by being Mash1/Olig2-positive and p27-regulated. Absence of p27 favors its proliferation and delays its start in differentiating. *In vivo*, the defect drives to a rise in neurogenesis and oligodendrogenesis. Their results suggested that p27 repress transcription factor sex-determining region Y-box containing gene-2 (Sox2) expression, which could be the molecular mechanism responsible for progenitors behavior. Their data propose that SEZ-produced neurons and oligodendrocytes derive from a biopotential common progenitor, which start in differentiating is p27-regulated via transcriptional repression of the multipotency factor Sox2, which assumes that p27 exerts functions in the neural progenitors, which are independent of its participation in the cell cycle regulation and depend on its action as transcriptional regulator, in a similar way to p21. Fariñas reported on the era of the cell revolution.[50]

7.3.22 ARE NEURONS REGENERATED? STEM CELL–NICHE INTERACTIONS IN ADULT BRAIN

Fariñas raised the following questions on neural-system aging and neurons regeneration.[51]

Q1. Are neurons regenerated?

Q2. How can the brain be improved?

Q3. How is the cognitive aging produced?

Q4. Can one take part in a more direct way in brain plasticity?

The SCs are found at specific locations within adult tissues, suggesting that the microenvironment/niche in which they reside is essential for regulated behavior and lifelong maintenance. In the adult mammalian brain, NSCs continuously generate neurons, which functionally integrate into circuits in restricted regions. Endogenous NSCs may represent a reservoir of cells that could be harnessed for brain repair, and cells that could become dysregulated in pathological states or during aging. The NSCs dynamically regulate their numbers and cell output under homeostasis and in response to injury integrating cell intrinsic mechanisms and niche signals. Fariñas combined morphological studies with the search for molecular mechanisms underlying the dynamic regulation of adult NSCs in their in-vivo niche.[52] The adult brain subependymal niche, located in the wall of the lateral ventricles, contains a relatively quiescent population of self-renewing astrocyte-like NSCs, which continually produce new neurons. Within the specialized microenvironment in which SCs reside, different elements play a role in their bahavior regulation (e.g., innervation, irrigation, cerebrospinal fluid of lateral ventricles). Subependymal NSCs have access to the compartments but the signaling pathways are still under investigation. She proposed hypothesis (H)/Qs on SC–niche interactions in the adult brain.

H1. Cyclin-dependent kinase inhibitor gene (p21)-deficient cultures lose potency (glycogenic phenotype) and enter senescence.

Q1. Are there angiocrine factors for quiescence?

Q2. Is endothelial nitric oxide synthase responsible for the cytostatic effect?

7.3.23 INTERACTION OF NEURAL STEM CELLS WITH THEIR MICROENVIRONMENT: HOME, SWEET HOME

Concept of SC, associated with regenerative medicine, is already part of people's dairy language.[53] In the name, people include cells of different organs [ESCs, induced pluripotent stem cells (iPSCs), adult SCs]. All can be expanded indefinitely in a culture plaque and instructed to differ in those cell types that could be useful to restore cells lost by wounds or degenerative processes, which offers undoubted hopes for cell therapy. Specifically, SCs present in practically all tissues and organs of adult mammals, including man, carry out the function of sustaining the continuous cell renewal of the tissues, even, in the nervous system. The adult SCs are in definite locations and the specific microenvironment or niche that surrounds them regulates their huge plasticity and persistence. Many signals that regulate SCs behavior in the niches are still unknown, and ever so much left to understand the basic biology of adult SCs exists. However, the knowledge derived from the cells' study could lead not only to improve the cultures that are performed nowadays but also to raise the possible controlled reactivation of endogen SCs, removing the need for transplant in regenerative medicine. Fariñas raised Qs on NSCs interaction with microenvironment.

Q1. What is a young scientist?

Q2. How must a young scientist be?

Q3. How do adult SCs work?

Q4. How do people understand themselves?

The understanding of how adult SCs work is also changing people's way of thinking about aging or cancer, that is, how people understand themselves.

Fariñas proposed additional questions and hypothesis.

Q5. Are there angiocrine factors for quiescence?

Q6. How do tissues function, maintain, and repair?

H1. Defining the association of position with cell state is critical for people's understanding of how tissues function, maintain, and repair.

Q7. Do people have only two endogen zones in the brain that they want to regenerate?

7.3.24 REMOTE SIGNALING AND STEM CELL-NICHE BIOLOGY IN THE ADULT BRAIN

Adult SCs are found at specific locations, and their behavior and lifelong maintenance are regulated by cell intrinsic factors, and signals from the microenvironment or niche in which they reside. However, SC niches are poorly characterized, because of the complexity of the interactions between SCs and their neighbors, and to the dynamic changes required for the continuous production of cells. In the adult brain SEZ, radial glia/astrocyte-like NSCs continually produce neurons and oligodendrocytes, via a population of rapidly diving transit-amplifying progenitor cells. In the adult SEZ, different elements (e.g., innervation, irrigation, cerebrospinal fluid of the brain lateral ventricles) appear to play important roles in the regulation of NSCs behavior, but the involved signaling pathways are under investigation. Increasing evidence indicates that immune cells and immune mediators modulate NSCs behavior. Effects on neurogenesis of pro-inflammatory cytokines that are produced under nonphysiological conditions (e.g., irradiation, inflammation, status epilepticus, stroke) were described. However, their effects appeared sometimes contradictory, suggesting potentially distinct effects depending on the involved cell or receptor type. Tumor necrosis factor α (TNFα), a pro-inflammatory cytokine, is a multifunctional protein with a broad range of activities in different systems. Fariñas group evaluated potential roles of TNFα and its receptors (TNFRs) in SEZ remodeling/regeneration, analyzing direct effects of TNFα on the proliferation/self-renewal of NSCs in culture, and assessing its relevance in different in-vivo scenarios where SEZ homeostasis is compromised.[54] They analyzed the role of both TNFRs via specific TNFR1 and TNFR2 agonists, and TNFR knock-out mice. They found that TNFα modulates proliferation, self-renewal, and the balance of symmetrical/asymmetrical divisions of NSCs, and each receptor mediates a distinct bioresponse.

7.3.25 SC-NICHE BIOLOGY IN BRAIN: A MARKER FOR CELLULAR SENESCENCE DETECTION?

Adult SCs are found at specific locations, and their behavior and lifelong maintenance are regulated by cell-intrinsic factors and signals from the microenvironment or niche in which they reside.[55] Astrocyte-like NSCs continually produce neurons and oligodendrocytes, in two discrete neurogenic niches of the adult mammalian brain: SGZ and SEZ. The potential of NSCs for brain repair is fueling the concept of SC niches as druggable targets for restorative therapies. However, SC niches are poorly characterized because of the complexity of SC–their neighbor interactions, and to the dynamic changes required for the continuous production of cells. In the adult SEZ, different elements (e.g., innervation, irrigating vasculature, cerebrospinal fluid) appear to play important roles in the regulation of NSC behavior, but the involved signaling pathways are still under investigation. Despite the presence of a blood/brain physical barrier, NSC behavior can be modulated by circulating factors, in ways that are far from understood. Increasing evidence indicates that remote lesions outside CNS can activate NSCs via the engagement of systemic innate immune responses, which are transduced to the brain. Therapeutic approaches for the activation of endogenous NSCs will necessarily require ways to deliver specific signals to neurogenic niches.

Cellular senescence is a state of permanent arrest of the cell cycle that cells undergo after certain number of divisions, or when they suffer stress or DNA damage. The accumulation of senescent cells with aging was related to age-associated pathologies, for example, neurodegenerative disorders (NDs).[56] The NSCs progressively lose their neurogenic potential with aging. However, if this is caused by a senescence process remains unknown.[57] A comprehensive understanding about senescence and its influence in SEZ neurogenic niche is needed. In order to achieve the proposal, the reliable detection of senescent cells becomes a priority. The most widely used method to monitor senescence is the detection of the lysosomal activity of senescence-associated β-galactosidase (β-Gal) (SA-β-Gal), an enzyme that becomes especially abundant in senescent cell. However, the marker can fail to differentiate properly between senescent and non-senescent cells, and it was proposed that the detection of another enzyme, a lysosomal hydrolase called α-fucosidase (α-Fuc), could be an alternative, which is a novel biomarker with the potential to become a sensitive and specific marker, in general, and, specially, when SA-β-Gal do not provide a reliable signal.[58] The α-Fuc enzyme activity is specifically increased in senescent cells, which can be observed via a histochemical labeling, which uses X-Fuc as substrate, rendering an intensive blue staining, which enabled Fariña group to detect cellular senescence *in vitro* in senescence-induced cells and tissue samples from aging-mice models, which could offer them the possibility of identifying those areas in the neurogenic niche where senescent cells are accumulated.[59]

Cellular senescence is a state of permanent arrest of the cell cycle. The accumulation of senescent cells with aging was related to age-associated pathologies, for example, NDs. The NSCs in SEZ progressively lose their proliferative capability and neurogenic potential with aging. To investigate the implication of cellular senescence in neurogenesis within the SEZ, a reliable method for senescence detection is needed. The activity of SA-β-Gal was the most widely used senescence marker. However, another lysosomal enzyme (α-Fuc) was proposed as a potential alternative method to monitor cellular senescence. Fariñas group compared side-by-side β-Gal and α-Fuc histochemical stainings in a well-established model of cellular senescence with SK-MEL-28 melanoma cell line under different conditions of pH, fixation, and cofactors in the staining solution.[60] Staining of β-Gal provided by far a more reliable labeling for senescence detection. They compared the efficiency of the two methods for the detection of cell senescence *in vivo*, by staining whole-mount preparations of SEZ of senescence-accelerated-prone 8 mice and their control senescence-accelerated-resistant mice. Senescent cells were only observed with β-Gal method.

7.3.26 IN SEARCH FOR VASCULAR SENESCENT CELLS

Cell senescence is an alternative cellular process to apoptosis in which the cell cycle is in permanent arrest.[61] Senescent cells present an active secretome, in which proinflammatory factors are released with consequently deleterious effects for the neighboring cells. Those cells accumulate with age, causing a reduction in tissue function that leads to the development of age-associated diseases, for example, NDs. In the adult mammalian brain, SEZ, lining the lateral wall of the lateral ventricle, represents the major reservoir of NSCs and continually generates neurons. Several signals are implicated in the regulation of NSC activity, for example, vasculature plays an important role providing a number of factors to the neurogenic niche. With age, the capacity for self-renewal and proliferation of NSCs declines, resulting in a loss of neurogenic potential in SEZ. However, it is unknown whether the phenomenon is influenced by cell senescence. Given the important role of the vasculature in sustaining the neurogenic niche, a process of senescence in the blood vessels, which are in close proximity to SEZ, could contribute to the depletion of NSCs. In order to explore the relationship, an on–off two-photon probe (AHGa) and a functionalized nanoparticle (GalNP) were utilized as tools for senescence detection. Both become activated by increased lysosomal (SA-β-Gal) activity compared to normal cells. On one hand, AHGa consists of a chemical compound that becomes fluorescent when it is hydrolyzed by β-Gal. On the other hand, GalNPs are capped with a galacto-oligosaccharide, which is hydrolyzed by β-Gal, allowing the release of their content, which device allows the detection or clearance of senescent cells when loaded with a fluorophore or a senolytic drug, respectively.

7.3.27 SIGNALS OF SUCCESS

Cancers are SC diseases, and pathways that are important in regulating these may represent novel targets for cancer therapy. Pathways active in pattern development in embryo play roles in control of SC fate decisions. Lewis and Maitland reviewed work that highlighted two pathways [Hedgehog/ Wingless (Wnt) signaling ones], in regulating SCs, and how pathways may be activated in prostate cancer (PC).[62]

7.3.28 STEM CELL-BASED NEUROPROTECTIVE AND NEURORESTORATIVE STRATEGIES

The SCs, special subset of cells derived from embryo/adult tissues, present characteristics of self-renewal, multiple differentiation lineages, plastic capability, and long-term maintenance. Reports suggested that NSCs, derived from adult hippocampus/SVZ, possess potential to develop transplantation strategies and screen candidate agents for neurogenesis/ protection/plasticity, in NDs. Hung *et al.* reviewed roles of NSCs/other SCs in neuroprotective/restorative therapies, for neurological/psychiatric diseases.[63] They showed that NSCs are involved in pathogenesis of several NDs, for example, depression, stroke, and Parkinson's disease. They discussed iPSC utilities, reprogramming from adult fibroblasts with ectopic expression of four embryonic genes. Understanding SCs biophysiology helps elucidating pathogenicity and developing treatments for NDs. In contrast to cell transplantation therapies, SC application provides drug discovery and small-molecule testing platform. High-throughput SC-based systems are used to discriminate neuroprotective candidate mechanisms in translation medical research for NDs.

7.3.29 STEM CELL CONCEPTS RENEW CANCER RESEARCH

Although uncontrolled proliferation is distinguishing property of tumor as whole, individual cells that make up this exhibit variation in properties, for example, morphology, proliferation kinetics, and ability to initiate tumor growth in transplant assays. Understanding heterogeneity molecular/cellular basis presents implications in therapeutic-strategy design. Mechanistic basis of tumor heterogeneity resulted uncertain; however, neoplasia is cellular hierarchy with cancer stem cells (CSCs) at apex. Dick historically reviewed hematology influence on development of SC concepts and cancer linkage.[64]

7.3.30 ANALYSIS OF MEMBRANE PROTEINS EXPRESSED IN PLURIPOTENT MURINE EMBRYONIC STEM CELLS

The ESCs are established from inner cell mass of preimplantation embryos, capable of self-renewal and exhibit pluripotency. Given properties, ESCs are expected to present therapeutic potential in regenerative medicine and as tool for in-vitro differentiation studies of SCs. Growth factors and extracellular matrix components regulate ESC progeny pluripotency/differentiation. Cell surface receptors that bind regulatory factors are crucial for SC regulation. In order to identify membrane proteins that are involved in regulation of PSCs, Intoh *et al.* purified/analyzed membrane proteins of murine ESCs, cultured with/out leukemia inhibitory factor, by quantitative proteomics two-dimensional polyacrylamide gel electrophoresis-based analysis, via fluorescently labeled proteins, and shotgun-based analysis with isotope-labeled peptides identified 338 proteins, for example, transmembrane, membrane-binding and extracellular ones, which were expressed specifically in pluripotent/differentiated murine ESCs.[65] Identified-protein functions revealed cell adhesion molecules, channels, and receptors, which are expected to play important roles in maintenance of murine ESC pluripotency. Membrane proteins that are expressed in pluripotent ESCs but not in differentiated cells, for example, Slc16a1/Bsg, could be useful for SC selection *in vitro*.

7.3.31 STEM CELL ELIMINATION

Holden and Jones argued that targeting CSCs might be key to effectively treating cancer and avoiding tumor recurrence.[66] Two theories exist to explain tumor formation/growth. (1) Stochastic model assumes that all cancer cells present low but equal tumor-forming capacity. (2) The CSC/tumor-initiating model suggests that small cell population exists, within tumor, responsible for its generation/maintenance, and they are capable of forming new tumors; all other cells in tumor lack capacity; cellular hierarchy exists within tumor with all cells ultimately derived from CSC small population (cf. Fig. 7.3).

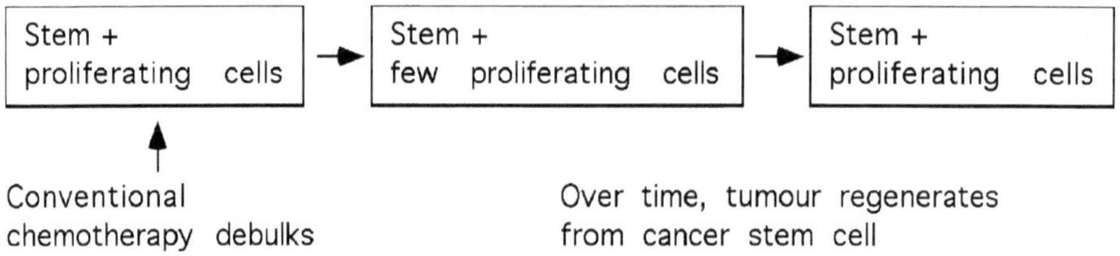

FIGURE 7.3 Conventional chemotherapy misses cancer stem cells.

Publications evidenced in support of CSC/tumor-initiating cell hypothesis (cf. Fig. 7.4).

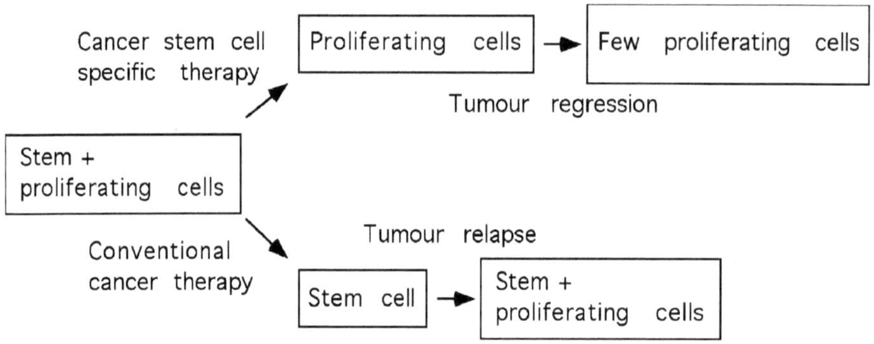

FIGURE 7.4 Ideal therapy for tumors.

7.3.32 TRAVELING CELLS GIVE FORM TO EMBRYOS AND PRODUCE CANCER METASTASIS

Theme in biology is the study of mechanisms that determine cell position during embrygenesis, that is, how from an only cell complete individual is formed, how cells *know* if they are going to form part of arm/lips and how they identify way that they must follow. Cells exist that are born in their definitive place but others, far from their final location and must travel great distances to reach their destination. Study of *traveling* cells and how they move is essential to understand how is generated organism/human body. Nieto-Toledano reported normal embryonic development in various species, and advances in knowledge of how migratory cells *know* that they must become detached, from their original place, and initiate their travel.[67] When processes do not work adequately, congenital malformations or pathologies during adult life are originated; for example, first step in cancer metastasis formation consists in that tumor cells are detached and migrate to colonize other organs, in form similar to what happens during embryonic development for tissue formation. Experiments showed that these are the same mechanisms that control cell movements in embryos/cancer. Knowledge of processes that control embryonic *traveling* cells helps to fight major dangerous side of cancer: metastasis.

7.3.33 STEM CELLS AND NEUROPROTECTION: UNDERSTANDING THE PLAYERS

Neuroprotective therapy use begs question of how treatments could affect preexisting SC populations, within host, and those introduced via cell-replacement therapy. Multiple mechanisms mediate SC responses to neuroprotectants, for example, host/donor age/gender, cellular lineage/differentiation status, and mitochondrial dynamics. Therapeutic sources for SCs are embryonic, somatic, or induced pluripotent with little known about gender, age, cell-type, and mitochondrial-dynamics effects. With advent of therapies to stimulate/recruit endogenous stem or transplant donor cells into damage areas, in hopes of recuperative regeneration of lost neurons, it is important to discuss mechanisms that dictate winning players in neuroprotection game. Pearce reviewed characteristics of renewing SCs that affect neuroprotection.[68]

7.3.34 PERMANENTLY BLOCKED STEM CELLS DERIVED FROM BREAST CANCER CELL LINES

The CSCs are thought to be resistant to standard chemotherapeutic drugs and inimical conditions of tumor microenvironment. Obtaining CSCs in sufficient quantities, and maintaining their undifferentiated state, were hurdles to their characterization and identification of pharmaceuticals that preferentially target cells. Sajithlal *et al.* described CSC-like population tagging from four human breast cancer (BC) cell lines with GFP, under control of

Oct3/4 SC-specific promoter.[69] Populations enriched in CSC expressed GFP; however, cells remained blocked in CSC-like state, and tended to be resistant to chemotherapeutic drugs and acidotic/hypoxic conditions. The CSC-like cells possessed other in-vitro CSC attributes and were able to generate reproducibly tumors in immunocompromised mice from 100 cells, which were comprised almost exclusively of pure CSCs. Ability of Oct3/4 promoter, to block CSC differentiation, underscores its potential general utility for obtaining highly purified CSC populations, although mechanism remains undefined; nonetheless, such stable cell lines should be extremely valuable tools, for studying basic questions pertaining to CSC biology and initial identification of CSC-specific chemotherapeutic agents, which can be verified in primary CSCs.

7.3.35 TOLL-LIKE RECEPTORS ROLE IN SYSTEMIC CANDIDA ALBICANS INFECTIONS: PROLIFERATION

Toll-like receptors (TLRs) constitute a family of pattern-recognition receptors (PRRs), which recognize molecular signatures of microbial pathogens, and function as sensors for infection, which induce activation of innate immune responses and subsequent development of adaptive immune responses.[70] The TLRs, mainly TLR2/4, are involved in host interaction with *Candida albicans* and play role in development of host immune responses during candidiasis. Recognition of *C. albicans* by TLRs on phagocytic cells activates intracellular signaling pathways, which trigger production of proinflammatory cytokines, which are critical for innate host defense and orchestrate adaptive response. Reactivity of T-helper (Th)-cell plays central role in regulating immune responses to *C. albicans*: Th1-response controls fungal infectivity, although proinflammatory (Th1) host response must be counterbalanced via Th2 and regulatory T (Treg) cells to ensure optimal, protective Th1 response. Subset of Th cells, Th17, played role in antifungal immunity and TLRs contribute to polarization toward proinflammatory Th17 response. *Candida albicans*–TLRs interaction is complex as (1) TLR2 functions as homodimer or TLR2–TLR1/6 heterodimers and collaborates with other non-TLR PRRs, in recognizing fungal ligands or triggering intracellular signaling pathways and (2) expression of fungal ligands is different at fungal cell surface, depending on morphotype (yeast cells or hyphae), which influences type of induced host immune response. Gil reported TLR participation in proliferation/differentiation of hematopoietic SCs and progenitors versus *C. albicans* (cf. Fig. 7.5).

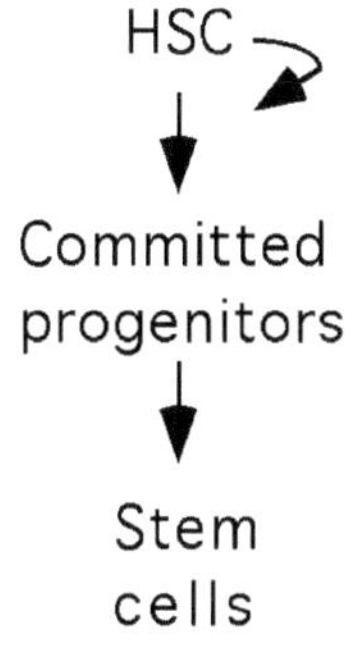

FIGURE 7.5 The continuous process of hematopoiesis: Increasing maturing.

7.3.36 DISSECTING CANCER STEM CELL THEORIES

If model is apt, it depends on cancer type and individual patient.[71] Theory of CSC, which holds that small number of tumorigenic cells are responsible for metastases, applies to only some cancers/patients; learning when CSC model is valid is step to understanding/treating cancer.

7.3.37 WATT FUN!

Fiona Watt probed individual SCs and determined genes/molecules, which direct them to differentiate or cause them to contribute to cancer.[72] Her hits are as follows: (1) determined that adhesion regulates differentiation/tissue assembly in mammalian epidermis and catalogued key pathways that govern effect;[73] (2) identified molecules that regulate exit from SC compartment in epidermis; (3) showed that integrins on cell surface of skin epithelia can be in/active and demonstrated that antibodies versus integrins could be used to obtain samples enriched in SCs; (4) discovered that nondividing, differentiated cells can contribute to benign tumor formation in skin;[74] (5) produced chemically patterned surfaces that provide *islands* on which single SCs can live, to identify environmental factors that regulate their behavior;[75] (6) as editor of *Journal of Cell Science*, she initiated interview series recording women experiences in science.

7.3.38 (RE)PROGRAMMING DIRECTOR

Unwilling to accept terminal-differentiation finality, Helen Blau-honed techniques that showcase cells flexibility to adopt different identities.[76] Her hits are as follows: (1) Via a technique involving cell fusion, she was the first to demonstrate that differentiated mammalian (human) cells retain their developmental plasticity and can be reprogrammed, to activate genes typical of completely different cell types. (2) Via the fusion technique combined with new short-interfering RNA (siRNA) and RNA-sequencing technologies, she showed that the mechanisms regulating reprogramming to iPSCs can be elucidated and identified one factor, key to the initiation of pluripotency in fused cells.[81] (3) She revealed how physical extracellular matrix parameters can impact muscle SC fate/function, with broad implications for SCs use in regenerative medicine. (4) She determined that mammalian differentiated cells, for example, cells of regenerative newts, can be dedifferentiated by transiently disabling tumor suppressors: the cells re-enter the cell cycle and make copies of themselves, providing a novel cell source for regenerative medicine. (5) She decided that Duchenne muscular dystrophy results from dystrophin deficiency in conjunction with shortened telomeres.

7.3.39 AN EPIGENETIC (EPI) PHENOMENON

While exploring the genetics of a rare type of tumor, Stephen Baylin discovered an epigenetic modification that occurs in almost every cancer (a finding he is helping bring to the clinic).[82] His hits are as follows: (1) Characterized the biology and genetics of a rare, heritable form of thyroid cancer. (2) Determined that methylation of cytosine (C) bases in the promoter region of genes plays a key role in human cancers. Identified a handful of tumor suppressor genes that are silenced by this DNA methylation and continues to study how these alterations work in concert with mutations to drive the progression of cancer. (3) Working with colleague J. Herman, he developed a technique, called methylation-specific polymerase chain reaction, for detecting and quantitating DNA methylation. (4) Coordinated efforts to catalog the genome-wide epigenetic changes that contribute to human cancers. (5) With research associate H. O'Hagan and graduate student W. Wang, he discovered that chronic inflammation can rapidly promote the widespread methylation of genes, potentially laying the groundwork for cancer.

7.3.40 HEMATOPOIETIC DIFFERENTIATION: COORDINATED DYNAMICAL PROCESS

The differentiation process, proceeding from SCs toward the different committed cell types, can be considered as a trajectory toward an attractor of a dynamic process, which view, taking into consideration the transcriptome and microRNA (miRNA) genome (miRNome) dynamics considered as a whole, instead of looking a few *master genes* driving the system, offers a novel perspective on phenomenon. Felli *et al.* investigated the *differentiation trajectories* of the hematopoietic system considering a genome-wide scenario.[83] They developed serum-free liquid suspension unilineage cultures of cord blood (CB) $CD34^+$ hematopoietic progenitor cells via erythroid (E), megakaryocytic (MK), granulocytic (G), and monocytic (Mo) pathways, which cultures recapitulate physiological hematopoiesis, allowing the analysis of almost pure unilineage precursors starting from initial differentiation of hierarchy of progenitor populations until terminal maturation. Analyzing the expression profile of protein-coding genes and miRNAs in unilineage CB E, MK, G, and Mo cultures, at sequential stages of differentiation and maturation, they observed a coordinated, fully interconnected, and scalable character of cell population behavior in both transcriptome and miRNome spaces reminiscent of an attractor-like dynamics. The miRNome and transcriptome spaces differed for a still not terminally committed behavior of miRNAs.

7.3.41 GENE EXPRESSION PROFILE OF STEM CELL PLURIPOTENTIALITY/DIFFERENTIATION

Understanding fundamental tumorigenesis mechanisms remains pressing problem in biology; to this end, stem-like cells with tumor-initiating potential became a central focus in cancer research. While CSC hypothesis presents a compelling model of self-renewal and partial differentiation, tumor–normal SCs relation remains unclear. Palmer *et al.* identified, in an unbiased fashion, messenger RNA (mRNA) transcription patterns associated with PSCs.[84] Via profile, they derived a quantitative measure of SC like gene expression activity. They showed how 189-gene signature stratifies a variety of SC, malignant and normal tissue samples by their relative plasticity and differentiation state within Concordia, a diverse gene expression database consisting of 3209 Affymetrix HGU133+ 2.0 microarray assays.

Furthermore, orthologous murine signature correctly orders a time course of differentiating embryonic mouse SCs. They showed how stem-like signature serves as a proxy for tumor grade in a variety of solid tumors, for example, brain, BC, lung, and colon.

7.3.42 THE IMPLICATIONS OF CANCER STEM CELLS FOR CANCER THERAPY

Surgery, radiotherapy, and chemotherapy are recognized as most effective anticancer therapies. Despite advances toward elucidating molecular mechanisms and developing clinical trials, cancer still remains major public health issue. Studies showed that CSCs, small subpopulation of tumor cells, can generate bulk populations of nontumorigenic cancer cell progeny via self-renewal/differentiation. As CSCs are proposed to persist in tumors as distinct population and cause relapse/metastasis, development of CSC-targeted therapeutic strategies holds hope for improving survival and life quality in cancer patients. Therapeutic innovations will emerge from understanding of CSCs bioenvironment which, however, are unexplored. Jiang et al. reviewed CSCs characteristics, evidences and development, and implications/challenges for cancer treatment.[85]

7.3.43 RECENT TRENDS IN STEM CELL RESEARCH ON CANCER THERAPEUTICS

Shanju *et al.* reviewed recent trends in SC research on cancer therapeutics.[86] The SCs are unique cells having the property of self-renewal characteristics and differentiation. Protein antigen markers in surface of cells are used to identify SC population. The CSCs or tumor-initiating cells (TICs) occur within tumor cells causing cancer. Cell surface antigen markers on SCs are not able to identify the specific cancer type. Further research of tumor suppressor genes, signaling pathways during SC differentiation would be able to combat cancer.

7.3.44 VARIATION BECAUSE OF DIFFERENTIATION

A hypothesis suggests tumors develop from CSCs that mature to enter a more differentiated state. Individual cells within a tumor might move along this continuum of differentiation at different rates, accounting for the genetic, epigenetic, and gene expression variation seen within and between tumors.

Lending credibility to this hypothesis, SC-like gene expression patterns were detected in early tumor cells.[87,88] This model could be seen either as complementary or antagonistic to the selection model, and more data are required to ascertain the exact contribution of each process to the heterogeneity seen in tumor samples.

7.3.45 AGING/CANCER: POTENTIAL REPROGRAMMING VIA MI-2/ NURD MODULATION

From nonconventional angles, Zhang argued *Don't put all eggs in one basket* and natural *self-organization* are among keys to unveil aging and cancer.[89] The former is behind phenomena, for example, asymmetry division, sharing interaction, chaotic and oscillatory gene expressions, sex, stemness, tumor, and aging damage heterogeneities; the latter is a core of system biology-based cell attractors theory, and evolution in this vision and reflections. He examined aging and cancer inevitabilities and postulated that the sharing in modern-life origin and multicellularity commencement is essential for cell normality, and the carcinogenesis is speciation. As for the nucleosome remodeling and histone deacetylase (Mi-2/NuRD), he explained the carcinogenesis alongside diseased multicellularity.[90] He extended it to mammalian target and rapamycin (mTOR) protein kinase and metabolism in aging and cancer (cf. Fig. 7.6). He proposed the modulation on them to reprogram aging and cancer.

Developmental program → Ageing program → Cancer stem-like program

FIGURE 7.6 Reprogramming of Mi-2/NuRD, mTOR and metabolism (3M) in aging and cancer.

7.4 RESULTS: STEM CELLS AND THEIR RELEVANCE TO PROSTATE CANCER INHIBITION

7.4.1 PROSTATE CANCER STEM CELLS: NEW HOPE FOR THERAPY?

The PC remains clinical/biologically difficult problem for translational researchers (cf. Table 7.1); with improved methods for selection and propagation/phenotypic analysis of critical tumor-initiating cells, real progress toward rational therapies is made.[91]

TABLE 7.1 Prostate Cancer Stem Cell Resistance to Conventional Cancer Treatment.

Therapy	Primary resistance	Secondary resistance
Surgery	Residual tumor	–
Androgen therapy	CSCs AR negative	Genomic instability
Radiotherapy	CSCs divide slowly	Expression of known radioresistance genes elevated in CSCs
Chemotherapy	CSCs divide slowly	Expression of drug resistance genes, e.g., ABC transporters in SCs
Immunotherapy/ vaccine	Normally directed versus common/differentiated cell antigens not expressed in CSCs	Genomic instability
Gene therapy	CSCs do not express common differentiated cell-targeting molecules	Genomic instability

AR, Androgen receptor; ABC, ATP-binding cassette; CSCs, cancer stem cells; SCs, stem cells.

7.4.2 ROLE OF ENHANCER OF ZESTE HOMOLOG 2 IN GROWTH OF PROSTATE CANCER STEM CELLS ISOLATED FROM LYMPH NODE CANCER OF PROSTATE (LNCAP) CELLS

Enhancer of zeste homolog 2 (EZH2) plays a crucial role in ESCs and SSCs for their proliferation and differentiation. However, the roles and underlying mechanisms of EZH2 in prostate cancer stem cells (PCSCs) remain unknown. Li et al. investigated EZH2 effects on PCSCs, which they isolated from human PC cell line LNCaP by fluorescence-activated cell sorting.[92] They compared EZH2 expression between PCSCs and non-PCSCs. They investigated the association between EZH2 function and PSCS growth via siRNA-mediated knockdown of EZH2. They investigated cell growth by 3-(4,5-dimethylthiazol-2-yl)-2,5-diphenyltetrazolium bromide and explored cell cycle/apoptosis of PCSCs by flow cytometric analysis. They determined the upstream pathway miRNA level via a luciferase reporter assay and examined the downstream pathway cycle regulators via reverse transcriptase-polymerase chain reaction. Their results showed that LNCaP cell line comprised a greater proportion of $CD44^{+}/133^{+}$ cells by comparison to cell line PC-3. The EZH2 was upregulated in PCSCs compared with non-PCSCs. Silence of EZH2 inhibited cell growth/cycle and promoted apoptosis progression. The EZH2 was an miRNA (miR)-101 direct target in

PCSCs and EZH2's mRNA levels were inversely correlated with miR-101 expression and cyclin E2 (a cell-cycle regulator) was suppressed by siRNAs targeting EZH2 (siEZH2). The EZH2 is essential for PCSC growth, partly via a negative regulation by miR-101 and positively regulating cyclin E2.

7.4.3 THE MOTHER OF ALL CELLS: THE PLURIPOTENT STEM CELLS

Torres proposed the following questions and answers on the mother of all cells, PSCs.[93]

Q1. Why is there no treatment for all diseases?

A1. Because of the lack of scientific knowledge on the biological mechanisms that are altered in the diseases.

Q2. In addition, what does happen if SCs are not accessible to extract them?

Q3. Why does the deficiency of gene *Gdap1* reduce the efficiency of reprogramming.

Q4. When does the fragmentation of mitochondria happen?

Q5. How does the fragmentation of mitochondria happen?

Q6. Could innate immune response be involved in the development of totipotent SCs?

During the process of reprogramming to iPSCs, somatic cells switch from oxidative to glycolytic metabolism, a transition associated with profound mitochondrial reorganization. Neither the importance of mitochondrial remodeling for cell reprogramming nor the molecular mechanisms controlling this process are well understood. Torres group showed that an early wave of mitochondrial fragmentation occurs upon expression of reprogramming factors.[94] Reprogramming-induced mitochondrial fission is associated with a minor decay in mitochondrial mass but not with mitophagy. The profission factor dynamin-related protein (Drp)1 is phosphorylated early in reprogramming and its knockdown/inhibition impairs mitochondrial fragmentation and generation of iPS cell colonies. Phosphorylation of Drp1 depends on extracellular signal-regulated kinase (Erk) activation in early reprogramming, which occurs, at least in part, because of downregulation of the mitogen-activated protein (MAP) kinase (MAPK) phosphatase (MKP) dual-specificity phosphatase (Dusp)6 (cf. Fig. 7.7). Their data indicate that mitochondrial fission controlled by an Erk-Drp1 axis constitutes an early and necessary step in the reprogramming process to pluripotency. They raised a Q on early Erk1/2 activation that promotes Drp1-dependent mitochondrial fission for cell reprogramming.

Q1. What works badly in an individual's ill cells that they give rise to disease sign?

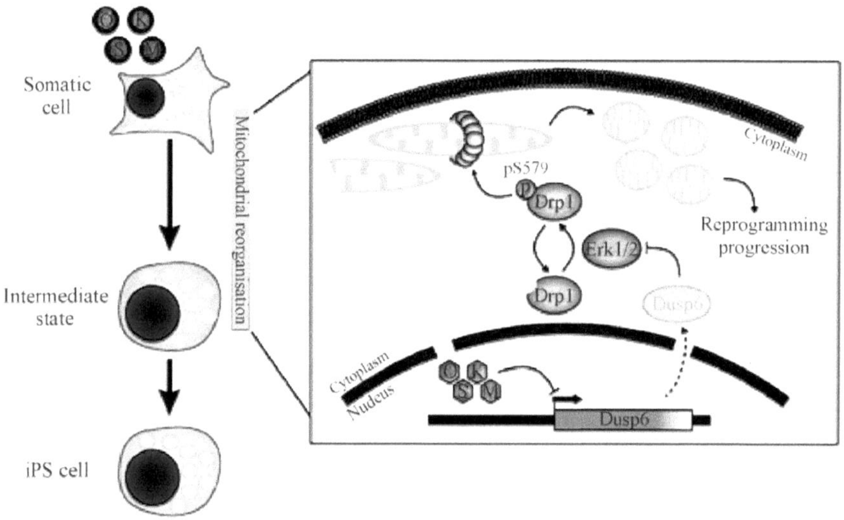

FIGURE 7.7 ERK signaling in activating Drp1 via phosphorylation at S579 during early reprogramming.

Takahashi and Yamanaka were the first to generate iPSCs by nuclear reprogramming of other cell types with defined transcription-factors combinations.[95] Raya proposed Q/As on possibilities and limitations of iPSC-based human disease modeling.[96]

Q1. Cancer predisposition?

Q2. Is induced reprogramming a laboratory-exclusive phenomenon?

A2. Yes, it does not exist in nature.

Q3. In regenerative therapy, are there alternatives to iPSC, for example, miRNAs?

A3. Yes, it could be done with anything that induces regeneration but one must find such a thing!

Q4. However, does the environment affect iPSCs?

A4. Yes, the environment must be taken into account.

Q5. Can it be reused to reprogram the cells that are already *in situ*?

A5. In *directed reprogramming* (*transdifferentiating*) parallel line, cells are differentiated without iPSCs.

Q6. Why is an iPSC line so expensive?

A6. It is a question of means and regulations.

Q7. Is there such biobank of allogenic cells?

A7. Yes, the patient will be immunodeficient; it is like a transplant but one is never without organs.

Q8. In Parkinson, Is it important to have polysaccharides around neurons?

A8. Sure! The extracellular matrix is important: one thing to do is to improve the culture systems.

Q9. When one adds four factors inducing pluripotency, what does happen in a cell that it takes a step back in differentiation?

A9. People are beginning to know that it forms the *core net*, which is a self-regulated net.

7.4.4 GENETICALLY SUPERVISED SELF-ORGANIZATION IN ENSEMBLES OF MOUSE EMBRYONIC STEM CELLS

Martínez-Arias proposed Q/A/H/Fs on genetically supervised self-organization in mouse-ESCs ensembles.[97]

Q1. How does one learn about genes, cells, and patterns?

A1. They are a function of genes, proteins, and forces.

H1. (Richard Feynman). *What I cannot create, I do not understand.*

F1. The ESCs have the capacity of developing embryos if they are placed in embryos.

F2. The cells are never measuring levels of proteins but rather correlations in the levels of proteins.

Q2. What have people learned?

A2. People have learned gene regulatory nets.

Q3. How can cells self-organize into an organism?

Q4. ...or do they need references, cues, coordinates?

Q5. How does the mammalian embryo break symmetry?

Q6. What if ... may be ... proteins are not controlled by function?

Q7. (Philosophical Q). How can one go from this to know what is common (categories, principles)?

A7. Principle is molecules and molecular interactions; the strategy varies.

Q8. Is there a signal or pattern in the environment recorded by the embryo if cells are coming?

A8. Yes, they are coming from the background.

Q9. Is there any metabolism or metabolite change in the medium?

A9. Yes, one needs O_2, and so on.

7.4.5 FROM IN VITRO TO IN VIVO, A QUANTUM JUMP FOR REGENERATIVE MEDICINE

Embryonic pluripotency can be recapitulated *in vitro* by a spectrum of PSC states captured by different synthetic niches.[98] Their distinct spatiotemporal characteristics provide an unprecedented tool toward the study of early human development and evolution. The newly unveiled ability of some SC types for crossing xeno-barriers will facilitate the generation of cross-species chimeric embryos from distant species, for example, humans. When combined with efficient zygote gene-editing technologies, xenogeneic human PSCs may also open new frontiers for regenerative translational medicine applications, for example, the possibility of generating human organs in animals via interspecies chimeric complementation.

7.4.6 BIOETHICAL CONSIDERATIONS

Technical challenges are not only issues facing human ESCs (hESCs). Skloot delved into bioethical issues surrounding development of Henrietta Lacks (HeLa) cells, which were derived without consent from woman dying of cervical cancer.[99] The hESCs demand scrutiny: issues are research business interest and information resources lack by population.[100–106] A thought experiment reported how a bioethicist[107] explores the soul of Venter's new life form (*synthetic cell*) and of his experiment.[108]

7.4.7 STORIES OF MOTHER AND DAUGHTER CELLS: HOW TO ORGANIZE PEOPLE'S BRAIN

Gil Sanz raised questions on mother/daughter-cells stories, and how to organize ours brain.[109]

Q1. How to organize people's brain?

Q2. How do we get to generate the great cell diversity from a population of similar NSCs?

Q3. How alterations in NSCs diversification are connected with neural disorders?

7.4.8 DISCOVERY OF SMALL-MOLECULE INHIBITORS OF TRANSCRIPTIONAL ACTIVITY

Orphan nuclear receptor tailless (TLX) (NR2E1) plays a critical role in the NSCs regulation and NSC-derived-brain tumors development. Data emerged implicating TLX in PC and BC. Inhibitors of TLX transcriptional activity present an impact on the treatment of several critical malignancies. However, TLX protein possesses a noncanonical ligand-binding domain (LBD), which lacks a ligand-binding pocket (conventionally targeted in case of nuclear receptors), which complicates the development of small-molecule inhibitors of TLX. A rational structure-based design approach was utilized to identify small molecules targeting Atro-box BS of human TLX LBD.[110] As a result of virtual screening of 7,000,000 molecular structures, 97 compounds were identified and evaluated in TLX-responsive luciferase reporter assay. Among them, three chemicals demonstrated 40–50% inhibition of luciferase-detected transcriptional activity of TLX orphan nuclear receptor at 35-μM dose. The identified compounds represent the first class of small-molecule inhibitors of TLX transcriptional activity identified via methods of computer-aided drug discovery.

7.4.9 RETT'S SYNDROME: A NOT MUCH FREQUENT AND FEMININE DISEASE

Rett's syndrome is a neural development, not much frequent disease caused by mutations in a gene of X chromosome (MECP2), which is an expression regulator of other genes and essential for the correct working of cells, which does that male fetuses, and which only present one copy of X chromosome, never be born if they suffer for mutations at MECP2. Rett's syndrome affects young girls, who are born apparently healthy and develop the disease when 6–18 months old. Symptoms that the devastating disease causes include a delay in development with loss of learned capacities, motor problems, mental disability, loss of the power of speech, and characteristics similar to autistic spectrum disorders. Agustín Pavón group researched, in a murine model of Rett's syndrome, how MECP2 lack affects adult neurogenesis, that is, the neurons generation that occurs after birth.[111] In mammals, including humans, the generation of new neurons is restricted to certain brain areas between which olfactory system is found. Their results suggest that neural maturing processes in olfactory system are found altered in the mice, as well as the function of olfactory neurons. They showed that, although the mice can distinguish different odors, they present a decayed interest in investigating the olfactory signals of their congeners, a social behavior that can resemble that observed in the persons that present autistic spectrum disorder. To know the bioprocesses that are altered in Rett's syndrome will help people to design better therapeutic strategies, in future.

7.4.10 DISCOVERY OF A FUNCTION OF AN ENZYME THAT COULD BE USED TO BRAIN TUMORS

Genomic imprinting is implicated in the control of gene dosage in neurogenic niches. Ferrón group addressed the importance of *Igf2* imprinting for murine adult neurogenesis in SVZ and in SGZ of the hippocampus *in vivo*.[112] In SVZ, paracrine IGF2 is a cerebrospinal fluid and endothelial-derived neurogenic factor requiring biallelic expression, with mutants presenting reduced activation of SC pool and impaired olfactory bulb neurogenesis. In contrast, *Igf2* is imprinted in the hippocampus, acting as an autocrine factor expressed in NSCs solely from the paternal allele. Conditional mutagenesis of *Igf2* in blood vessels confirms that endothelial-derived IGF2 contributes to NSC maintenance in SVZ but not in SGZ, which is regulated by the biallelic expression of IGF2 in the vascular compartment. They indicated that a regulatory decision to imprint or not is a functionally important mechanism of transcriptional dosage control in adult neurogenesis. Ten–eleven translocation (TET) proteins catalyze DNA hydroxylation, playing an important role in demethylation of DNA in mammals. Although hydroxymethylation levels are high in the mouse brain, the potential role of TET proteins in adult neurogenesis is unknown. They showed that a non-catalytic action of TET3 is essentially required for the maintenance of NSC pool, in the adult SVZ niche, preventing premature differentiation of NSCs into non-neurogenic astrocytes, which occurs via direct binding of TET3 to the paternal transcribed allele of the imprinted gene small nuclear ribonucleoprotein-associated polypeptide N (*Snrpn*), contributing to transcriptional repression of the gene.[113] They identified *Bmp2* as an effector of SNRPN-mediated astrocytic terminal differentiation. They described a novel mechanism of control of an imprinted gene in the regulation of adult neurogenesis via an unconventional role of TET3.

7.5 FINAL REMARKS

From the present results and discussion, the following final remarks can be drawn.

(1) Acute myeloid leukemia enables research on different-oncogene biofunction in leukemic transformation/progression and allows devising healing strategies, for example, gene/antisense therapies, and so on, aimed at leukemic SC.

(2) Chain net, in SVZ, carries neuronal precursors from caudal SVZ to rostral parts of adult brain. It should be determined what guides migration via SVZ net, which other vertebrate brains contain pathway array and biosignificance of such an extensive neuronal-precursor net, many destined for single brain region: olfactory bulb.

(3) Despite robust germinal capacity, no evidence of cells migrating in chains exists along SVZ or olfactory peduncle to bulb, which is explained by a long distance that separates olfactory bulbs from cerebrum in man or micro-osmatic capabilities. Findings raise fate question of cells born in human SVZ. Identification of large, anatomically discrete population of proliferating, multipotent human astrocytes leads to understanding role that human neural SCs act in tumorigenic demyelination/degeneration.

(4) Planar polarity of ciliated ependymal cells is essential for formation of chemorepulsive-factor gradients, which guide neuroblast migration in adult brain. Polarized ciliated cells contribute important vectorial information for body plan development. Polarized epithelia and motile cilia in brain serve as directional-information conveyors for neuronal migration.

(5) Postnatal neurogenesis is important for learning/memory, and plasticity induction, linked to neuronal maturation timing. Hypothesis is that medial migratory stream, and other potential escape pathways

from SVZ, supplies interneurons to regions of developing human brain as mechanism of postnatal plasticity is delayed. Although function of recipient cortical domain, ventromedial prefrontal cortex is unknown in children, region in adult human brain is activated during specific cognitive tasks, for example, spatial conceptualization and emotional processing of visual cues. Ventromedial prefrontal cortex is focally inactivated in patients with advanced Alzheimer's disease. Beyond functional implications, developmental study of human SVZ shows major period of neurogenesis migration, which extends into postnatal life but is limited to early childhood, which holds implications for understanding neonatal neurological diseases, for example, germinal matrix hemorrhages and perinatal hypoxic/ischemic injuries, each potentially altering SVZ neurogenesis, and its apparent downstream cortical targets at formative stages of human development. Detection of migratory route for immature neurons, within infant human brain, highlights mechanisms via which increased regional complexity is achieved during brain evolution.

(6) Adult neural SCs allows understanding how SCs behave, in absence of dynamic patterning mechanisms that control embryonic neural development; SC bioapproach enriches understanding basic neural development, while providing exciting avenues to injury/illness treatment.

(7) Proliferative potential of neural SCs residing in adult brain is subject to intrinsic regulation, and knowledge is needed on intracellular mechanisms of cycling control, if one pursues reactivation of latent neural SC populations to engage endogenous neurogenesis, for brain-disease treatment. Embryonic neural SCs escape cellular checkpoints, and methods to monitor genetic stability of ex-vivo expanded neural SCs are necessary before therapeutic intervention.

(8) It was identified pigment epithelium-derived factor as ependymal and endothelial-derived molecule, with actions on SC renewal in nervous system. Factor receptor is unknown but understanding pathways, which are activated by it, results fundamental for determining mechanisms, which are implicated in self-renewal, and their interplay with other endogenous signaling systems. Understanding signals that regulate vascular/neural SC behavior, in coordinated way, results critical for development of brain-tumor treatments. Factor is beneficial in SC-based therapies, degenerative diseases, and cancer.

(9) Analysis in cell pair assays/populations, which differ in Notch activity levels, indicated that Notch asymmetry/symmetry at SC division is functionally important for replication. It is unclear how asymmetric distribution of activated Notch is produced in adult neural SCs, although it depends on regulated partitioning of effectors, for example, Numb, which inhibits Notch signaling and whose segregation is dictated by cell-polarity determinants. Data showed that asymmetry/symmetry is modulated by extrinsic cues/microenvironments, and neurovascular niches modify SC fate decisions that determine their persistence/response to nonphysiological demand.

(10) Apoptosis/elimination is essential for development/maintenance of functional nervous systems. Effort was made to understand molecular mechanisms; however, less attention was given to by-products. During regressive processes, degenerated/excess cellular debris is made that need to be eliminated. Reports evidenced for link between defective clearance of apoptotic cells and autoimmunity development. Hypothesis is that defects in neural waste clearance, during development, predisposes organism to autoimmune attack on nervous system later in life.

(11) Human gene *DIRK1A* was associated with Down syndrome neurological defects. Although transient ectopic expression of Dyrk1A, in neural progenitor cells of developing fetal cortex, results in premature differentiation, moderate increase in Dyrk1A levels has no effects on adult neural SC behavior; dose-dependent effects of Dyrk1A, in human pathology and rodent neural progenitor cells, support that Dyrk1A dose maintenance is important to preserve normal neurogenesis during development and in adult.

(12) Data support DLK1 role in neurogenic continuum, initiating at early postnatal period and being maintained over aging of an animal's lifetime. Compromised neurogenesis is driven by early increase followed by depletion of SC pool. In postnatal SVZ, soluble DLK1 derived from niche astrocyte signals via DLK1 membrane-bound form in neural SCs to regulate their number with continued requirement for DLK1 in their maintenance with age. Findings indicate that neural SCs, and niche astrocytes, are distinguished shortly after birth by expression of DLK1 membrane-bound/secreted isoforms and that differential processing of single gene confers distinct functional properties to both. Epigenetically regulated

selective absence of imprinting results in biallelic dosage of *Dlk1*, which is required for normal neurogenesis; biallelic expression of imprinted genes in specific cell types constitutes regulatory event in developmental program; possibility that genomic imprinting loss/gain be used as dynamic developmental mechanism, to control gene dosage, implicates for imprinting function/evolution understanding, and indicates that epigenetic mechanisms that control process in somatic lineages be adaptable to environmental niche in which they act.

(13) Signaling complexities in carcinogenesis/development overwhelm at stage; however, main player identity is known at least for Hedgehog/Wnt pathways. Cancer therapies directed versus pathways are likely to attack different cell populations, for example, CSCs, from those targeted by therapies, which seek to eliminate rapidly dividing cells, and those with differentiated phenotype; for example, agent effectiveness to block androgen response axis in PC. While initially effective, their long-term success is questionable. Therapies versus targets based not on growth rate but cancer biobasis should offer cure.

(14) SC development provides neurological/psychiatric-disease treatment. (a) Understanding biopathology of neural SCs helps elucidating pathophysiology of neurological/psychiatric diseases, for example, depression, Parkinson's disease, or ischemic stroke. Knowledge helps developing neuroprotective/restorative therapies. (b) Neural SCs provide platform to clarify mechanism and test drug efficacy. (c) Embryonic/induced pluripotent SC development makes cell transplantation therapies promising, in ischemic stroke or degenerative diseases; problems exist, for example, teratoma formation from implanted SCs or viability/differentiation ability of implanted cells. (d) Collaborative efforts of basic/translational research are needed. (e) SC-based neuroprotective/restorative strategies preserve potential to develop transplantation strategies and screen candidate agents for neurogenesis/protection/plasticity in degenerative diseases.

(15) SC concepts lie at heart of primary cancer therapies, which showed first success in curing patients with acute leukemia. Interest renewal, in SCs/cancer linkage, leads to understanding how tumor cellular context affects cancer-specific (epi)genetic pathways and drives therapeutic advances.

(16) Hypothesis is that PSCs present functional/specific surface membrane proteins, given sensitivity to extracellular conditions. Receptors/cell adhesion molecules, which are specifically expressed in pluripotent murine embryonic SCs, are successfully identified, suggesting that ligand–receptor signaling is crucial for mouse embryonic SC pluriponency; other pluripotency-specific membrane proteins, for example, transporters/channels, are identified. Small-molecule transport across plasma membrane regulates pluripotency of murine embryonic SCs. Membrane proteins, identified in pluripotent mouse embryonic SCs, are useful as pluripotency-specific markers or surface antigens for SC purification; they are involved not only in pluripotency maintenance but also in somatic cell reprogramming, whereby combination of embryonic SC-specific factors reprograms embryonic/somatic fibroblasts, to induce PSCs. Research uncovers membrane-protein roles in regulation of embryonic SCs.

(17) Despite the rarity of CSCs, and so on, with development of therapies to target them, interest exists in disease SCs and potential to remove tumors without recurrence.

(18) Heterogeneity/variety identification/characterization of progenitors, and their determinants, is necessary for understanding their role in neuroprotection. Cellular identification of neuronal progenitors with limited proliferative capacity subtypes, and their specific stimuli, is crucial for developing mechanistically based neuroprotective therapies that target optimal cell type. To move forward, in developing neuroprotectants, requires understanding SCs maximizing therapies and minimizing risks. Links in SC function are mediated by age, gender, telomerase, and mitochondrial factors. Interrelated characteristics are misunderstood in all SC types. To develop, neuroprotectants emphasis needs to be on understanding the factors and how they influence SC renewal/neurogenesis; understanding the factors provides insight into approaches, applications, and targets for SC therapies; how factors influence/deregulate functional pathways mediates more efficacious neuroprotective targets.

(19) Undifferentiated CSCs from many diseases were difficult to isolate/maintain, limiting characterization/extent to which they can be used to identify pharmaceuticals to which they are selectively susceptible. Finding that CSCs, from four different BC cell lines,

is maintained in undifferentiated state *in vitro/vivo* that provides opportunities. Cell lines possess properties that are distinct from those of their more abundant *transit amplifying cell*-like counterparts, for example, differences in their cell surface phenotypes, differential sensitivities to cytotoxic agents and environmental stresses, in-vitro growth properties, tumor-initiating capacities, and transcriptional profiles. Despite similarities, differences among CSC lines likely reflect first cell of original primary tumors, inherent growth biodifferences, and therapy histories of patients from whom lumps were obtained. Cell lines should provide homogeneous populations of CSCs, derived from naturally occurring and spontaneously arising human BC, which serve as powerful tools. Agents, which inhibit specifically CSCs, require verification in large number of disease SC lines and primary illness SCs. Understanding differences among CSC lines allows for their bioadaptive behavior knowledge.

(20) Consistent with their roles, the transcriptome system can be considered as the state space of a cell population, while the continuously evolving miRNA space corresponds to the tuning system necessary to reach the attractor. The behavior of miRNA machinery could be of great relevance not only for the promise of reversing the differentiated state but even for tumor biology.

(21) Core stemness gene expression signature represented a quantitative measure of SC-associated transcriptional activity. Signature intensity correlated to relative plasticity level and differentiation across human tissues. The fact that signature intensity be capable of differentiating histological grade for a variety of human malignancies suggested potential therapeutic/diagnostic implications.

(22) Gaps exist but self-organization insights in system biology point of aging/cancer suggest that both are inevitable and reversible or preventable. Breaking *Don't put all eggs in one basket* and *self-organization* promote aging and cancer. Cancer may hijack host system with/out switching their original functions. Inhibitors for Mi-2/NuRD and its functional partner chromatin remodeling system, mTOR could separately reprogram and reverse aging/cancer. However, it is tempting to find out effective combination rapamycin/mTOR inhibitors, vitamin C, and histone deacetylase inhibitors of low-dosage and lesser side effect in such reversals. Simplified genome-wide expression partners could be developed as ensemble markers for aging-related diseases or cancer.

(23) Prostate cancer remains clinical/biologically difficult problem for translational researchers. Invoking CSC mechanism, confounding features of common tumor can be explained with respect to therapy resistance. With improved methods for selection, propagation, and phenotypic analysis of critical tumor-initiating cells, real progress toward rational therapies can be made. SC phenotype feature is differentiation/malignancy separable nature.

(24) The EZH2 is essential for PCSCs growth, partly via regulating cyclin E2, and EZH2 expression is regulated by miR-101. Considering EZH2 role in PCSCs and the relation between EZH2 and miR-101, introduction of miR-101 to silence EZH2 could be a potential therapeutic strategy for prostate cancer.

(25) Recruitment of young neurons to the primate hippocampus decreases rapidly during the first years of life, and neurogenesis in the dentate gyrus does not continue, or is rare, in adult humans. The early decline in hippocampal neurogenesis raises questions about how the function of the dentate gyrus differs between humans and other species in which adult hippocampal neurogenesis is preserved.

(26) The role of genomic imprinting in gene dosage control is not fully understood. It is interesting to consider whether biallelic expression evolved from the imprinted state or represent preservation of an ancestral mode of regulation maintained because the mitogen demand in the SVZ renders the evolutionary cost of functional haploidy too high in the tissues. It is possible that imprinted expression, in the SGZ, represents an evolved mechanism to limit insulin-like growth factor-2 dosage, as a means to prevent overstimulation of the autocrine signaling axis. Regulation of adult neurogenesis, via the selective absence of insulin-like growth factor-2 imprinting, is reminiscent of recent work showing that selective biallelic expression of another imprinted gene, *Dlk1*, is required for adult neurogenesis. These findings suggest that genomic imprinting could be used as a mechanism of gene dosage control in particular developmental contexts and raises questions about the evolution, adaptability, and flexibility of imprinting as an epigenetically regulated process that could respond to intrinsic environmental cues.

(27) Adult niches preserve the undifferentiated property of their resident SCs, but contribution of epigenetic mechanisms to the process

remained largely unexplored. These data elucidated the action of TET3, preventing terminal differentiation and exhaustion of neural SCs by transcriptional repression of *Snrpn* gene independently of its methylation. It was described how the regulation acts in concert with niche signals (e.g., bone morphogenetic protein, Noggin) to modulate the proper differentiation of the neural SC pool.

Further work will deal with MEK/ERK pathway overactivation in liver tumors and resistance to transforming growth factor-β-induced death.

ACKNOWLEDGMENTS

The authors thank support from Fundacion Universidad Catolica de Valencia San Vicente Martir (Project No. 2019-217-001UCV).

KEYWORDS

- **cancer stem cell**
- **chemotherapy**
- **tumor microenvironment**
- **neural stem cell**
- **stem-cell-based strategy**
- **differentiation**
- **cellular senescence**

REFERENCES

1. Wilmut, I.; Sullivan, G.; Chambers, I. The Evolving Biology of Cell Reprogramming. *Phil. Trans. R. Soc. Lond., Ser. B* **2011,** *366*, 2183–2197.
2. Torrens, F; Castellano, G. Modelling of Complex Multicellular Systems: Tumour-Immune Cells Competition. *Chem. Centr. J.* **2009,** *3* (Suppl. I), 75-1–1.
3. Torrens, F.; Castellano, G. Information Theoretic Entropy for Molecular Classification: Oxadiazolamines as Potential Therapeutic Agents. *Curr. Comput.-Aided Drug Des.* **2013,** *9*, 241–253.
4. Torrens, F.; Castellano, G. Molecular Classification of 5-Amino-2-aroylquinolines and 4-Aroyl-6,7,8-trimethoxyquinolines as Highly Potent Tubulin Polymerization Inhibitors. *Int. J. Chemoinf. Chem. Eng.* **2013,** *3* (2), 1–26.
5. Estrela, J. M.; Mena, S.; Obrador, E.; Benlloch, M.; Castellano, G.; Salvador, R.; Dellinger, R. W. Polyphenolic Phytochemicals in Cancer Prevention and Therapy: Bioavailability versus Bioefficacy. *J. Med. Chem.* **2017,** *60*, 9413–9436.

6. Torrens, F.; Castellano, G. Molecular Classification of Antitubulin Agents with Indole Ring Binding at Colchicine-Binding Site. In *Molecular Insight of Drug Design*; Parikesit, A. A., Ed.; InTechOpen: Vienna, 2018; pp 47–67.
7. Torrens, F.; Castellano, G. Molecular Classification of 2-Phenylindole-3-carbaldehydes as Potential Antimitotic Agents in Human Breast Cancer Cells. In *Theoretical Models and Experimental Approaches in Physical Chemistry: Research Methodology and Practical Methods*; Haghi, A. K., Thomas, S., Praveen, K. M., Pai, A. R., Eds.; Apple Academic–CRC: Waretown, NJ, in press.
8. Torrens, F.; Castellano, G. Cancer and Hypotheses on Cancer. In Molecular Chemistry and Biomolecular Engineering: Integrating Theory and Research with Practice; Pogliani, L., Torrens, F., Haghi, A. K., Eds.; Apple Academic–CRC: Waretown, NJ, in press.
9. Torrens, F.; Castellano, G. Precision Personalized Medicine from Theory to Practice: Cancer. In Green Chemistry and Biodiversity: Principles, Techniques, and Correlations; Aguilar, C. N., Ameta, S. C., Haghi, A. K., Eds.; Apple Academic–CRC: Waretown. NJ, in press.
10. Torrens, F.; Castellano, G. AIDS Destroys Immune Defences: Hypothesis. *New Front. Chem.* **2014,** *23*, 11–20.
11. Torrens-Zaragozá, F.; Castellano-Estornell, G. Emergence, Spread and Uncontrolled Ebola Outbreak. *Basic Clin. Pharmacol. Toxicol.* **2015,** *117* (Suppl. 2), 38.
12. Torrens, F.; Castellano, G. 2014 Spread/Uncontrolled Ebola Outbreak. **New Front. Chem**. **2015, 24**, 81–91.
13. Torrens, F.; Castellano, G. Ebola Virus Disease: Questions, Ideas, Hypotheses and Models. *Pharmaceuticals* **2016,** *9*, 14-6–6.
14. Torrens, F.; Castellano, G. Clinical Translational Research: Cancer, Diabetes, and Cardiovascular Disease. In *Physical Biochemistry, Biophysics, and Molecular Chemistry: Applied Research and Interactions*; Torrens, F., Mahapatra, D. K., Haghi, A. K., Eds.; Apple Academic–CRC: Waretown, NJ, in press.
15. Torrens, F.; Castellano, G. *Primary Health, Smoking Habits and Immunotherapy in Cancer*. In *Environmental Technology and Engineering Techniques: Basic Concepts and Health Interventions*; Khalaf, M. N., Smirnov, M. O., Kannan, P., Haghi, A. K., Eds.; Apple Academic–CRC: Waretown, NJ, in press.
16. Torrens, F.; Castellano, G. Epigenomics, Epithelial Plasticity, Clinical Genetics and Rare Diseases. In *Environmental Technology and Engineering Techniques: Basic Concepts and Health Interventions*; Khalaf, M. N., Smirnov, M. O., Kannan, P., Haghi, A. K., Eds.; Apple Academic–CRC: Waretown, NJ, in press.
17. Torrens, F.; Castellano, G. Predictive in Silico Global Metabolism Analysis: Disease Mechanism. In *Processing and Characterization of Green Engineered and Sustainable Materials*; Haghi, A. K., Ed.; Apple Academic–CRC: Waretown, NJ, in press.
18. Torrens, F.; Castellano, G. Intracellular Proteolysis and Neoplasia: Ubiquitin-Proteasome System, Cell Cycle and Cancer. In *Natural Pharmaceuticals and Green Microbial Technology: Health, Environment and Resource Recovery*; Mahapatra, D. K.; Shinde, R. S., Haghi, A. K., Eds.; Apple Academic–CRC: Waretown, NJ, in press.
19. Torrens, F.; Castellano, G. Vitamin-D, Analogues, Drugs and Their Relevance to Cancer Inhibition. In *Natural Pharmaceuticals and Green Microbial Technology: Health, Environment and Resource Recovery*; Mahapatra, D. K., Shinde, R. S., Haghi, A. K., Eds.; Apple Academic–CRC: Waretown, NJ, in press.

20. Karley, D.; Gupta, D.; Tiwari, A. Biomarkers: The Future of Medical Science to Detect Cancer. *J. Mol. Biomark. Diagn.* **2011,** *2*, 118-1–7.
21. Bonnet, D.; Dick, J. E. Human Acute Myeloid Leukaemia Is Organized as a Hierarchy That Originates from a Primitive Hematopoietic Cell. *Nat. Med.* **1997,** *3*, 730–737.
22. Watson J. D.; Crick, F. C. Molecular Structure of Nucleic Acids. A Structure for Deoxyribose Nucleic Acid. *Nature (London)* **1953,** *171*, 737–738.
23. Watson, J. D.; Crick, F. C. Genetical Implications of the Structure of Deoxyribose Nucleic Acid. *Nature (London)* **1953,** *171*, 964.
24. Meselson, M.; Stahl, F. W. The Replication of DNA in *Escherichia coli*. *Proc. Natl. Acad. Sci. U.S.A.* **1958,** *44*, 671–682.
25. Cairns, J. Mutation Selection and the Natural History of Cancer. *Nature (London)* **1975,** *255*, 197–200.
26. Lewis, R. The Meselson-Stahl Experiment Lives On. *Scientist* **2004,** *18* (22), 56–56.
27. Doetsch, F.; Alvarez-Buylla, A. Network of Tangential Pathways for Neuronal Migration in Adult Mammalian Brain. *Proc. Natl. Acad. Sci. U.S.A.* **1996,** *93*, 14895–14900.
28. Reillo, I.; de Juan Romero, C.; García-Cabezas, M. Á.; Borrell, V.; A Role for Intermediate Radial Glia in the Tangential Expansion of The Mammalian Cerebral Cortex. *Cereb. Cortex* **2011,** *21*, 1674–1694.
29. Kelava, I.; Reillo, I.; Murayama, A. Y.; Kalinka, A. T.; Stenzel, D.; Tomancak, P.; Matsuzaki, F.; Lebrand, C.; Sasaki, E.; Schwamborn, J. C.; Okano, H.; Huttner, W. B.; Borrell, V. Abundant Occurrence of Basal Radial Glia in the Subventricular Zone of Embryonic Neocortex of a Lissencephalic Primate, the Common Marmoset *Callithrix jacchus*. *Cereb. Cortex* **2012,** *22*, 469–481.
30. Eriksson, P. S.; Perfilieva, E.; Björk-Eriksson, T.; Alborn, A. M.; Nordborg, C.; Peterson, D. A.; Gage, F. H. Neurogenesis in the Adult Human Hippocampus. *Nat. Med.* **1998,** *4*, 1313–1317.
31. García-Verdugo, J. M. Personal Communication.
32. Sanai, N.; Tramontin, A. D.; Quiñones-Hinojosa, A.; Barbaro, N. M.; Gupta, N.; Kunwar, S.; Lawton, M. T.; McDermott, M. W.; Parsa, A. T.; García-Verdugo, J. M.; Berger, M. S.; Alvarez-Buylla, A. Unique Astrocyte Ribbon in Adult Human Brain Contains Neural Stem Cells But Lacks Chain Migration. *Nature (London)* **2004,** *427*, 740–744.
33. Sawamoto, K.; Wichterle, H.; Gonzalez-Perez, O.; Cholfin, J. A.; Yamada, M.; Spassky, N.; Murcia, N. S.; García-Verdugo, J. M.; Marin, O.; Rubenstein, J. L. R.; Tessier-Lavigne, M.; Okano, H.; Alvarez-Buylla, A. New Neurons Follow the Flow of Cerebrospinal Fluid in the Adult Brain. *Science* **2006,** *311*, 629–632.
34. Sanai, N.; Nguyen, T.; Ihrie, R. A.; Mirzadeh, Z.; Tsai, H. H.; Wong, M.; Gupta, N.; Berger, M. S.; Huang, E.; Garcia-Verdugo, J. M.; Rowitch, D. H.; Alvarez-Buylla, A. Corridors of Migrating Neurons in the Human Brain and Their Decline During Infancy. *Nature (London)* **2011,** *478*, 382–386.
35. Paredes, M. F. James, D.; Gil-Perotin, S.; Kim, H.; Cotter, J. A.; Ng, C.; Sandoval, K.; Rowitch, D. H.; Xu, D.; McQuillen, P. S.; Garcia-Verdugo, J. M.; Huang, E. J.; Alvarez-Buylla, A. Extensive Migration of Young Neurons into the Infant Human Frontal Lobe. *Science* **2016,** *354*, 81–81.
36. Sorrells, S. F.; Paredes, M. F.; Cebrian-Silla, A.; Sandovall, K.; Qi, D.; Kelley, K. W.; James, D.; Mayer, S.; Chang, J.; Auguste, K. I.; Chang, E. F.; Gutierrez, A. J.; Kriegstein, A. R.; Mathern, G. W.; Oldham, M. C.; Huang, E. J.; Garcia-Verdugo, J. M.; Yang, Z.;

Alvarez-Buylla, A. Human Hippocampal Neurogenesis Drops Sharply in Children to Undetectable Levels in Adults. *Nature (London)* **2018,** *555*, 377–381.
37. Paredes, M. F.; Sorrells, S. F.; Cebrian-Silla, A.; Sandoval, K.; Qi, D.; Kelley. K. W.; James, D.; Mayer, S.; Chang, J.; Auguste, K. I.; Chang, E. F.; Gutierrez Martin, A. J.; Kriegstein, A. R.; Mathern, G. W.; Oldham, M. C.; Huang, E. J.; Garcia-Verdugo, J. M.; Yang, Z.; Alvarez-Buylla, A. Does Adult Neurogenesis Persist in the Human Hippocampus? *Cell Stem Cell* **2018,** *23*, 780–781.
38. Anderson, D. J. Stem Cells and Pattern Formation in the Nervous System: The Possible versus the Actual. *Neuron* **2001,** *30*, 19–35.
39. Ferrón, S.; Mira, H.; Franco, S.; Cano-Jaimez, M.; Bellmunt, E.; Ramírez, C.; Fariñas, I.; Blasco, M. A. Telomere Shortening and Chromosomal Instability Abrogates Proliferation of Adult But Not Embryonic Neural Stem Cells. *Development* **2004,** *131*, 4059–4070.
40. Ramírez-Castillejo, C.; Sánchez-Sánchez, F.; Andreu-Agulló, C.; Ferrón, S. R.; Aroca-Aguilar, J. D.; Sánchez, P.; Mira, H.; Escribano, J.; Fariñas, I. Pigment Epithelium-Derived Factor Is a Niche Signal for Neural Stem Cell Renewal. *Nat. Neurosci.* **2006,** *9*, 331–339.
41. Andreu-Agulló, C.; Morante-Redolat, J. M.; Delgado, A. C.; Fariñas, I. Vascular Niche Factor PEDF Modulates Notch-Dependent Stemness in the Adult Subependymal Zone. *Nat. Neurosci.* **2009,** *12*, 1514–1523.
42. Wu, H. H.; Bellmunt, E.; Scheib, J. L.; Venegas, V.; Burkert, C.; Reichardt, L. F.; Zhou, Z.; Fariñas, I.; Carter, B. D. Glial Precursors Clear Sensory Neuron Corpses During Development via Jedi-1, An Engulfment Receptor. *Nat. Neurosci.* **2009,** *12*, 1534–1541.
43. Fariñas, I. Personal Communication.
44. Ferrón, S. R.; Pozo, N.; Laguna, A.; Aranda, S.; Porlan, E.; Moreno, M.; Fillat, C.; de la Luna, S.; Sánchez, P.; Arbonés, M. L.; Fariñas, I. Regulated Segregation of Kinase Dyrk1A during Asymmetric Neural Stem Cell Division Is Critical for EGFR-Mediated Biased Signalling. *Cell Stem Cell* **2010,** *7*, 367–379.
45. Oh, E. C.; Katsanis, N. Imprinting in the Brain. *Nature (London)* **2011,** *475*, 299–300.
46. Ferrón, S. R.; Charalambous, M.; Radford, E.; McEwen, K.; Wildner, H.; Hind, E.; Morante-Redolat, J. M.; Laborda, J.; Guillermot, F.; Bauer, S. R.; Fariñas, I.; Ferguson-Smith, A. C. Postnatal Loss of *Dlk1* Imprinting in Stem Cells and Niche Astrocytes Regulates Neurogenesis. *Nature (London)* **2011,** *475*, 381–385.
47. Fariñas, I. Personal Communication.
48. Porlan, E.; Morante-Redolat, J. M.; Marqués-Torrejón, M. Á.; Andreu-Agulló, C.; Carneiro, C.; Gómez-Ibarlucea, E.; Soto, A.; Vidal, A.; Ferrón, S. R.; Fariñas, I. Transcriptional Repression of Bmp2 by p21$^{Waf1/Cip1}$ Links Quiescence to Neural Stem Cell Maintenance. *Nat. Neurosci.* **2013,** *16*, 1567–1575.
49. Fariñas, I. Personal Communication.
50. Fariñas, I. Personal Communication.
51. Fariñas, I. Envejecimiento Saludable. In *Proceedings of the XXXVIII Congreso de la Sociedad Española de Bioquímica y Biología Molecular, València, Spain, 2015*; SEBBM: València, Spain, 2015; p 4.
52. Fariñas, I. Stem Cell-Niche Interactions in the Adult Brain. In *Proceedings of the XXXVIII Congreso de la Sociedad Española de Bioquímica y Biología Molecular, València, Spain, 2015*; SEBBM: València, Spain; 2015; pp 14.
53. Fariñas, I. Personal Communication.
54. Fariñas, I. Personal Communication.

55. Fariñas, I. Personal Communication.
56. Muñoz-Espín, D.; Serrano, M. Cellular Senescence: From Physiology to Pathology. *Nat. Rev. Mol. Cell Biol.* **2014,** *15*, 482–496.
57. Acosta, J. C.; Banito, A.; Wuestefeld, T.; Georgilis, A.; Janich, P.; Morton, J. P.; Athineos, D.; Kang, T. W.; Lasitschka, F.; Andrulis, M.; Pascual, G.; Morris, K. J.; Khan, S.; Jin, H.; Dharmalingam, G.; Snijders, A. P.; Carroll, T.; Capper, D.; Pritchard, C.; Inman, G. J.; Longerich, T.; Sansom, O. J.; Aznar Benitah, S.; Zender L.; Gil, J. A Complex Secretory Program Orchestrated by the Inflammasome Controls Paracrine Senescence. *Nat. Cell Biol.* **2013,** *15*, 978–990.
58. Hildebrand, D. G.; Lehle, S.; Borst, A.; Haferkamp, S.; Essmann, F.; Schulze-Osthoff, K. α-Fucosidase as a Novel Convenient Biomarker for Cellular Senescence. *Cell Cycle* **2013,** *12*, 1922–1927.
59. Rojas, S.; Alfonso, M.; Sirerol, M. S.; Fariñas, I.; Costero, A.; Martínez, R. Personal Communication.
60. Rojas, S.; Palop, M. J.; Sirerol, M. S.; Fariñas, I.; Martínez, R.; Costero, A. *A Novel Marker for Cellular Senescence Detection?* In *Sensors and Molecular Recognition*; Costero Nieto, A. M., Parra Álvarez, M., Gaviña Costero, P., Gil Grau, S., Eds.; Universitat de València: València, Spain, 2018; Vol 12, pp 16-1–4.
61. Rojas, S.; Mantikas, K.; Lozano, B.; Alfonso, M.; Palop, M. J.; Sirerol, M. S.; Fariñas, I.; Martínez, R. Personal Communication.
62. Lewis, J. L.; Maitland, N. J. Signals of Success. *Eur. Biopharm. Rev.* **2005,** *2005* (1), 58–65.
63. Hung, C. W.; Liou, Y. J.; Lu, S. W.; Tseng, L. M.; Kao, C. L.; Chen, S. J.; Chiou, S. H.; Chang, C. J. Stem Cell-Based Neuroprotective and Neurorestorative Strategies. *Int. J. Mol. Sci.* **2010,** *11*, 2039–2055.
64. Dick, J. E. Stem Cell Concepts Renew Cancer Research. *Blood* **2008,** *112*, 4793–4807.
65. Intoh, A.; Kurisaki, A.; Yamanaka, Y.; Hirano, H.; Fukuda, H.; Sugino, H.; Asashima, M. Proteomic Analysis of Membrane Proteins Expressed Specifically in Pluripotent Murine Embryonic Stem Cells. *Proteomics* **2009,** *9*, 126–137.
66. Holden, K. A.; Jones, K. Stem Cell Elimination. *Eur. Biopharm. Rev.* **2009,** *2009* (49) 32–37.
67. Nieto-Toledano, M. Á. Personal Communication.
68. Pearce, V. Stem cells and Neuroprotection: Understanding the Players. *Int. J. Mol. Sci.* **2010,** *11*, 3288–3297.
69. Sajithlal, G. B.; Rothermund, K.; Zhang, F.; Dabbs, D. J.; Latimer, J. J.; Grant S. G.; Prochownik, E. V. Permanently Blocked Stem Cells Derived from Breast Cancer Cell Lines. *Stem Cells* **2010,** *28*, 1008–1018.
70. Gil, M. L.; Gozalbo, D. Role of Toll-Like Receptors in Systemic *Candida albicans* Infections. *Front. Biosci.* **2009,** *14*, 570–582.
71. Dutton, G. Dissecting Cancer Stem Cell Theories. *Gen. Eng. Biotechnol. News* **2010,** *3* (3), 1–47.
72. Hopkin, K. Watt Fun! *Scientist* **2011,** *25* (1), 44–46.
73. Watt, F. M.; Jordan, P. W.; O'Neil, C. H. Cell Shape Controls Terminal Differentiation of Human Epidermal Keratinocytes. *Proc. Natl. Acad. Sci. U.S.A.* **1988,** *85*, 5576–5580.
74. Arwert, E. N.; Lal, R.; Quist, S.; Rosewell, I.; van Rooijen, N.; Watt, F. M. Tumor Formation Initiated by Nondividing Epidermal Cells via an Inflammatory Infiltrate. *Proc. Natl. Acad. Sci. U.S.A.* **2010,** *107*, 19903–19908.

75. Connelly, J. T.; Gautrot, J. E.; Trappmann, B.; Tan, D. W. M.; Donati, G.; Huck, W. T. S.; Watt, F. M. Actin and Serum Response Factor Transduce Physical Cues from the Microenvironment to Regulate Epidermal Stem Cell Fate Decisions. *Nat. Cell Biol.* **2010,** *12*, 711–718.
76. Hopkin, K. (Re)programming Director. *Scientist* **2012,** *26* (10), 52–54.
77. Blau, H. M.; Chiu, C. P.; Webster, C. Cytoplasmic Activation of Human Nuclear Genes in Stable Heterocaryons. *Cell* **1983,** *32*, 1171–1180.
78. Chiu C. P.; Blau, H. M. Reprogramming Cell Differentiation in the Absence of DNA Synthesis. *Cell* **1984,** *37*, 879–887.
79. Chiu C. P.; Blau, H. M. 5-Azacytidine Permits Gene Activation in a Previously Noninducible Cell Type. *Cell* **1985,** *40*, 417–424.
80. Blau, H. M.; Pavlath, G. K.; Hardeman, E. C.; Chiu, C. P.; Silberstein, L.; Webster, S. G.; Miller S. C.; Webster, C. Plasticity of the Differentiated State. *Science* **1985,** *230*, 758–766.
81. Yamanaka, S.; Blau, H. M. Nuclear Reprogramming to a Pluripotent State by Three Approaches. *Nature (London)* **2010,** *465*, 704–712.
82. Baylin, S. B.; Beaven, M. A.; Engelman, K.; Sjoerdsma, A. Elevated Histaminase Activity in Medullary Carcinoma of the Thyroid Gland. *N. Engl. J. Med.* **1970,** *283*, 1239–1244.
83. Felli, N.; Cianetti, L.; Pelosi, E.; Carè, A.; Liu, C. G.; Calin, G. A.; Rossi, S.; Peschle, C.; Marziali, G.; Giuliani, A. Hematopoietic Differentiation: A Coordinated Dynamical Process towards Attractor Stable States. *BMC Syst. Biol.* **2010,** *4*, 85-1–14.
84. Palmer, N. P.; Schmid, P. R.; Berger, B.; Kohane, I. S. A Gene Expression Profile of Stem Cell Pluripotentiality and Differentiation Is Conserved across Diverse Solid and Hematopoietic Cancers. *Genome Biol.* **2012,** *13*, R71-1–40.
85. Jiang, W.; Peng, J.; Zhang, Y.; Cho W. C. S.; Jin, K. The Implications of Cancer Stem Cells for Cancer Therapy. *Int. J. Mol. Sci.* **2012,** *13*, 16636–16657.
86. Shanju, S.; Sangeetha, K.; Chandran, K. N.; Satheesan, B. Recent Trends in Stem Cell Research on Cancer Therapeutics. *Res. J. Biotechnol.* **2013,** *8* (4), 86–89.
87. Ben-Porath, I.; Thomson, M. W.; Carey, V. J.; Ge, R.; Bell, G. W.; Regev, A.; Weinberg, R. A. An Embryonic Stem Cell-Like Gene Expression Signature in Poorly Differentiated Aggressive Human Tumors. *Nat. Genet.* **2008,** *40*, 499–507.
88. Schoenhals, M.; Kassambara, A.; De Vos, J.; Hose, D.; Moreaux, J.; Klein, B. Embryonic Stem Cell Markers Expression in Cancers. *Biochem. Biophys. Res. Commun.* **2009,** *383*, 157–162.
89. Zhang, Y. Ageing and Cancer: Breaking the *Don't Put All Eggs in One Basket* and Natural *Self-organisation*, and Their Potential Reprogramming via Modulation of Mi-2/NuRD, mTOR Kinase and Metabolism. *Enz. Eng.* **2013,** *2* (1), 109-1–8.
90. Zhang, Y.; Moriguchi, H. Chromatin Remodeling System, Cancer Stem-Like Attractors, and Cellular Reprogramming. *Cell. Mol. Life Sci.* **2011,** *68*, 3557–3571.
91. Holmes, S. Prostate Cancer Stem Cells: New Hope for Therapy? *Innov. Pharm. Technol.* **2011,** *2011* (38), 20–24.
92. Li, K.; Liu, C.; Zhou, B.; Bi, L.; Huang, H.; Lin, T.; Xu, K. Role of EZH2 in the Growth of Prostate Cancer Stem Cells Isolated from LNCaP cells. *Int. J. Mol. Sci.* **2013,** *14*, 11981–11993.
93. Torres, J. Personal Communication.

94. Prieto, J.; León, M.; Ponsoda, X.; Sendra, R.; Bort, R.; Ferrer-Lorente, R.; Raya, A.; López-García, C.; Torres, J. Early ERK1/2 Activation Promotes DRP1-Dependent Mitochondrial Fission Necessary for Cell Reprogramming. *Nat. Commun.* **2016,** *7,* 11124-1–13.
95. Takahashi, K.; Yamanaka, S. Induction of Pluripotent Stem Cells from Mouse Embryonic and Adult Fibroblast Cultures by Defined Factors. *Cell* **2006,** *126*, 663–676.
96. Raya, Á. Personal Communication.
97. Martínez-Arias, A. Personal Communication.
98. Izpisúa Belmonte, J. C. Personal Communication.
99. Skloot, R. *The Immortal Life of Henrietta Lacks*; Crown: New York, NY, 2010.
100. Oyama, S., Griffiths, P. E., Gray, R. D., Eds. *Cycles of Contingency: Developmental Systems and Evolution*; Massachusetts Institute of Technology: Cambridge, MA, 2001.
101. Takahashi, K.; Tanabe, K.; Ohnuki, M.; Narita, M.; Ichisaka, T.; Tomoda K.; Yamanaka, S. Induction of Pluripotent Stem Cells from Adult Human Fibroblasts by Defined Factors. *Cell* **2007,** *131*, 861–872.
102. Yu, J.; Vodyanik, M. A.; Smuga-Otto, K.; Antosiewicz-Bourget, J.; Frane, J. L.; Tian, S.; Nie, J.; Jonsdottir, G. A.; Ruotti, V.; Stewart, R.; Slukvin, I. I.; Thomson, J. A. Induced Pluripotent Stem Cell Lines Derived from Human Somatic Cells. *Science* **2007,** *318*, 1917–1920.
103. Chung, Y.; Klimanskaya, I.; Becker, S.; Li, T.; Maserati, M.; Lu, S. J.; Zdravkovic, T.; Ilic, D.; Genbacev, O.; Fisher, S.; Krtolica, A.; Lanza, R. Human Embryonic Stem Cell Lines Generated without Embryo Destruction. *Cell Stem Cell* **2008,** *2*, 113–117.
104. Isasi, R. M.; Knoppers, B. M. Governing Stem Cell Banks and Registries: Emerging Issues. *Stem Cell Res.* **2009,** *3*, 96–105.
105. Webber, P. European Patent Law: Stemmed Potential. *Eur. Biopharm. Rev.* **2012,** *2012* (58), 28–30.
106. Grinnell, F. *Everyday Practice of Science: Where Intuition and Passion Meet Objectivity and Logic*; Oxford University: New York, NY, 2009.
107. Kaebnick, G. Is the *Synthetic Cell* about Life? *Scientist* **2010,** *24* (7), 27.
108. Venter, J. C. Synthesizing life. *Scientist* **2011,** *25* (10), 60.
109. Gil Sanz, C. Personal Communication.
110. Dueva, E.; Singh, K.; Kalyta, A.; LeBlanc, E.; Rennie, P. S.; Cherkasov, A. Computer-Aided Discovery of Small Molecule Inhibitors of Transcriptional Activity of TLX (NR2E1) Nuclear Receptor. *Molecules* **2018,** *23*, 2967-1–10.
111. Agustín Pavón, C. Personal Communication.
112. Ferron, S. R.; Radford, E. J.; Domingo-Muelas, A.; Kleine, I.; Ramme, A.; Gray, D.; Sandovici, I.: Constancia, M.; Ward, A.; Menheniott, T. R.; Ferguson-Smith, A. C. Differential Genomic Imprinting Regulates Paracrine and Autocrine Roles of IGF2 in Mouse Adult Neurogenesis. *Nat. Commun.* **2015,** *6*, 8265-1–12.
113. Montalbán-Loro, R.; Lozano-Ureña, A.; Ito, M.; Krueger, C.; Reik, W.; Ferguson-Smith, A. C.; Ferrón, S. R. TET3 Prevents Terminal Differentiation of Adult NSCs by a Non-catalytic Action at *Snrpn*. *Nat. Commun.* **2019,** *10*, 1726-1–14.

CHAPTER 8

MEK/ERK Pathway Overactivation in Liver Tumors: TGF-β Death

FRANCISCO TORRENS[1*] and GLORIA CASTELLANO[2]

[1]*Institut Universitari de Ciència Molecular, Universitat de València, Edifici d'Instituts de Paterna, P. O. Box 22085, E-46071 València, Spain*

[2]*Departamento de Ciencias Experimentales y Matemáticas, Facultad de Veterinaria y Ciencias Experimentales, Universidad Católica de Valencia San Vicente Mártir, Guillem de Castro-94, E-46001 València, Spain*

[*]*Corresponding author. E-mail: torrens@uv.es*

ABSTRACT

Real-time cell electronic sensing is a tool for label-free detection and cell-based parameters. Cancerous cells present varying migratory/invasive behaviors specific to chemoattractant stimuli/Matrigel concentration. Cellular invasion/migration for real-time cell analyzer is a versatile device for invasive/migration analysis and a unique platform capable of quantitatively measuring invasive/migratory behavior of cells in real time, without exogenous labels, which enables identification of migratory/invasive time points, which will help increase throughput by shortening assay time, minimize number of endpoint assays, and assist to determine optimal time points for inhibitor studies. Analysis using xCELLigence reduces hands-on cell-manipulation steps leading to physiologically relevant results. Upregulation of nicotinamide adenine dinucleotide phosphate hydride oxidase four, by transforming growth factor-β, is required for its pro-apoptotic activity in hepatocytes. Impairment of transforming factor-β-induced response might confer apoptosis resistance in hepatocellular carcinoma cells. Overactivation of pathway mitogen-activated protein/extracellular signal-regulated kinase, in liver tumor cells, confers resistance to transforming growth factor-β-induced cell death, via impairing nicotinamide adenine dinucleotide phosphate hydride oxidase four upregulation, which is required for efficient mitochondrial-dependent apoptosis. In vivo murine study corroborated that silencing nicotinamide adenine dinucleotide phosphate hydride oxidase four increases tumorigenic capacity. Human hepatocellular carcinoma downregulates oxidase. Silencing oxidase induces epithelial–mesenchymal transition mechanisms, which brings on migration.

8.1 INTRODUCTION

Setting the scene: mitogen-activated protein kinase (MAPK)/extracellular signal-regulated kinase (ERK) kinase (MEK)/ERK pathway overactivation in liver tumors and resistance to transforming growth factor (TGF)-β-induced death. Hepatocyte model systems are useful for diverse applications (cf. Fig. 8.1).

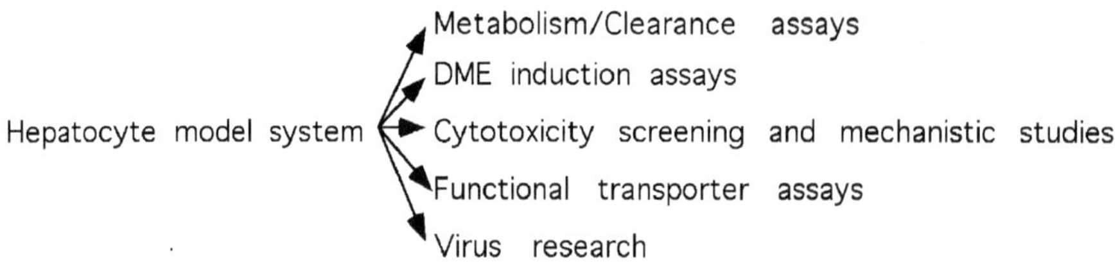

FIGURE 8.1 Diverse applications of hepatocyte model systems.
DME, drug-metabolizing enzyme.

The TGFs were originally named based on their abilities to induce a transformed phenotype in non-neoplastic rat kidney fibroblasts.[1] Dobolyi et al. reviewed the neuroprotective functions of TGF-β proteins.[2] Merck Millipore reported automated live cell imaging of cell migration across a microfluidic-controlled chemoattractant gradient.[3]

Earlier publications informed about the modeling of complex multicellular systems and tumor–immune cells competition,[4] information theoretic entropy for molecular classification of oxadiazolamines as potential therapeutic agents,[5] molecular classification of 5-amino-2-aroylquinolines and 4-aroyl-6,7,8-trimethoxyquinolines as highly potent tubulin polymerization inhibitors,[6] polyphenolic phytochemicals in cancer prevention, therapy, and bioavailability vs. bioefficacy,[7] molecular classification of antitubulin agents with indole ring binding at colchicine-binding site,[8] molecular classification of 2-phenylindole-3-carbaldehydes as potential antimitotic agents in human breast cancer cells,[9] cancer, its hypotheses,[10] and precision personalized medicine from theory to practice and cancer.[11] Studies covered the following areas: how human immunodeficiency virus/acquired immunodeficiency syndrome destroys immune defences, hypothesis,[12] 2014 emergence, spread, and uncontrolled Ebola outbreak,[13,14] Ebola virus disease, questions, ideas, hypotheses, and models,[15] clinical translational research, cancer, diabetes, cardiovascular disease,[16] primary health, smoking habits, and immunotherapy in cancer,[17] epigenomics, epithelial plasticity, clinical genetics, and rare diseases,[18] predictive in silico global metabolism analysis, disease mechanism,[19] intracellular proteolysis, neoplasia, ubiquitin-proteasome system, cell cycle and cancer,[20] vitamin-D, analogues, drugs, their relevance to cancer inhibition,[21] and stem cells, and their connection with cancer and prostate disease.[22] This report reviews state of the art of MEK/ERK pathway overactivation in liver tumors and resistance to TGF-β-induced death.

8.2 METHOD

Label-free detection was developed for cell-based screening; unlike classic detection, which uses fluorescence, radioisotope, luminescence, or light absorption, label-free detection measures cell function without a labeled molecule, whose advantages are: simple homogeneous assay format, noninvasive measurement, lesser interference with cell function, kinetic measurement, reduced assay time, etc. Solly et al. applied electrical impedance detection in a real-time (RT)-cell electronic sensing (CES™) system for cell-based assays.[23] Cell growth rate measured by RT-CES was comparable to actual cell number counted manually; proliferation, cytotoxicity, cytoprotection, growth inhibition, and apoptosis data generated by RT-CES correlated with classic methods.

Bird and Kirstein showed that for invasive/migration assays, cellular invasion/migration (CIM)-Plate® 16 used with Instrument xCELLigence® RT Cell Analyzer (RTCA) Dual Plate (DP, Roche) allows monitoring cellular responses in RT without exogenous labels via impedance-based technology. Each well in CIM-Plate 16 is a modified Boyden chamber. The CIM-Plate 16 provides kinetic cell-response profile throughout experiment detailing invasive/migration onset/rate. It is important to understand underlying biomolecular mechanisms; xCELLigence detects cell RT invasive/migration without labels.[24] Cell invasion is intrusion on/

destruction of adjacent tissues, especially for cancer cells. Cell migration is its movement from one area to another in response to chemical signal and important in physio/pathological processes, for example, embryonic development, cell differentiation, wound healing, immune response, inflammation, and metastasis; conventional methods for migration detection are endpoint/label-based assays. In Boyden chamber method, cells placed in inserts invade/migrate via artificial microporous membrane into lower chambers containing given chemoattractant. Drawbacks exist: migrated cells must be stained with chemical dyes/labeled with fluorescent molecules; membrane is removed from insert; and stained/labeled cells are counted manually using microscope/evaluated using spectrometer, which is a labor-intensive process that limits throughput and may alter gene expression profiles and lead to intra-assay variability because of efficiency differences.

In contrast, RTCA provides kinetic information about cell migration dynamically recording migration/invasion without labeling cells and, thus, improving assay; it uses CIM, which features microelectronic sensors integrated onto underside of microporous polyethylene terephthalate membrane of Boyden-like chamber. As cells migrate from upper chamber via membrane into bottom one in response to chemoattractant, they contact/adhere to electronic sensors on membrane underside resulting in impedance increase, which correlates to increasing numbers of migrated cells on membrane underside, and cell index (CI) reflecting impedance changes are automatically/continuously recorded by RTCA; cell migration activity is monitored via CI profile. They performed migration assay protocol used on RTCA with foetal bovine serum (FBS) as chemoattractant, which is optimized for cancer cell lines HT1080/HeLa. (1) Cells should be grown in FBS-free medium for 4–24 h. (2) For invasion experiments, CIM top chamber is coated with Matrigel™ (BD) before adding medium (with/without FBS) to bottom chamber. (3) Placing top onto bottom chamber and snapping both together assemble CIM. (4) Medium free of FBS is placed in top chamber to hydrate/pre-incubate membrane for 1 h in CO_2 incubator at 37°C. (5) Cells are trypsinized, pelleted, and re-suspended in FBS-free medium. (6) Once CIM equilibrated, it is placed in RTCA and background CI is measured. (7) The CIM is removed from RTCA and cells are added to top chamber. (8) The CIM is placed in RTCA and migration is monitored every 2 min. They analyzed densities of HeLa cells in 10% FBS ab/presence in bottom chamber. Cell migration kinetics were recorded by RTCA for 18 h, which were distinctly/easily detected by automated RT monitoring. Low/high seeding densities were quantitatively monitored/reflected by CI and correlated with manual cell count, determined at assay end by fixing/staining cells on top-chamber membrane. They monitored migration/invasion kinetics (in Matrigel ab/presence, respectively) in response to 10% FBS, in whose absence they observed little/no migration. Profiles of CI showed that cell invasion/migration were affected in dose-dependent manner by Matrigel concentration. The RT monitoring while assessing migration quantified invasion rate/onset. Delay of CI signal was because of time required for cells to invade via Matrigel and was concentration dependent.

8.3 RESULTS

8.3.1 UPREGULATION OF NADPH OXIDASE NOX4 BY TGF-β FOR PRO-APOPTOTIC ACTIVITY

TGF-β induced apoptosis in hepatocytes via oxidative stress (OS). Carmona-Cuenca et al. analyzed role of different nicotinamide adenine dinucleotide phosphate hydride (NADPH) oxidase (Nox) isoforms, in intracellular signaling induced by TGF-β in hepatocytes, to explore whether mechanism is altered in liver tumor cells (cf. Fig. 8.2a).[25] Primary cultures of rat/human hepatocytes HepG2/3B were used in in vitro studies to analyze TGF-β response. The TGF-β-induced apoptosis in rat hepatocytes does not require Rac (subfamily of ρ-family of GTPases)-dependent Noxs. The TGF-β upregulates Rac-independent Nox4, which correlates with pro-apoptotic activity. Regulation of Nox4 occurs at transcriptional level and is counteracted by intracellular survival signals. Short interfering ribonucleic acid (RNA) (siRNA)-targeted knockdown of Nox4 attenuates Nox activity, cysteine-aspartic protease (caspase) activation and cell death in rat hepatocytes. They observed NOX4 upregulation by TGF-β in human hepatocytes, coincident with apoptosis; in human hepatocellular carcinoma (HCC) cell lines, they observed NOX4 upregulation by TGF-β in only cells that are sensitive to its cytotoxic effect, for example, Hep3B. The siRNA-targeted knockdown of Nox4 in cells impairs TGF-β-induced apoptosis.

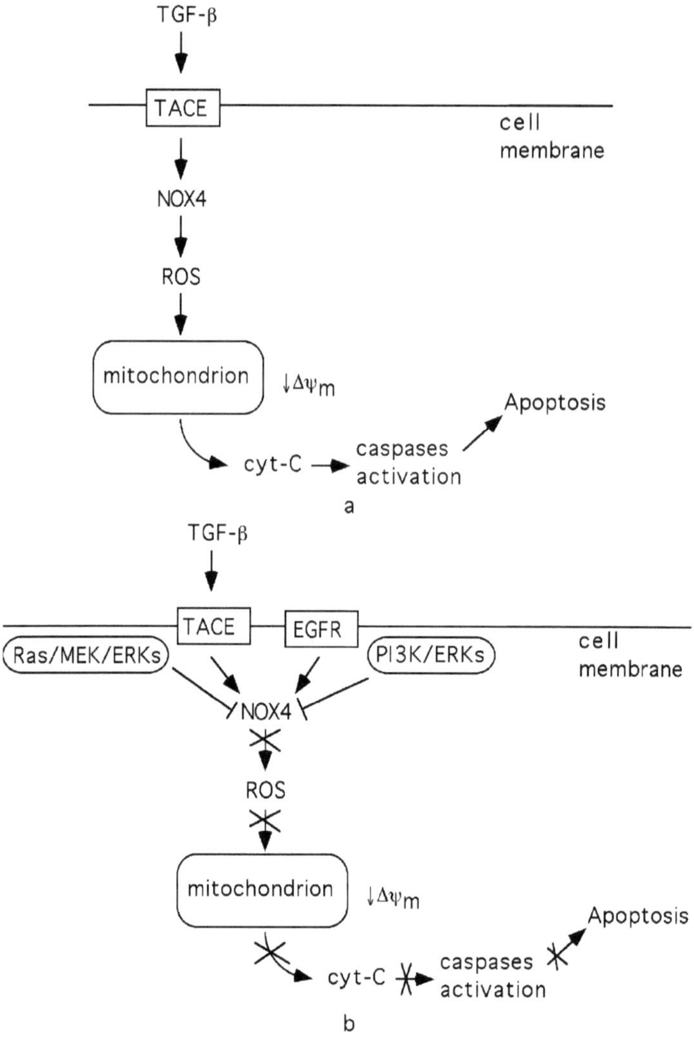

FIGURE 8.2 (a) Normal hepatocite and (b) MEK/ERK pathway overactivation in liver tumor.

8.3.2 *MEK/ERK PATHWAY OVERACTIVATION IN LIVER TUMORS: TGF-β-INDUCED DEATH*

The TGF-β induced apoptosis in hepatocytes, being considered liver tumor suppressor; however, many human HCC cells escape from pro-apoptosis gaining response to cytokine in malignancy terms. Caja et al. reported that

TGF-β-induced apoptosis in hepatocytes requires NOX4 upregulation, which mediates reactive oxygen species (ROS) production (Fig. 8.2b).[26] The TGF-β-induced NOX4 expression is inhibited by anti-apoptotic signals, for example, pathways phosphatidylinositol-3-phosphate kinase (PI3K) or MAPK/ERK. Their aim was to analyze whether resistance to TGF-β-induced apoptosis, in HCC cells, is related to NOX4-upregulation impairment because of survival-signals overactivation. Inhibition of pathway MEK/ERK in HepG2 cells, which are refractory to TGF-β pro-apoptosis, sensitizes them to cell death via mitochondrial-dependent mechanism, coincident with increased levels of B-cell lymphoma (BCL) type-2 (2)-like protein 11 (BIM)/BCL-2-modifying factor (BMF), decreased levels of BCL-extralarge (X_L)/myeloid leukaemia cell differentiation protein (MCL1) and BCL-2-associated X-protein (BAX)/homologous antagonist/killer (BAK) activation. Regulation of BMF, BCL-X_L, and MCL1 occurs at messenger (m)RNA level, whereas BIM regulation happens post-transcriptionally. Production of ROS and glutathione (GSH) depletion are observed in only cells treated with TGF-β/PD98059 (MEK1 inhibitor), which correlates with NOX4 upregulation. Targeting NOX4 knockdown impairs ROS increase and all mitochondrial-dependent apoptotic features by mechanism, which is upstream from BIM, BMF, BCL-X_L, and MCL1 levels regulation.

8.3.3 REAL-TIME ANALYSIS OF CELL RESPONSE TO MITOGENIC/ MOTOGENIC STIMULI

Fabregat performed RT analysis of cell response to mi/motogenic stimuli with xCELLigence.[27] One-chamber wells of xCELLigence allow adhes/ proliferation (0–3/>8 h) studies and two-chamber wells, migration (0 h), which is especially useful in metastasis/cancer. In adhes/proliferation experiments, HCCs from different patients were examined. (1) To elucidate mechanism: HCC cells either proliferate or do not die. (2) To analyze whether cells are able to migrate/invade: To study whether they cross extracellular matrix to discriminate invasion mechanism. (3) To determine cellular adhesion parameters. Incubating cells with TGF-β, CI decays to 70% of control: TGF-β stops growth for 7–8 h; then cells somewhat grow/ proliferate. Silencing NOX4 with siRNA decreases adhesion and increases proliferative capacity of HCC cells compared to control. Oxidase NFH is enzyme responsible for TGF-β effect. Silencing NOX4 decreases adhesion and confers proliferative advantage to HCC cells, which is not caused by apoptotic decay and was corroborated with other methods. Adhesion study to different substrates showed that silencing NOX4, HCC cells aggregate less to diverse extracellular matrixes. Silencing stable NOX4 decreases number of focal adhesions on HCC cells. The NOX4, which produces cell death, at the same time controls cell adhesion, a different mechanism. Cell response to adhesion consists in silencing NOX4 decreasing three integrins. In migration model, cells generate metalloproteinases that cross extracellular matrix. Migration studies show adhesion decay of HCC cells, which could cause matastasis provided migration capacity. The NOX4 not only causes cell death but also decreases adhesion and increases migratory capacity. In vivo studies in athymic nude mouse corroborate that silencing NOX4 increases HCC-cells proliferation, which rises tumorigenic capacity. Concentration of NOX4 is lesser in HCC cells than in control.

8.3.4 HEPATIC FIBROSIS, AUTOPHAGY, AND CARCINOGENESIS: DANGEROUS RELATIONS

Genetic models in which autophagic genes (*Atg5*, *Atg7*, *Vps45*) were inactivated develop the common features of hepatomegaly and tumorigenesis by unknown mechanism(s). Hippo tumor suppressor pathway plays a role in organ size and growth control since liver-specific deletions of its components (e.g., *Mst1/2*, *Sav1*, *Mer*) or overexpression of Yap (Hippo pathway effector) lead to hepatomegaly and tumorigenesis. Hippo pathway dysregulation was implicated in HCC. Friedman exploited a hepatocyte-specific promoter, Olig1, to drive Cre recombinase expression in the livers of mice with floxed alleles of Atg7.[28] OligCRE/Rosa26 reporter mice showed *cre* recombination in hepatocytes but not endothelial cells, cholangiocytes, or stellate cells. Olig1Cre:Atg7$^{f/f}$ conditional knock-out mice displayed hepatomegaly with a 3-fold increase in liver-to-body weight ratio and highly risen serum AST/ALT levels. Hepatocytes rose proliferative activity (Ki67) (12.8 vs. 76.5 Ki67^{+} cells/field, $n = 3$) at 3 months of age. Risen nuclear staining of hepatocyte Yap/Taz existed consistent with Yap/Taz activation and decayed Yap inactivation by the core components of Hippo pathway. The findings were complemented by results of: (1) risen protein levels of Yap/Taz; (2) decays in Yap/Taz inactivating Hippo pathway members (Mer, Sav1, pLats, pMob) in whole liver lysates; (3) risen mRNA expression of Yap/Taz targets (*Ctgf*, *Birc5*, *Itgb2*, *Afp*, *Areg*) by quantitative real-time polymerase chain reaction (qRT-PCR) and gene arrays; (4) risen nuclear Yap in AML12-shAtg7 cells by staining and immunoblotting, and risen proliferation by [^{3}H]-thymidine incorporation, which could be inhibited by Yap/Tead inhibitor verteporfin; (5) inhibition of hepatocyte proliferation in vivo by verteporfin in Olig1Cre:Atg7$^{f/f}$ as assessed by Ki67 staining; (6) development of multinodular HCCs in Olig1CRE:Atg7$^{f/f}$ and AlbCRE:Atg7$^{f/f}$ mice at 8 months. They described a hepatocyte-specific promoter in the liver and uncovered a link between hepatic autophagy, fibrosis, cancer, and Hippo tumor suppressor pathway. Their findings presented implications for illuminating the basis of hepatic growth regulation, liver disease, and tumorigenesis.

8.4 FINAL REMARKS

From the results and discussion, the following final remarks can be drawn.

1. Real-time cell electronic sensing is a tool for label-free detection and cell-based parameters. Cancerous cells present varying migratory/invasive behaviors, specific to chemoattractant stimuli/Matrigel concentration. Cellular invasion/migration for real-time cell analyzer is a versatile device, for invasive/migration analysis, and a unique platform capable of quantitatively measuring invasive/migratory behavior of cells in real time, without exogenous labels, which enables identification of migratory/invasive time points, which will help increase throughput by shortening assay time, minimize number of endpoint assays, and assist determine optimal time points for inhibitor studies. Analysis using xCELLigence reduces hands-on cell-manipulation steps leading to physiologically relevant results.
2. Upregulation of nicotinamide adenine dinucleotide phosphate hydride oxidase four, by TGF-β, is required for its pro-apoptotic activity in hepatocytes. Impairment of TGF-β-induced response might confer apoptosis resistance in hepatocellular carcinoma cells.
3. Overactivation of pathway mitogen-activated protein/extracellular signal-regulated kinase, in liver tumor cells confers resistance to TGF-β-induced cell death, via impairing nicotinamide adenine dinucleotide phosphate hydride oxidase four upregulation, which is required for efficient mitochondrial-dependent apoptosis.
4. In vivo murine study corroborated that silencing nicotinamide adenine dinucleotide phosphate hydride oxidase four increases tumorigenic capacity. Human hepatocellular carcinoma downregulates oxidase. Silencing oxidase induces epithelial–mesenchymal

transition mechanisms, which bring on migration. Silencing oxidase increases mesenchyme.

ACKNOWLEDGMENTS

The authors thank support from Fundacion Universidad Catolica de Valencia San Vicente Martir (Project No. 2019-217-001UCV).

KEYWORDS

- **hepatocarcinogenesis**
- **apoptosis**
- **NOX4**
- **NADPH oxidase**
- **real-time**
- **label-free cellular analysis**
- **mitogenic and motogenic stimuli**
- **hepatic fibrosis**

REFERENCES

1. Roberts, A. B.; Anzano, M. A.; Lamb, L. C.; Smith, J. M.; Sporn, M. B. New Class of Transforming Growth Factors Potentiated by Epidermal Growth Factor: Isolation From Non-Neoplastic Tissues. *Proc. Natl. Acad. Sci. U. S. A.* **1981,** *78*, 5339–5343.
2. Dobolyi, A.; Vincze, C.; Pál, G.; Lovas, G. The Neuroprotective Functions of Transforming Growth Factor Beta Proteins. *Int. J. Mol. Sci.* **2012,** *13*, 8219–8258.
3. Millipore, M. Automated Live Cell Imaging of Cell Migration Across a Microfluidic-Controlled Chemoattractant Gradient. *BioTechniques* **2013,** *54* (5), Advertorial–1-2.
4. Torrens, F.; Castellano, G. Modelling of Complex Multicellular Systems: Tumour–Immune Cells Competition. *Chem. Central J.* **2009,** *3* (Suppl. I), 75–1-1.
5. Torrens, F.; Castellano, G. Information Theoretic Entropy for Molecular Classification: Oxadiazolamines as Potential Therapeutic Agents. *Curr. Comput.-Aided Drug Des.* **2013,** *9*, 241–253.
6. Torrens, F.; Castellano, G. Molecular Classification of 5-Amino-2-Aroylquinolines and 4-Aroyl-6,7,8-Trimethoxyquinolines as Highly Potent Tubulin Polymerization Inhibitors. *Int. J. Chemoinf. Chem. Eng.* **2013,** *3* (2), 1–26.

7. Estrela, J. M.; Mena, S.; Obrador, E.; Benlloch, M.; Castellano, G.; Salvador, R.; Dellinger, R. W. Polyphenolic Phytochemicals in Cancer Prevention and Therapy: Bioavailability *versus* Bioefficacy. *J. Med. Chem.* **2017,** *60*, 9413–9436.
8. Torrens, F.; Castellano, G. Molecular Classification of Antitubulin Agents with Indole Ring Binding at Colchicine-Binding Site. In *Molecular Insight of Drug Design*; Parikesit, A. A., Ed.; InTechOpen: Vienna, 2018; pp 47–67.
9. Torrens, F.; Castellano, G. Molecular Classification of 2-Phenylindole-3-Carbaldehydes as Potential Antimitotic Agents in Human Breast Cancer Cells. In *Theoretical Models and Experimental Approaches in Physical Chemistry: Research Methodology and Practical Methods*; Haghi, A. K., Thomas, S., Praveen, K. M., Pai, A. R., Eds.; Apple Academic–CRC: Waretown, NJ, in press.
10. Torrens, F.; Castellano, G. Cancer and Hypotheses on Cancer. In Molecular Chemistry and Biomolecular Engineering: Integrating Theory and Research with Practice; Pogliani, L., Torrens, F., Haghi, A. K., Eds.; Apple Academic–CRC: Waretown, NJ, in press.
11. Torrens, F.; Castellano, G. Precision Personalized Medicine from Theory to Practice: Cancer. In Green Chemistry and Biodiversity: Principles, Techniques, and Correlations; Aguilar, C. N., Ameta, S. C., Haghi, A. K., Eds.; Apple Academic–CRC: Waretown. NJ, in press.
12. Torrens, F.; Castellano, G. AIDS Destroys Immune Defences: Hypothesis. *New Front. Chem.* **2014,** *23*, 11–20.
13. Torrens-Zaragozá, F.; Castellano-Estornell, G. Emergence, Spread and Uncontrolled Ebola Outbreak. *Basic Clin. Pharmacol. Toxicol.* **2015,** *117* (Suppl. 2), 38–38.
14. Torrens, F.; Castellano, G. 2014 Spread/Uncontrolled Ebola Outbreak. *New Front. Chem.* **2015,** 24, 81–91.
15. Torrens, F.; Castellano, G. Ebola Virus Disease: Questions, Ideas, Hypotheses and Models. *Pharmaceuticals* **2016,** *9*, 14–6-6.
16. Torrens, F.; Castellano, G. Clinical Translational Research: Cancer, Diabetes, and Cardiovascular Disease. In *Physical Biochemistry, Biophysics, and Molecular Chemistry: Applied Research and Interactions*; Torrens, F., Mahapatra, D. K., Haghi, A. K., Eds.; Apple Academic–CRC: Waretown, NJ, in press.
17. Torrens, F.; Castellano, G. Primary Health, Smoking Habits and Immunotherapy in Cancer. In *Environmental Technology and Engineering Techniques: Basic Concepts and Health Interventions*; Khalaf, M. N., Smirnov, M. O., Kannan, P., Haghi, A. K., Eds.; Apple Academic–CRC: Waretown, NJ, in press.
18. Torrens, F.; Castellano, G. Epigenomics, Epithelial Plasticity, Clinical Genetics and Rare Diseases. In *Environmental Technology and Engineering Techniques: Basic Concepts and Health Interventions*; Khalaf, M. N., Smirnov, M. O., Kannan, P., Haghi, A. K., Eds.; Apple Academic–CRC: Waretown, NJ, in press.
19. Torrens, F.; Castellano, G. Predictive in Silico Global Metabolism Analysis: Disease Mechanism. In *Processing and Characterization of Green Engineered and Sustainable Materials*; Haghi, A. K., Ed.; Apple Academic–CRC: Waretown, NJ, in press.
20. Torrens, F.; Castellano, G. Intracellular Proteolysis and Neoplasia: Ubiquitin-Proteasome System, Cell Cycle and Cancer. In *Natural Pharmaceuticals and Green Microbial Technology: Health, Environment and Resource Recovery*; Mahapatra, D. K., Shinde, R. S., Haghi, A. K., Eds.; Apple Academic–CRC: Waretown, NJ, in press.
21. Torrens, F.; Castellano, G. Vitamin-D, Analogues, Drugs and Their Relevance to Cancer Inhibition. In *Natural Pharmaceuticals and Green Microbial Technology: Health,*

Environment and Resource Recovery; Mahapatra, D. K., Shinde, R. S., Haghi, A. K., Eds.; Apple Academic–CRC: Waretown, NJ, in press.
22. Torrens, F.; Castellano, G. Stem Cells and Their Connection with Cancer: Prostate Disease. In *Natural Pharmaceuticals and Green Microbial Technology: Health, Environment and Resource Recovery*; Mahapatra, D. K., Shinde, R. S., Haghi, A. K., Eds.; Apple Academic–CRC: Waretown, NJ, in press.
23. Solly, K.; Wang, X.; Xu, X.; Strulovici, B.; Zheng, W. Application of Real-Time Cell Electronic Sensing (RT-CES) Technology to Cell-Based Assays. *Assay Drug Dev. Technol.* **2004,** *2*, 363–372.
24. Bird, C.; Kirstein, S. Real-Time, Label-Free Monitoring of Cellular Invasion and Migration With the xCELLigence System. *Nat. Meth.* **2009,** *6*, v–vi.
25. Carmona-Cuenca, I.; Roncero, C.; Sancho, P.; Caja, L.; Fausto, N.; Fernández, M.; Fabregat, I. Upregulation of the NADPH Oxidase NOX4 by TGF-Beta in Hepatocytes is Required for Its Pro-Apoptotic Activity. *J. Hepatol.* **2008,** *49*, 965–976.
26. Caja, L.; Sancho, P.; Bertran, E.; Iglesias-Serret, D.; Gil, J.; Fabregat, I. Overactivation of the MEK/ERK Pathway in Liver Tumor Cells Confers Resistance to TGF-β-Induced Cell Death Through Impairing Up-Regulation of the NADPH Oxidase NOX4. *Cancer Res.* **2009,** *69*, 7595–7602.
27. Fabregat, I. Análisis en Tiempo Real de la Respuesta Celular a Estímulos Mitogénicos y Motógenos: El Sistema xCELLigenge. *Seminario Científico Real Time PCR, Real Time Cell Analysis y Secuenciación Masiva de DNA: Aplicaciones*, Lightcycler University No. 50, Lightcycler University: València, Spain, 2012.
28. Friedman, S. Personal Communication.

CHAPTER 9

Chalcone Scaffold Bearing Natural Antigout Agents

DEBARSHI KAR MAHAPATRA[1*], VIVEK ASATI[2], and SANJAY KUMAR BHARTI[3]

[1]*Department of Pharmaceutical Chemistry, Dadasaheb Balpande College of Pharmacy, Nagpur 440037, Maharashtra, India*

[2]*Department of Pharmaceutical Chemistry, NRI Institute of Pharmacy, Bhopal 462021, Madhya Pradesh, India*

[3]*Institute of Pharmaceutical Sciences, Guru Ghasidas Vishwavidyalaya (A Central University), Bilaspur 495009, Chhattisgarh, India*

[*]*Corresponding author. E-mail: mahapatradebarshi@gmail.com*

ABSTRACT

Gout is a form of an arthritic syndrome characterized by severe chronic pain, discomfort, swelling, and redness in the joint present in the big toe as a result of monosodium urate (MSU) crystals accumulation. This chapter focuses on the ability of the natural and semisynthetic chalcone compounds such as 3,5,2,4-tetrahydroxychalcone, 4-hydroxyderricin, hesperidin methylchalcone, isobavachalcone, okanin, polyhydroxylated chalcones, quercetin chalcone, sappanchalcone, tetrahydroxychalcone, *trans*-chalcone, xanthoangelol, xanthoangelol B, and xanthoangeleol F in expressing antigout activity by completely suppressing the active disease proliferating enzyme, xanthine oxidase (XO), reducing the pro-inflammatory components, and suppressing the activation of nuclear factor kappa-light-chain-enhancer of activated B cells (NF-κβ). This chapter will provide unparallel information of chalcone scaffold bearing natural and semisynthetic molecules having pharmacotherapeutic perspectives. However, at present, these molecules are at nascent stages and only preclinical studies have been done so far and a large section of clinical studies are still pending. The structural, chemical, biological, pathophysiological, and miscellaneous aspects have an overall pharmaceutical interest, which will motivate the active researchers in the exploration and development of antigout agents for treating the associated symptoms.

9.1 INTRODUCTION

Gout is a form of an arthritic syndrome that has affected millions of people worldwide. It is often characterized by severe chronic (weeks to months) pain, discomfort, burning sensation, swelling, akinesis, and redness in the joint present in the big toe as well as ankles, elbows, fingers, knees, wrists, etc. due to monosodium urate (MSU) crystals accumulation.[1,2]

Uric acid, the purine component and the prime reason for the proliferation of gouty symptoms at its peak level in the body, results due to high intake of meat, tea, seafood, alcoholic beverages, sweetened drinks, and streak.[3] Apart from diet, ethnicity (more in African–American population), obesity, age (enhancement in disease augmentation), sex (more in men), medical conditions (abnormal kidney function), certain medicines, family history (more chances in the next generation), surgical conditions, low thyroid hormone levels, traumatic situations, etc., play a dominant role in the proliferation of serum uric acid levels.[4]

In the natural way of treatment, particularly the nonpharmacological approach, an absolute control of alcohol, adequate intake of fluid, reduction in body weight, avoiding intake of purine-rich foods, controlled intake of dairy products, etc. serve primarily in the management of gouty symptoms.[5] In contrast to it, drugs, such as allopurinol (ALP), febuxostat, probenecid, and lesinurad, are often prescribed with naproxen, ibuprofen, corticosteroids, and colchicine.[6] However, these marketed products have low pharmacodynamics attributes and demonstrate prominent adverse effects.[7]

For decades, these old, ranged drugs are into practice. Although from the last 25 years, the disease is fast progressing due to lifestyle and other prevailing factors and the role of therapy in the complete eradication of the disease is quite suspicious.[8] In the modern research trends, a peak shift toward products of natural origin has been seen along with fascination toward a product made through tailored approach.[9] The prescription pattern has been revolutionized from traditional combinations to unexplored classes of drugs. The classes of natural products like chalcones have recently gained attention by the modern-day researchers in the treatment of gout with a greater safety characteristic.[10]

This chapter broadly highlights the ability of numerous chalcone (propene-1-one) scaffold bearing compounds of both natural and semisynthetic origins in the management of gout symptoms and the control of serum uric acid level. From the data available from several pharmaceutical, medical, and research databases, the information regarding the mode of modulation of various therapeutic targets (xanthine oxidase (XO), NFκβ, etc.) and patents have been comprehensively depicted.

9.2 CHALCONES

Benzylideneacetophenone or 1,3-diphenyl-2-propene-1-one or Chalcone are scaffolds of natural origin. They comprise of two aromatic rings linked together by a bridge consists of three-carbon α, β-unsaturated carbonyl system.[11] They are regarded as the flavonoid and isoflavonoid precursor and as open-chain intermediate in the aurone synthesis of flavones. They undergo interconversion to flavonoid in the presence of an acid, whereas, in the presence of a base, it forms flavonone. They serve as a template for the elucidating flavanone, flavonoid, tannin, and chromanochromane structures.[12] They act as Michael's acceptors in Michael addition reactions. These molecules or preferable called chromophores due to their bright colors were first synthesized in the laboratory by Kostanecki and Tambor from acetophenone and benzaldehyde in the 19th century and termed them "chalcone."[13]

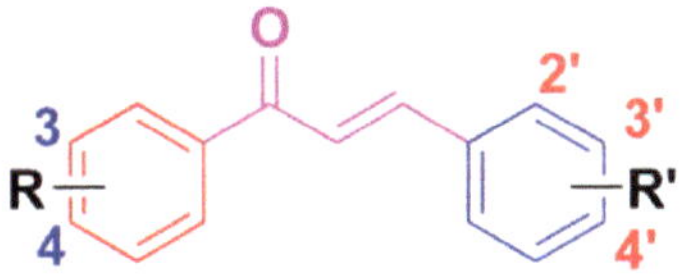

(1)

The low molecular weight ligands having chalcone scaffold have been reported to exhibit remarkable therapeutic effect in the treatment/management of anxiety, bacterial infections, cancer, diabetes, epilepsy, fungal infections, gout, HIV, hypertension, inflammation, leishmaniasis, malaria, metastasis, oxidation, protozoal infections, reduced immune response, sleep disorder, thrombosis, trypanosomiasis, tuberculosis, ulcer, etc.[14,15] Nonpharmacologically, they are known for their specific applications as artificial sweetener, brightening agent, catalyst, chemosensor, fluorescent polymers, fluorescent whitening agent, insecticide, polymerization agent, scintillator, etc.[16]

In the modern era, these compounds have gained attention and popularity in the scientific community due to their simple chemistry, ease of computational studies, availability of 10 replaceable hydrogens, and wide attribute in modulating the therapeutic targets.[17] They serve as a template for the synthesis of pharmacologically active six-membered and five-membered scaffolds such as benzodiazepine, benzothiazepine, benzoxazepine, isoxazole, pyrazole, pyrazoline, pyrimidine, thiadiazole, etc.[18] Traditionally, chalcones are synthesized by Claisen-Schmidt reaction; however, Carbonylated Heck coupling reaction, Suzuki-Miyaura reaction, Sonogashira isomerization coupling, Julia-Kocienski reaction, Friedel-Crafts reaction, Direct crossed-coupling reaction, one-pot reactions, solid acid catalyst mediated reactions, solvent-free reactions, microwave-assisted reactions, etc. are miscellaneous ways for high-yield synthesis of the diversely substituted benzylideneacetophenone scaffold bearing molecules with multifarious applications.[19]

9.3 CHALCONES AS ANTIGOUT AGENTS

From the methanol extract of *Caesalpinia sappan*, a therapeutically active chalcone molecule "sappanchalcone" **(2)** (Fig. 9.1) was isolated by a group of active researchers from Vietnam. The obtained natural compound expressed noteworthy concentration-dependent in vitro XO inhibition with an IC_{50} value of 3.9 μM in a competitive manner (Ki = 2.6 μM).[20]

Similarly, another research group studied the inhibitory potential of sappanchalcone **(2)** along with the standard drug ALP. In this study, sappanchalcone **(2)** demonstrated analogous XO inhibitory activity of ALP with an IC_{50} value of 2.5 μM. Inspired from the aforementioned results, the caffeoyl-substituted chalcones **(3)** were designed synthetically, which presented alike inhibition of the therapeutic target with IC_{50} of 2.5 μM.[21]

2',4'-dimethoxy-4,5',6'-trihydroxychalcone **(4)**, a prevalent phytochemical isolated from the aqueous extract of *Perilla frutescens* leaves, displayed micromolar concentration inhibition against the prime antigout target XO. The chalcone molecule showed nearly similar in vitro inhibition of the target as compared to the positive control compound, ALP. The aqueous extract had flavanone and aurone components along with chalcone, but only the chalcone had amazing enzyme inhibitory activity, which is reflective of the fact that open-chain chalcones in flavones pathway have more efficacy in comparison to the closed ring compounds.[22]

In the successive pathway of research, numerous chalcones have been isolated from the stem, root bark, root, and leaf of *Angelica keiskei* and investigated for their XO inhibitory perspectives. The well-known chalcone molecules—xanthoangelol (**5**, IC_{50} = 8.1 μM), 4-hydroxyderricin (**6**, IC_{50} = 54.3 μM), xanthoangeleol B (**7**, IC_{50} = 20.3 μM), isobavachalcone (**8**, IC_{50} = 27.1 μM), and xanthoangelol F (**9**, IC_{50} = 34.6 μM) expressed marvelous antigout activity through a mixed-type mechanism.[23]

Trans-chalcone **(1)**, the flavonoid precursor and hesperidin methyl chalcone **(10)** presented a complete attenuation of gouty conditions in rats at 30 mg/kg per oral dose. The simplest form of chalcone and the hesperidin-based chalcone exhibited an absolute reduction of oxidative stress by nitrite radical and superoxide anion downregulation. The expression of anti-inflammatory activity was mediated through reducing the pro-inflammatory factors such as interleukins, TGF-β, and TNF-α. The in vivo MSU-induced hyperalgesia was prevented by the acute reduction in MSU-induced infiltration of LysM-eGFP+ cells. Inhibition of NLRP3 inflammasome components mRNA expression and nuclear factor kappa-light-chain-enhancer of activated B cells (NF-κβ) activation were miscellaneous modes for edema reduction and expression of antigout activity. The inhibition of aforementioned components ultimately results in prevention of synovitis and downregulation of pro-caspase-1, ASC, and pro-IL-1β levels, which additionally provides relief from the deposition of MSU.[24,25]

Several natural and semisynthetic chalcones containing some hydroxyl groups have been observed as antigout agents capable of inhibiting the primary enzyme target XO. Numerous polyhydroxylated (methylenedione, tri-hydroxyl, tetra-hydroxyl, and penta-hydroxyl) chalcones **(11–15)** from nature presented in vitro XO inhibition with IC_{50} values < 1.5 μM. The chalcone compounds exhibited complete inhibition of the target enzyme with tri-hydroxylated molecules being recognized as the most potent series.[26,27]

Analogously, the XO inhibitory potential of a hydroxyl-containing nonpurine chemical 3,5,2,4-tetrahydroxychalcone **(16)** has been reported by Niu et al. with IC_{50} value of 22.5 μM in a noncompetitive manner (Ki value of 17.4 μM). In hyperuricemic mice model, the compound showed a significant dose-dependent decrease in hepatic XO and serum uric acid levels after in vivo intragastric administration with a high level of safety at a dose of 5 g/kg b.w.[28]

An incredible antigout activity has been perceived for a natural compound 2',3,4,4',6'-pentahydroxychalcone (quercetin chalcone, **17**) for the immense inhibition of the prime therapeutic target XO in a dose-dependent manner. In the year 1995, Throne Research Inc. Ltd. received a patent for the application of this orally administered chalcone compound. The pharmacological potential of the compound is based on the inhibition of MSU-induced NF-κβ activation and MSU-induced hyperalgesia.[29]

U.S. patent was credited to Taiwan National University for exploring the noteworthy XO inhibitory (EC_{50} value of 0.074 μM) potential of okanin **(18)**, a major component found in acacia extract.[30,31]

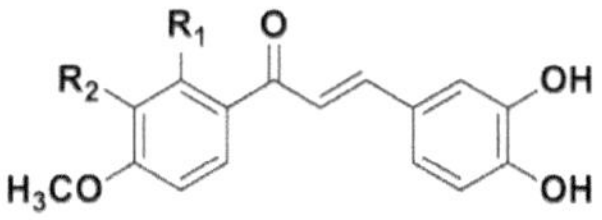

(2) $R_1 = OCH_3$, $R_2 = H$; **(3)** $R_1 = OH$, R_2 = Caffeoyl

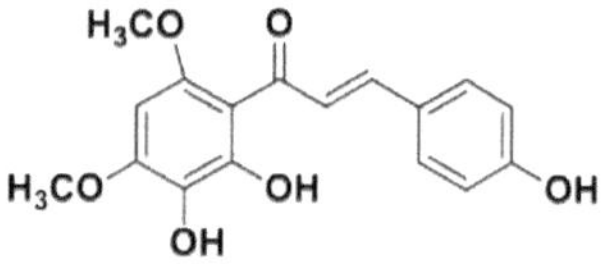

(4)

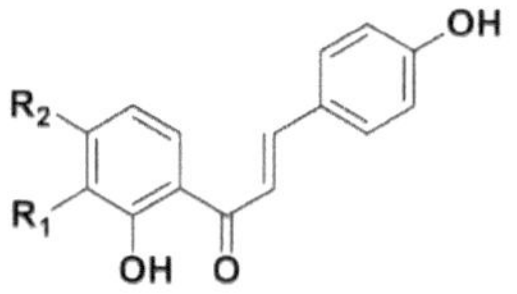

(5) R_1 = 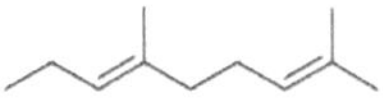, $R_2 = OH$; **(6)** R_1 = 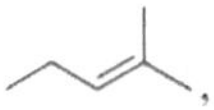, $R_2 = OCH_3$;

(7) R_1 = 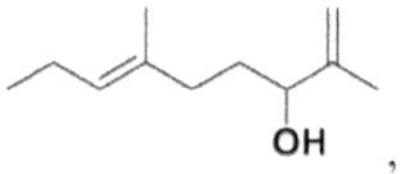

, $R_2 = OH$; **(8)** R_1 = 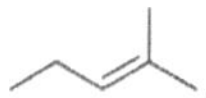, $R_2 = OH$;

(9) R_1 = 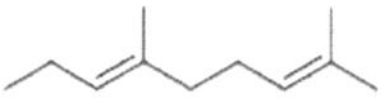, $R_2 = OCH_3$

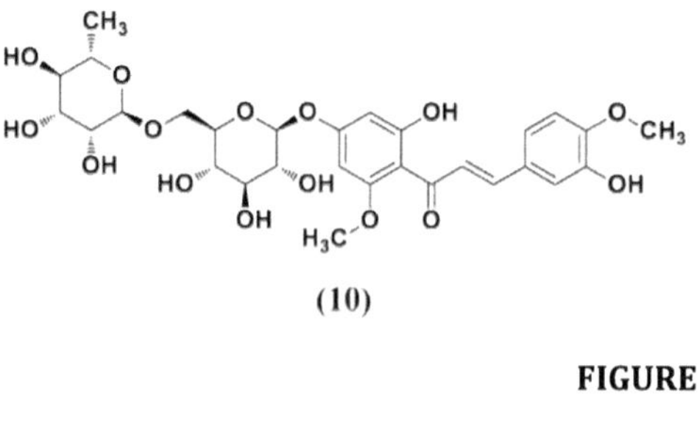

(10)

FIGURE 9.1 *(Continued)*

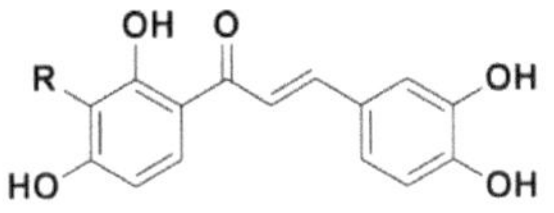

(11) R = OH; **(12)** R = H

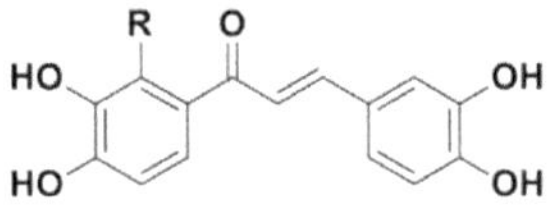

(13) R = H; **(14)** R = OH

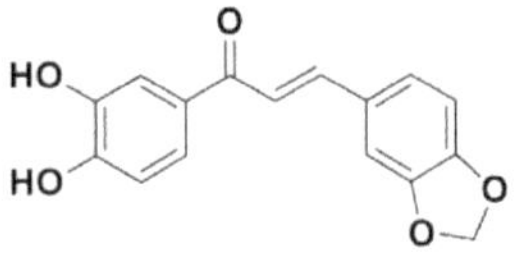

(15)

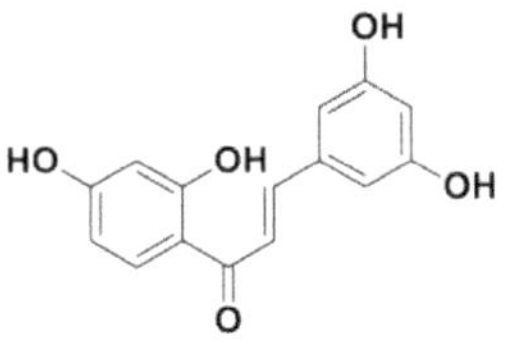

(16)

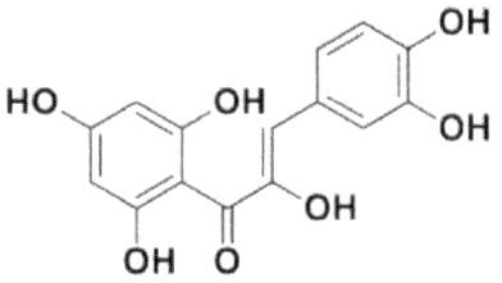

(17)

FIGURE 9.1 Chalcone compounds with notable antigout activity.

9.4 CONCLUSION

The chalcones obtained from natural and semisynthetic sources have been seen to produce strikingly high antigout activity in comparison to the standard marketed formulation allopurinol by inhibiting the active enzyme XO that simultaneously reduces the serum uric acid level and hepatic XO levels. In addition to it, the molecules exhibit biological activity by suppressing the activation of NF-κβ and preventing the formation of pro-inflammatory factors. The number(s), position(s), and type(s) of substituents play a key role in the modulation of the therapeutic target by forming possible hydrogen bonding/Van der Waals interactions. This chapter will provide unparalleled information of chalcone scaffold bearing natural and semisynthetic molecules having pharmacotherapeutic perspectives. However, at present, these molecules are at nascent stages and only preclinical studies have been done so far and a large section of clinical studies are still pending. The structural, chemical, biological, pathophysiological, and miscellaneous aspects have an overall pharmaceutical interest, which will motivate the active researchers in the exploration and development of antigout agents for treating the associated symptoms.

KEYWORDS

- **chalcone**
- **inhibition**
- **gout**
- **xanthine oxidase**
- **inflammation**
- **NF-κβ**
- **mono sodium ureate**

REFERENCES

1. Reinders, M. K.; Tim, L. New Advances in the Treatment of Gout: Review of Pegloticase. *Ther. Clin. Risk Manag.* **2010,** *6*, 543–550.
2. Robinson, P. C.; Taylor, W. J.; Merriman, T. R. Systematic Review of the Prevalence of Gout and Hyperuricaemia in Australia. *Inter. Med. J.* **2012,** *42* (9), 997–1007.
3. Terkeltaub, R. A. Gout. *New Eng. J. Med.* **2003,** *349* (17), 1647–1655.
4. Harrold, L. New Developments in Gout. *Curr. Opin. Rheumatol.* **2013,** *25* (3), 304–309.
5. Campano, G. D.; Holmes, E. W. Hyperuricemia and Gout. *Med. Clin. North Am.* **1986,** *70* (2), 419–436.
6. Kim, K. Y.; Schumacher, H. R.; Hunsche, E.; Wertheimer, A. I.; Kong, S. X. A Literature Review of the Epidemiology and Treatment of Acute Gout. *Clin. Ther.* **2003,** *25* (6), 1593–1617.
7. Singh, J. A.; Reddy, S. G.; Kundukulam, J. Risk Factors for Gout and Prevention: A Systematic Review of the Literature. *Curr. Opin. Rheumatol.* **2011,** *23* (2), 192–202.
8. Emmerson, B. T. The Management of Gout. *New Eng. J. Med.* **1996,** *334* (7), 445–451.
9. Schlesinger, N. Management of Acute and Chronic Gouty Arthritis. *Drugs* **2004,** *64* (21), 2399–2416.
10. Terkeltaub, R. Update on Gout: New Therapeutic Strategies and Options. *Nature Rev. Rheumatol.* **2010,** *6* (1), 30–38.
11. Mahapatra, D. K.; Bharti, S. K.; Asati, V. Anti-Cancer Chalcones: Structural and Molecular Target Perspectives. *Eur. J. Med. Chem.* **2015,** *98*, 69–114.
12. Mahapatra, D. K.; Asati, V.; Bharti, S. K. Chalcones and Their Therapeutic Targets for The Management of Diabetes: Structural and Pharmacological Perspectives. *Eur. J. Med. Chem.* **2015,** *92*, 839–865.
13. Mahapatra, D. K.; Bharti, S. K.; Asati, V. Chalcone Scaffolds as Anti-Infective Agents: Structural and Molecular Target Perspectives. *Eur. J. Med. Chem.* **2015,** *101*, 496–524.
14. Mahapatra, D. K.; Bharti, S. K. Therapeutic Potential of Chalcones as Cardiovascular Agents. *Life Sci.* **2016,** *148*, 154–172.
15. Mahapatra, D. K.; Bharti, S. K.; Asati, V. Chalcone Derivatives: Anti-Inflammatory Potential and Molecular Targets Perspectives. *Curr. Top Med. Chem.* **2017,** *17* (28), 3146–3169.

16. Mahapatra, D. K.; Bharti, S. K.; Asati, V.; Singh, S. K. Perspectives of Medicinally Privileged Chalcone Based Metal Coordination Compounds for Biomedical Applications. *Eur. J. Med. Chem.* **2019,** *174*, 142–158.
17. Mahapatra, D. K.; Asati, V.; Bharti, S. K. An Updated Patent Review of Therapeutic Applications of Chalcone Derivatives (2014–Present). *Exp. Opin. Ther. Pat.* **2019,** *29* (5), 385–406.
18. Mahapatra, D. K.; Asati, V.; Bharti, S. K. Recent Therapeutic Progress of Chalcone Scaffold Bearing Compounds as Prospective Anti-Gout Candidates. *J. Crit. Rev.* **2019,** *6* (1), 1–5.
19. Mahapatra, D. K.; Bharti, S. K.; Asati, V. Recent Perspectives of Chalcone Based Molecules as Protein Tyrosine Phosphatase 1B (PTP-1B) Inhibitors. In *Medicinal Chemistry with Pharmaceutical Product Development*; Mahapatra, D. K., Bharti, S. K., Eds., 1st ed.; Apple Academic Press: New Jersey; 2019, pp 235–251.
20. Nguyen, M. T.; Awale, S.; Tezuka, Y.; Le Tran, Q.; Kadota, S. Xanthine Oxidase Inhibitors From the Heartwood of Vietnamese Caesalpinia Sappan. *Chem. Pharm. Bull.* **2005,** *53* (8), 984–988.
21. Bui, T. H.; Nguyen, N. T.; Dang, P. H.; Nguyen, H. X.; Nguyen, M. T. Design and Synthesis of Chalcone Derivatives as Potential Non-Purine Xanthine Oxidase Inhibitors. *SpringerPlus* **2016,** *5* (1), 1789.
22. Liu, Y.; Hou, Y.; Si, Y.; Wang, W.; Zhang, S.; Sun, S.; Liu, X.; Wang, R.; Wang, W. Isolation, Characterization, and Xanthine Oxidase Inhibitory Activities of Flavonoids From the Leaves of Perilla Frutescens. *Nat. Prod. Res.* **2018,** *26*, 1–7.
23. Kim, D. W.; Curtis-Long, M. J.; Yuk, H. J.; Wang, Y.; Song, Y. H.; Jeong, S. H.; Park, K. H. Quantitative Analysis of Phenolic Metabolites From Different Parts of Angelica Keiskei by HPLC–ESI MS/MS and Their Xanthine Oxidase Inhibition. *Food Chem.* **2014,** *153*, 20–27.
24. Staurengo-Ferrari, L.; Ruiz-Miyazawa, K. W.; Pinho-Ribeiro, F. A.; Fattori, V.; Zaninelli, T. H.; Badaro-Garcia, S.; Borghi, S. M.; Carvalho, T. T.; Alves-Filho, J. C.; Cunha, T. M.; Cunha, F. Q. Trans-Chalcone Attenuates Pain and Inflammation in Experimental Acute Gout Arthritis in Mice. *Front Pharmacol.* **2018,** *9*, 1123.
25. Ruiz-Miyazawa, K. W.; Pinho-Ribeiro, F. A.; Borghi, S. M.; Staurengo-Ferrari, L.; Fattori, V.; Amaral, F.; Teixeira, M. M.; Alves-Filho, J. C.; Cunha, T. M.; Cunha, F. Q.; Casagrande, R. Hesperidin Methylchalcone Suppresses Experimental Gout Arthritis in Mice by Inhibiting NF-kB Activation. *J. Agri. Food Chem.* **2018,** *66* (25), 6269–6280.
26. Hofmann, E.; Webster, J.; Do, T.; Kline, R.; Snider, L.; Hauser, Q.; Higginbottom, G.; Campbell, A.; Ma, L.; Paula, S. Hydroxylated Chalcones with Dual Properties: Xanthine Oxidase Inhibitors and Radical Scavengers. *Bioorg. Med. Chem.* **2016,** *24* (4), 578–587.
27. Beiler, J. M.; Martin, G. J. The Inhibition of Xanthine Oxidase by Flavonoids and Related Compounds. *J. Biol. Chem.* **1951,** *192* (2), 831–834.
28. Niu, Y.; Zhu, H.; Liu, J.; Fan, H.; Sun, L.; Lu, W.; Liu, X.; Li, L. 3, 5, 2′, 4′-Tetrahydroxychalcone, a New Non-Purine Xanthine Oxidase Inhibitor. *Chem. Biol. Interact* **2011,** *189* (3), 161–166.
29. Birdsall, T. C.; Czap, A. F.; Inventor; Thorne Research Inc. Ltd., Assignee. Quercetin Chalcone and Methods Related Thereto. U.S. Patent US5977184A. September 1999.
30. Chang, S. T.; Tung, Y. T.; Inventor; National Taiwan University. Acacia Extracts and Their Compounds on Inhibition of Xanthine Oxidase. U.S. Patent US20100310688A1. June 2009.

31. Chang, S. T.; Tung, Y. T.; Inventor; National Taiwan University. Use of Acacia Extracts and Their Compounds on Inhibition of Xanthine Oxidase. U.S. Patent US8414936B2. June 2009.

CHAPTER 10

Double-Pole Electrode in New Devices and Processes for Direct Electrochemical Oxidation of Blood Inside the Blood Vessel

V. A. RUDENOK*

Izhevsk State Agricultural Academy, g. Izhevsk, Russia

E-mail: rudenva@ramber.ru

ABSTRACT

Electrochemical bipolar polarization of a solid-state electrode is realized in the case of a special electrode switching circuit, when an electrode without an external electrical contact is placed in the electrochemical cell between two electrodes connected to an electric current source. When the current flows between the first pair of electrodes, the power lines of the current interact with the third electrode, causing "induced" polarization in it. This phenomenon is used here for the development of measurement technology.

It is most effective for the development of a polarograph scheme with a solid-state electrode. The polarograph scheme, proposed in the paper, contains two electrochemical cells: polarographic and electrochemical. In the first one, a polarized bipolar wire platinum electrode is mounted, which is equipped with a movable contact to measure the current value of the potential; in the second, the working electrode and the reference electrode. The potential measured on the wire electrode is fed to the potentiostat input. At the potentiostat output, a current signal is synchronously generated. The current from the potentiostat output passes through the working electrode, so that the value of the working electrode potential coincides with the potential of the wire electrode at the point of contact with its movable contact.

A device for quantitative measurement of porosity of galvanic cathode coatings is also proposed. The calibration dependence links the results of the contact measurements of porosity and current corrosion flowing in microgalvanic vapor in the coating pores. The potential of the test sample, corroding in stationary conditions, is fed to the potentiostat input; the current at its output corresponds to the corrosion current density of the part.

Then, we consider a device for the organism detoxification and treatment by electrochemical synthesis of sodium hypochlorite and elemental hydrogen in the blood stream directly in the blood vessel from the components that make up the blood, on the surface of the wire electrode introduced into the blood vessel, and polarized by a pair of overhead electrodes.

10.1 INTRODUCTION

Various application options of a special polarization type of metal electrode in the absence of direct electrical contact with a current source are considered.

The technology of bipolar electrolysis is widely applied in large-scale synthesis, for instance, in the production of hydrogen and heavy water by the alkaline solution electrolysis in the electrolysers of filter-press type.[1] The solution in seal housing is split by multiple vertical flat-parallel electrodes, isolated from each other. Polarization emerges on the plate surface when the current flows through the electrolyte solution, contacting with the pair of end electrodes linked to the current source. However, this method is not practicable so far in instrumentation engineering. Several possible options of such application are discussed in the subsequent section.

10.1.1 POLAROGRAPH WITH EXTENDED WIRE ELECTRODE

Polarographic analysis is widely used in modern laboratories and it is distinguished with the capability to simultaneously find the concentration of all substances in the solution. However, the method is ecologically dangerous since it is connected with the necessity to use metal mercury.

In this paper, we propose the device for polarographic analysis of solutions composition, eliminating its use. The current force is measured in the points corresponding to the potentials of electrochemically active ions in the solution, interacting with individual areas of the platinum electrode. In the working solution, the electrode is polarized to the potential equaled to the potential in the given point of the platinum electrode, bipolarly polarized in another cell. The measuring electrode is located parallel to the solution surface and it is polarized by the current flowing between the electrodes located on both sides of the first electrode. As a result, in the areas of the electrode, facing these additional electrodes, potentials, different in sign, emerge. Since the wire electrode is used as a measuring electrode, the induced potential is uniformly distributed along the wire axis. This element is used as an active element of the analytical instrument, which allows defining the solution composition by constructing a diagram "current—potential."[10]

Device description. The device for measuring the distribution of potentials and current along the platinum wire electrode contains experimental cell 2 (Fig. 10.1) comprising electrode 7 being investigated, and electrochemical cell 1 containing working electrode 10 and additional electrode 11. The cells are connected by salt bridge 13. The investigated electrode 7, equipped with the device for uniform movement of reference electrode 9 along its surface, is placed in the second cell.

Reference electrode 9 is linked to the terminal "reference electrode" at the potentiostat input. Reference electrode 12, connected with voltmeter 4, is mounted in electrochemical cell 1. Self-recorder 6 registers the potential of working electrode 10; self-recorder 5 registers the current force of its polarization.

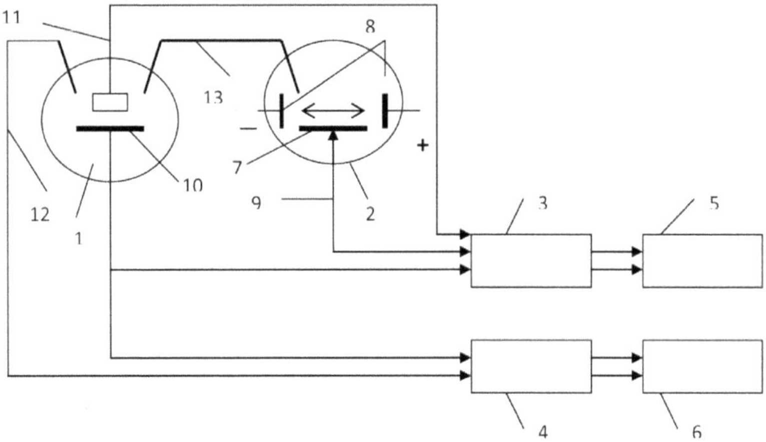

FIGURE 10.1 The device for measuring the distribution of potentials and current along the platinum wire electrode.
1—electrochemical cell; 2—experimental cell; 3—potentiostat; 4—voltmeter; 5 and 6—self-recorders; 7—extended platinum (investigated) electrode; 8—additional electrodes; 9—reference electrode; 10—platinum electrode; 11—additional platinum electrode; 12—silver-chloride electrode; 13—salt bridge. Electrode 10 is connected to potentiostat input "working electrode"; 11—to input "additional electrode"; 9—to input "reference electrode."

During the measurement, platinum wire electrode 7 is strained horizontally in the investigated solution near its surface and is polarized by the bipolar scheme. The polarization takes place with the help of two electrodes 8, mounted at the ends of electrode 7.

Reference electrode 9 slides with its capillary along the surface of the wire electrode.

The current via electrode 7 in the point of its touching by electrode 9 corresponds to the speed of electrochemical process established in this point at the given potential. Electrode 9 is connected to terminal "reference electrode" at the potentiostat input. It is seen that electrode 7 works as the reference electrode.

The potentiostat operates in the mode of maintaining the potential value of electrode 10 equaled to the potential measured by electrode 9. Then, the polarization current density of electrode 10 by the potentiostat to the set potential will correspond to the density of the current flowing in the touching point of the surface of electrode 7 by electrode 9. The potential of electrode 7 in its touching point by electrode 9 equals the potential of electrode 10 formed by the polarization current at the potentiostat output. Then the polarization current of electrode 10 at the given moment corresponds to the current density in the touching point of electrode 7 by electrode 9.

Self-recorder 5 registers the polarization current at the potentiostat output. The potential in this point is registered by voltmeter 4 and self-recorder 6. Such measuring scheme allows obtaining the graph "potential current force" (process polarogram).

The polarograph is useful to study processes in aqueous solutions of organic and inorganic substances.[3,4]

10.1.2 DEVICE FOR QUANTITATIVE MEASUREMENT OF POROSITY OF GALVANIC COATINGS

The majority of metal coatings are inclined to form layers with increased porosity. A coating with increased electrochemical stability in comparison with the basis metal stability under the action of aggressive atmosphere loses its protective ability, thus, resulting in corrosive decomposition of the basis in the coating pores. It is important to conduct quantitative measurements of porosity of galvanic coatings. It mostly refers to chromium coatings, since their porosity is connected with chromium inclination to spontaneous cracking. It is especially important to measure the coating porosity in hard-to-reach places of parts.

Microgalvanic cells operate in the pores of the porous coating. The current flowing in the coating pores polarizes the coating. The greater the current force in the pores, the more the surface electrochemical potential shifts to the potential of the basis metal. The author of this paper developed the technique to measure the protective ability of coatings by potentiostatic method.

Device description. The device for quantitative measurement of coating porosity was designed in accordance with the proposed technique. The main unit of the device for quantitative measurement of porosity is the device called "potentiostat" that is widely used in electrochemical measurements.[5] The proposed device for quantitative measurement of porosity comprises two blocks: measurement block and control block. The measurement block contains two elements: the investigated part and electrochemical cell. The second block—potentiostat with the function of maintaining the potential of the working electrode in the electrochemical cell equaled to the potential of the coating metal.

Device design. The coating characteristics were measured on the internal surface of the body cylindrical part. The vertical part was filled with the measuring solution. The electrochemical cell filled with the same solution was fixed near the upper cut of the body, and chromium and platinum electrodes were placed inside it. The part was connected to the terminal "reference electrode" on the potentiostat input. The chromium electrode was connected to the terminal "working electrode" and platinum one—to the terminal "reference electrode." That way, the difference between the potentials of the investigated body part and working chromium electrode was supplied to the potentiostat input. At the output, the potentiostat maintained the current between the chromium and additional electrodes of such force and polarity for the difference between the potentials supplied to the input to be maintained as zero. Consequently, the polarization potential value of the working chromium electrode is controlled by the potential spontaneously placed on the internal surface of the body part.[6] And the potential value of the part chromium surface is entirely defined by the corrosion current density spontaneously formed in the chromium pores as a result of operation of shortcut microgalvanic couple "steel basis—chromium coating." The area of the working electrode is 1 cm^2; therefore, the force of its polarization current before the investigated part potential numerically equals the density of corrosion current in the pores of the investigated coating.

The current change is displayed on the computer screen connected to the potentiostat. The exponential initial part of the graph reflects the process of iron activation at the pore bottom under the influence of corrosion solution. Then, the current reaches the saturation part, thus, characterizing the equilibrium between the oxygen diffusion rate and the rate of release of corrosion products from the pore channel. The current force in this part is accepted as the corrosion current density at the pore bottom.

Obviously, the task of quantitative measurement of porosity comes to the establishment of the link between the measured density of corrosion current and actual area of pores of the given coating. Consequently, it is necessary to make a calibration curve of the dependence of corrosion current density of the metal basis on the area of pore surface. The calibration curve was made using steel plates with chromium coating with different porosity on them. The pore area in the coating was found by the standard method of filter paper application. The traces of indicator interaction with iron ions, coinciding with the pore section area, were seen on the filter paper. Their total area was found by scanning the dry sheet with further scan processing with the help of the program developed by us. The program allows measuring the total share of spot surface in the coating. Further, the corrosion current density of each plate was measured. The data obtained were used to plot the diagram "current density—porosity."

10.1.3 USE OF BIPOLAR POLARIZATION FOR CAPILLARY ELECTROLYSIS IN ELECTROLYTE FLOW

Monooxygenase system of human liver removes toxic substances from the organism by oxidizing them with molecular oxygen catalyzed by special detoxicating ferment—cytochrome P-450.[7,8] During the blood electrolysis, the products decomposing bacteria and their toxins are formed in the electrochemical cell.[9] However, during the electrolysis, the blood corpuscles are adsorbed by the surface of metal electrode and the system electrical conductivity decreases to zero. The electrolysis stops.

The described device models the liver detoxicating functions in the process of direct electrochemical oxidation of blood inside the blood vessel. For the first time, the authors managed to bypass the problem of blood corpuscles sticking onto the surface of electrodes[11,12] by the electrolysis in the blood stream inside the blood vessel. The direct electrochemical oxidation of blood is performed by its direct electrolysis inside the blood vessel by the electrode bipolar polarization. During the bipolar polarization of the wire electrode, the polarization potential on one of its ends (anode) corresponds to the process of chlorine oxidation with the formation of hypochlorite, and on the second one (cathode)—to the potential of hydrogen ion reduction to molecular hydrogen.

Wire electrode is polarized by the bipolar scheme with the help of attachable electrodes in such a way that its one end is polarized cathodically, and the second one—anodically. At the same time, oxidation proceeds on the surface of one part of wire electrode, and reduction—on the second one.

10.2 CONCLUSIONS

For the first time, the following devices were developed and tested: polarograph with extended solid-state electrode; device for quantitative measurement of galvanic coating porosity in hard-to-reach places of an actual part; and device for direct electrochemical oxidation of blood inside the blood vessel.

KEYWORDS

- **bipolar electrochemical polarization**
- **electrochemical oxidation**
- **blood**
- **detoxification**
- **through porosity**
- **corrosion processes in the coating pores**

REFERENCES

1. Fedotiev, N. P., et al. *Applied Electrochemistry*. "Chemistry": Leningrad, 1967; pp 346–351.
2. Rudenok, V. A.; Bakhchisaraitsyan, N. G.; Kruglikov, S. S. Method of Measuring the Basis Corrosion Rate in the Pores of Cathodic Galvanic Coating. Patent No 1356726 MPK G01N27416. March 30, 1990, bulletin No 12.
3. Elistratova, K. N.; Smirnov, K. S.; Grigoryan, N. S.; Kharlamov, V. I. Corrosion-Resistant Galvanic Coatings on Titanium//Galvanic Engineering and Surface Treatment. **2008,** *XVI* (3), 24–29.

4. Kasatkin, V. E.; Solodkova, L. N.; Kondrashev, Yu. V. Potentiostats of IPC Series. Practical Application in Electrochemical Methods of Research. Part 1. Analyzer of Organic Additives "Korian-3." *Galvanic Engineering and Surface Treatment* **2011,** *19* (2), 27–34.
5. Solodkova, L. N.; Kudryavtsev, V. N. Method of Detecting Organic Additives and Their Action Efficiency During Zinc Electric Deposition From Zincate Electrolytes. *Galvanic Engineering and Surface Treatment* **1993,** *2* (2), 46–51.
6. Rudenok, V. A. Method of Measuring Distribution of Potentials and Intensity of Processes Flowing Along the Investigated Electrode, and Device for Its Implementation. Patent of Russia No 2599161, 2015, bulletin No 32.
7. Petrosyan, E. A. Pathogenetic Principles and Substantiation of Suppurative Surgical Infection Treatment By the Method of Indirect Electrochemical Oxidation: Extended Abstract of Dissertation of Doctor of Medical Sciences, Leningrad, 1991.
8. Vasiliev, Yu. B, Sergienko, V. I.; Grinberg, V. A.; Martynov, A. K. Removal of Toxins From the Organism with the Help of Electrochemical Oxidation. *Iss. Med. Chem.* **1991,** *37* (2), 74–78.
9. Yao, S. J.; Wolfson, S. K. U.S. Patent 3878564, April 22, 1975.
10. Rudenok, V. A.; Marasinskaya, E. I.; Zakomyrdin, A. A. Organism Detoxication Method and Device for the Method Implementation. Patent 2229300, Russian Federation, MPK 7A61K 33/14. No 2002120848/14, May 27, 2004, bulletin No 15, p 5.
11. Sergienko, V. I.; Martynov, A. K.; Khapilov, N. A. Author's certificate 1074493 USSR, 1984, bulletin No 7.
12. Vasiliev, Yu. B.; Grinberg, V. A.; Guseva, E. K.; Chechkov, A. A. *Electrochemistry* **1986,** *22*.

CHAPTER 11

Advanced Development in the Synthesis of Biologically Active Heterocyclic 1, 2, 3, 4 Tetrahydropyrimidine-2-Ones

KALPANA N. HANDORE[1*], VASANT V.CHABUKSWAR[1], and S. B. SHARMA[2]

[1]Nowrosjee Wadia College, Chemistry Department, University of Pune, Pune, India

[2]Modern Education Society's College of Engineering, Affiliated to Savitribai Phule Pune University, Bund Garden Road, Pune 411001, India

[*]Corresponding author. E-mail: kalpanahandore@gmail.com

ABSTRACT

This paper discusses atom economy, applications in combinatorial chemistry, and diversity-oriented synthesis. Such heterocycles show chiral center, which have a wide scope of pharmacological properties including antiviral, antitumor, antibacterial, and anti-inflammatory activities. The pyrimidine ring gives good antimicrobial activity and potent antitubercular activity. Recently, functionalized dihydropyrimidin (DHPM) analogs have emerged as active antihypertensive agents. Thus, the synthesis of these heterocyclic compounds is very important in present areas of study. The search for more suitable preparation of dihydropyrimidinone continues today also. The first protocol to prepare the compounds of this type was presented by Biginelli in 1893 and it involved a three-component, one-pot condensation. A major drawback to Biginelli's original was poor-to-moderate yields. This review explains collectively the synthesis of dihydropyrimidinones in various conditions using a variety of catalysts and some solid supported catalysts. This review also gives an application of Biginelli's reaction and its modifications and enantioselective synthesis.

11.1 INTRODUCTION

Hantzsch,[1] Knoevenagel,[2] and Biginelli reactions[3] have some similarity as each of these employs aldehyde and acetoacetic ester (active methylene compound). The earliest of these seems to be the discovery of the Hantzsch reaction, which was reported in 1881, wherein Hantzsch heated acetoacetic ester, an ammonia source, and an aldehyde, to obtain the now well-known dihydropyridines or Hantzsch pyridines[1].

A decade later the Italian chemist, P. Biginelli, reacted similar two components in equimolar ratio viz. acetoacetic ester and aldehyde, along with a third component as urea, in acidic alcoholic solution to obtain a new compound, the now well-known 3,4-dihydropyrimidin-2(1*H*)-ones or Biginelli compounds,[3,4] which are obvious aza-analogues of the Hantzsch dihydropyridines. Biginelli did not detect any Hantzsch dihydropyridines as byproducts.[5,6]

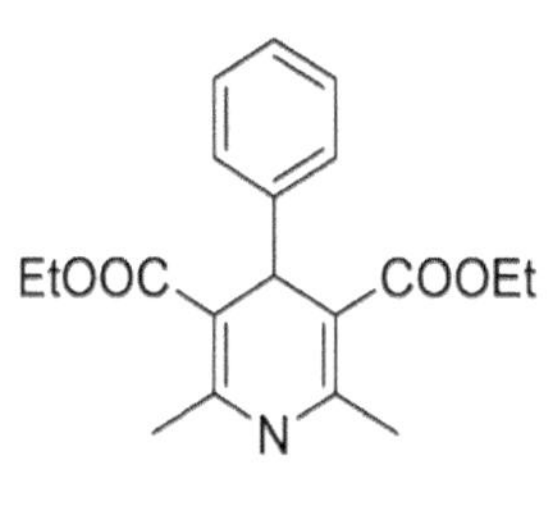

Hantzsch pyridines

(Dihydropyridine)

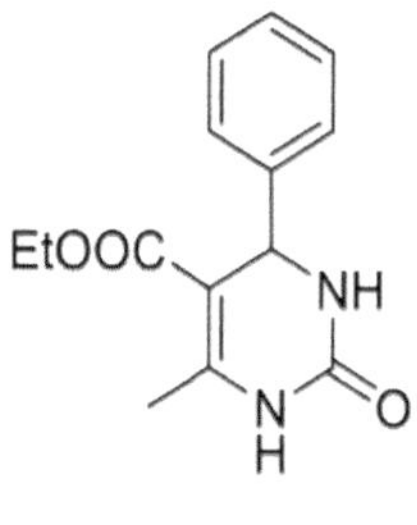

Biginelli compound

(Dihydropyrimidine)

Biginelli compounds, that is, 3,4-dihydropyrimidin-2(1H)-ones (DHPMs) possess interesting biological applications (Kappe, 1993; Snider and Shi, 1993; Overman et al., 1995; Kappe, 2000a, 2000b). The untapped potentials of DHPMs were explored because of their apparent structural similarities with clinically important dihydropyridines (DHPs) of nifedipine type (Rovnyak et al., 1995).

11.2 DISCOVERY

In 1893, after Hantzsch's discovery, the Italian chemist Pietro Biginelli (University of Florence) for the first time reported on the acid catalyzed cyclocondensation of ethyl aceto acetate (EAA), benzaldehyde, and urea. The reaction was carried out by simply heating the mixture of the three components in ethanol with catalytic amount of HCl at reflux temperature. The product of this novel one-pot, three-component synthesis precipitated on cooling; the reaction mixture was identified as 3,4-dihydropyrimidin-2-(1H)-ones (Scheme 11.1).

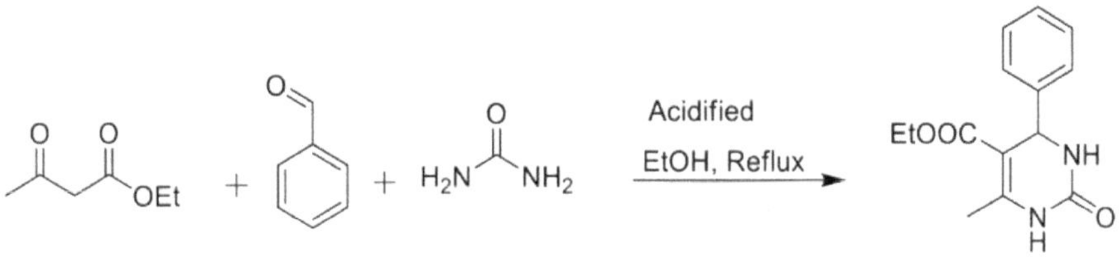

SCHEME 11.1 Acid catatalyzed dihydropyrimidinones synthesis.

The Biginelli reaction, one of the most useful multicomponent reactions, offers an efficient way to access multifunctionalized 3,4-dihydropyrimidin-2-(1H)-ones (DHPMs) and related heterocyclic compounds. The achievement of making multiple bonds in a one-pot, multicomponent coupling reaction promotes a sustainable synthetic approach to new molecule discovery. Compounds containing the dihydropyrimidinones moiety inherently asymmetric molecules; and the influence of the absolute configuration at the stereogenic center at C_4 Pyrimidinone and thione scaffold plays an important role in many biological processes, and hence they are used in the synthesis of drugs. Much effort was directed toward developing highly efficient asymmetric Biginelli reaction owing to the exhibition of a wide range of biological activities in DHPMs. 3,4-dihydropyrimidin 2-(1H)-ones and their sulfur analogues have been reported to have remarkable pharmacological properties (Fig. 11.1). Some of them have antibacterial, antitumor, antitubercular antifungal, antioxidant, antimalerial activity, anti-HIV, sedative, anticonvulsant, and anti-inflammatory activity. They are also associated with nucleic acid, antibiotic, and anticancer drug, and some are calcium channel modulators, as shown in Figure 11.2.

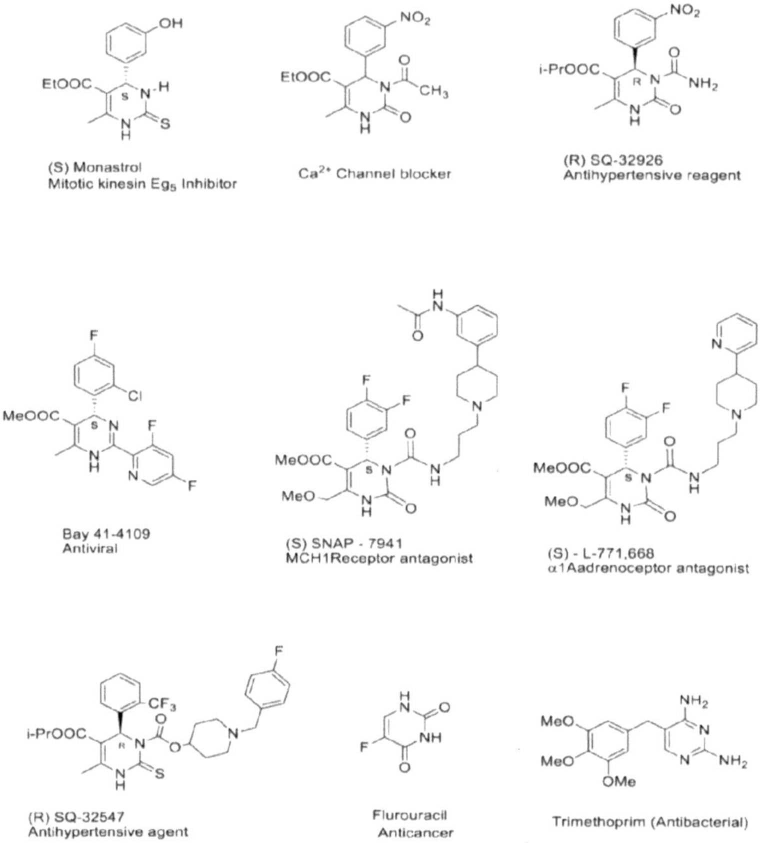

FIGURE 11.1 Biginelli compounds.

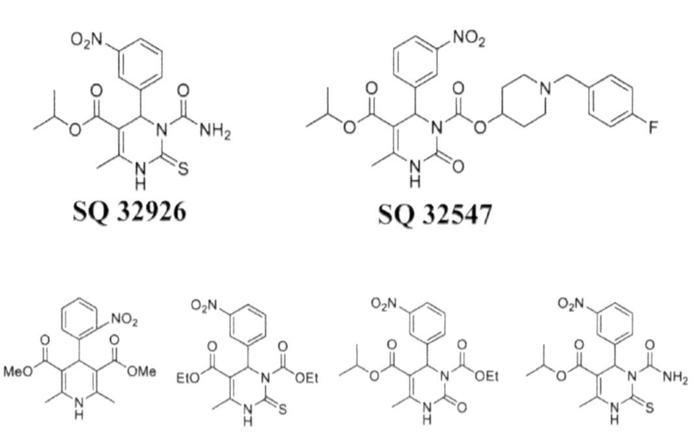

FIGURE 11.2 Structures of DHPM-based calcium channel modulators.

11.3 MECHANISTIC STUDIES

There was curiosity of researchers after the discovery of the reaction. There were studies to find out expected pathway for this reaction. Different research groups have discussed the mechanism of the reaction, which is as follows:

11.3.1 *FOLKERS AND JOHNSON (1933)*

In 1933, Folker and Johnson reported that one of the three intermediates was present in this reaction,[7] which includes bisureide 1, which was formed by condensation reaction between aldehyde and urea followed by subsequent attack of resultant imine with another equivalent of urea. Also, 3-ureido ethyl acrylate 2 arose from condensation reaction between keto ester and urea. Finally, the reaction of the keto ester and aldehyde delivered the aldol adduct 3 (Scheme 11.2)

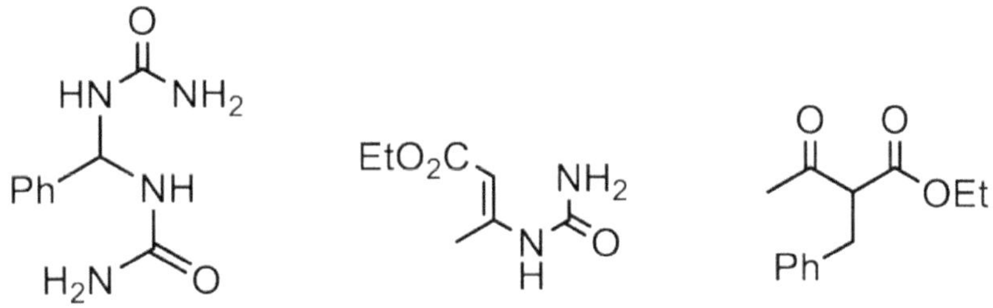

1 2 3

SCHEME 11.2 Intermediates proposed by Folkers and Johnson.

Folkers and Johnson, in 1933, studied the mechanism of the Biginelli reaction, which has been investigated by several research groups and over the past decades has been the subject of some debate. Recently, Kappe clearly demonstrated the formation of *N*-acyliminium ion as the key intermediate, resulting from the acid-catalyzed condensation of benzaldehyde and urea. Interception of this iminium ion by ethyl acetoacetate produces open-chain ureide, which subsequently cyclizes to the dihydropyrimidine (Scheme 11.3) on the basis of ^{1}H and ^{13}CNMR and trapping experiments.

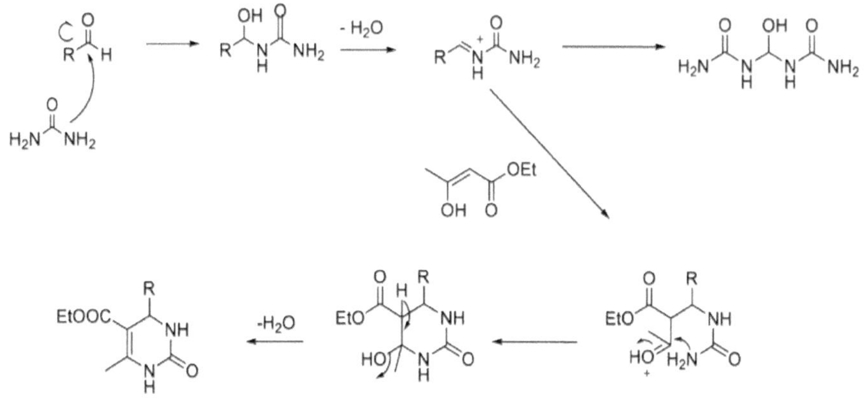

SCHEME 11.3 The mechanism of the Biginelli reaction proposed by Kappe.

11.3.2 SWEET AND FISSEKIS (1973)

Forty years after the initial process, Sweet and Fissekis proposed a more detailed pathway involving carbonuim ion species.[8] According to them, the first step is the aldol condensation between EAA and benzaldehyde to give aldol adduct, which, on dehydration, gives a carbenuim ion species that is in equilibrium with enone. Nucleophilic attack of urea on carbonuim ion gives ureide. Intermolecular cyclization produced a hemiaminal, which underwent dehydration to dihydropyrimidinone (Scheme 11.4).

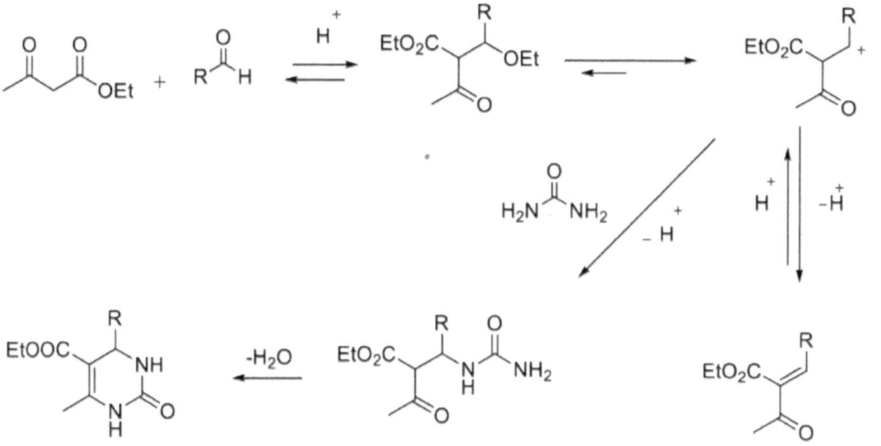

SCHEME 11.4 The mechanism of the Biginelli reaction proposed by Sweet and Fissekis.

11.3.3 ATWAL AND O'REILLY (1987)

Atwal and et al.[9–11] gave a proposal with poor yield of the dihydropyrimidinone for aliphatic aldehyde and aldehyde having a slightly hindered carbonyl function, that is,. orthosubstituents. They suggested that reaction takes place in two steps, the first step is the formation of unsaturated carbonyl compound via Knoevenagel condensation and the second step is the base catalyzed addition of urea (Scheme 11.5).

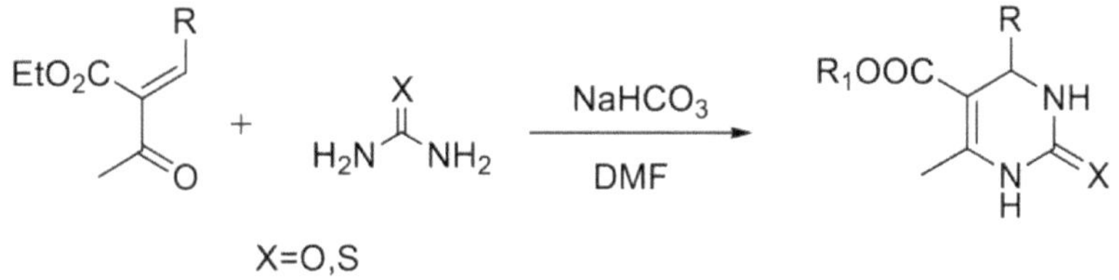

SCHEME 11.5 Formation of 3,4 dihydropyrimidinones/via Knoevenagel condensation.

There are very few studies on mechanistic aspect of this reaction. After Kappe's mechanistic proposal, this reaction attracted the attention of a large number of researchers around the world as far as large number of publications are concerned, which just modified the catalyst development. Most of the research papers were flooded with catalyst modification and mention Kappe's mechanism, which was based on spectroscopic evidences, which is given in the subsequent section.

11.3.4 O. KAPPE (1997)

On the basis of spectral techniques, C^{13} and $^{1}HNMR$, Kappe further re-examined the mechanism of this multicomponent reaction.[12] The first step involves the nucleophilic attack of urea on electron deficient carbon of aldehyde, which gives N-acyliminium ion intermediates by dehydration. He suggested that the first step is rate limiting. In the next step, active methylene compound adds to this intermediate in a Michael fashion. Kappe found that, in this reaction, dihydropyrimidinones were always formed in minor quantities,[6] which had not been observed by earlier research groups (Scheme 11.3). Saloutina and co-workers[13] used $CF_3COCH_2CO_2Et$ in place of acetoacetic ester; they isolated the intermediate, which, on dehydration with *p*-toluenesulfonic acid, gives DHPM.

11.3.5 CEPANEC (EXPERIMENTAL EVIDENCES 2007)

Cepanec studied Lewis acid catalyzed reaction mechanism by using $SbCl_3$ under these conditions.[14] It was found that reaction proceeds via intermediate 3-ureido-crotonates and not by iminuim ion proposed by Kappe (Scheme 11.6).

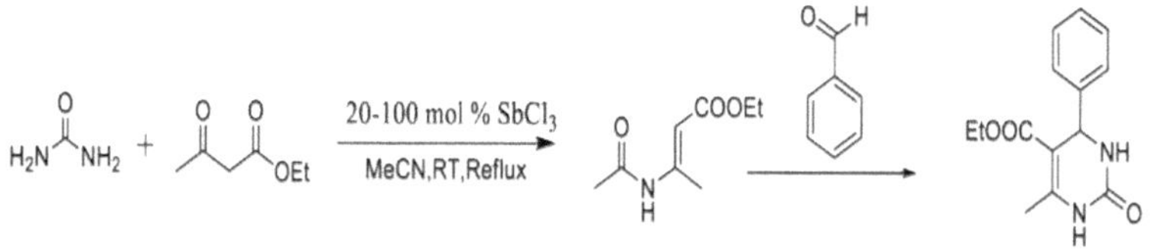

SCHEME 11.6 Formation of 3,4 dihydropyrimidinone via 3-ureido-crotonates.

11.3.6 JIAN-HUA ZHOU (DFT STUDY—2008)

The density functional theory (DFT) study reported that via condensation of urea, benzaldehyde and ethylacetoacetate were investigated under classical condition[15] and they have confirmed Kappe's proposal.

11.3.7 DE SOUZA (2009)

De Souza et al. investigated Biginelli reaction[16] and concluded in favor of Folkers and Johnson proposal (Scheme 11.7). They based their conclusions on DFT calculations and used a mass spectrometer having accessories for various ionization.

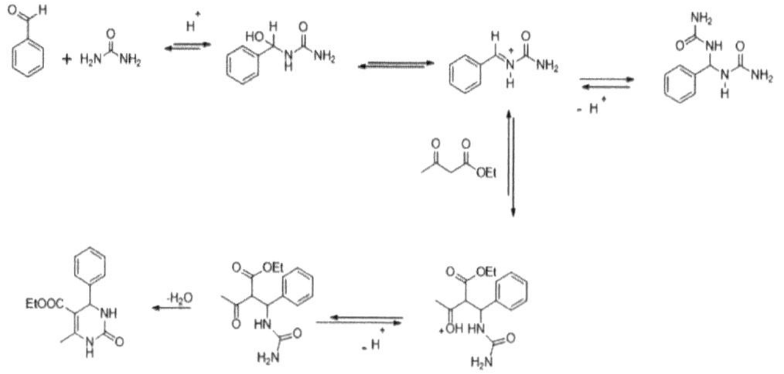

SCHEME 11.7 Formation of 3,4 dihydropyrimidinone via iminium ion intermediate.

11.3.8 SHUN-JUN JI 2010 (BASE CATALYZED)

A large number of investigations have been on the use of acidic catalysts, and only very few papers describe basic catalysts (Scheme 11.8). Very recently, Chinese workers[17] have described the use of strong bases and proposed different pathways, as shown below

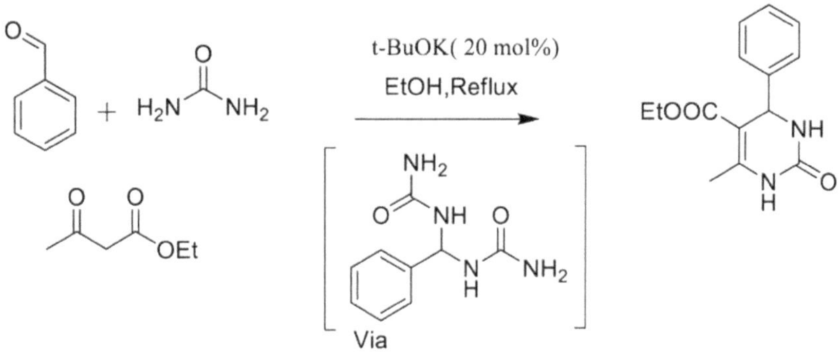

SCHEME 11.8 Base catalyzed synthesis of 3,4-dihydropyrimidinone.

All the aforementioned processes and mechanism reinvestigations with each catalysts used seems to be a healthy trend in this reaction in contrast to earlier reports casually mentioning that our reaction seems to follow Kappe mechanism. In mechanism advancements, yet another mechanism report is there using hexaaquo-Al (III) BF_4.[20]

11.3.9 LITVIC 2010 (BRØNSTED ACID)

The mechanism of synthesis of dihydropyrimidinones via 'ureido-crotonate' formation a reported by Cepanec which is in contrast to acylimino intermediate suggested by Kappe[18] (Scheme 11.9).

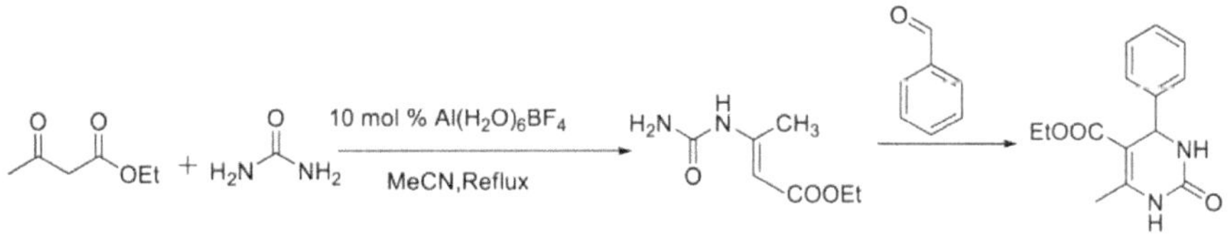

SCHEME 11.9 Acid catalyzed synthesis of 3,4-dihydropyrimidinone.

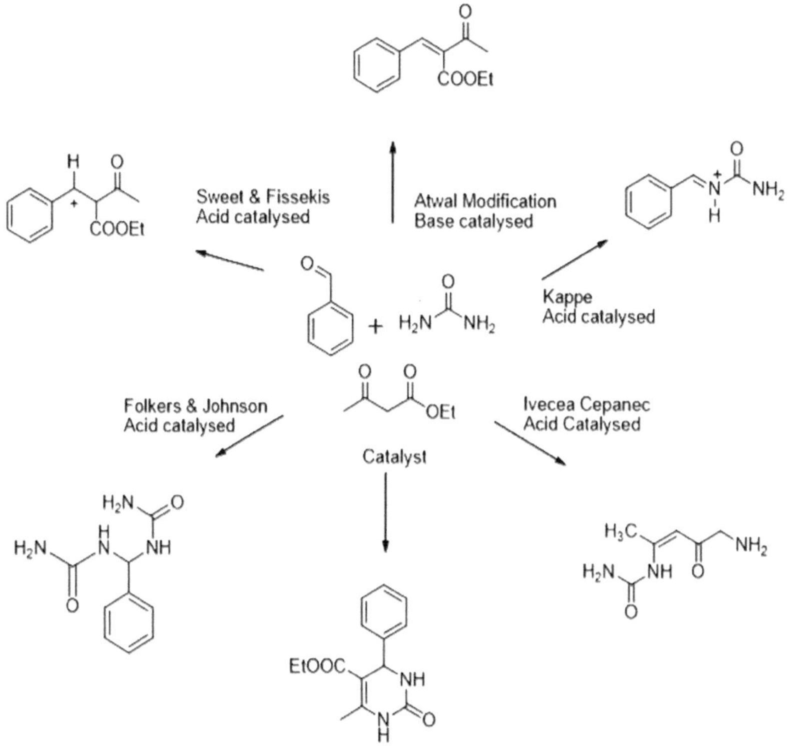

SCHEME 11.10 Investigated intermediates by various researchers.

11.4 MODIFIED BIGINELLI CONDENSATION

From the 19th century to mid-1970s, very few papers were published concerning this reaction. However, from mid-1970s, the utility of this reaction has grown rapidly, especially because some DHPM posses significant therapeutic and pharmacological properties. These synthetic interests have included both methodological improvement of the original Biginelli reaction conditions and total synthesis of natural products. As a result, new synthetic methods for the efficient preparation of these heterocyclic compounds are of great importance. Very recently, several modified and improved procedures for the one-pot synthesis of dihydropyrimidinones have been reported, but many reported methods have drawbacks such as long reaction time, harsh reaction condition, toxic catalysts, unstablity of catalyst, difficulty in separation, and reusability of catalyst.

In 1995, Wipf and Cunningham provided the first synthesis of DHPM using solid-phase resin-bound urea formation of DHPM and cleavage from the resin with TFA gives N-1 substituted product. Wipf and Cunningham were the first to report on the adaptation of this reaction to solid-phase techniques. Their approach consisted of the fixation of γ-aminobutyric acid-derived urea on Wang resin, followed by condensation with excesses of β-keto esters and aldehydes in the presence of a catalytic amount of HCl. Subsequent cleavage of the product by 50% trifluoroacetic acid provided DHPMs in high yield and excellent purity (Scheme 11.11).

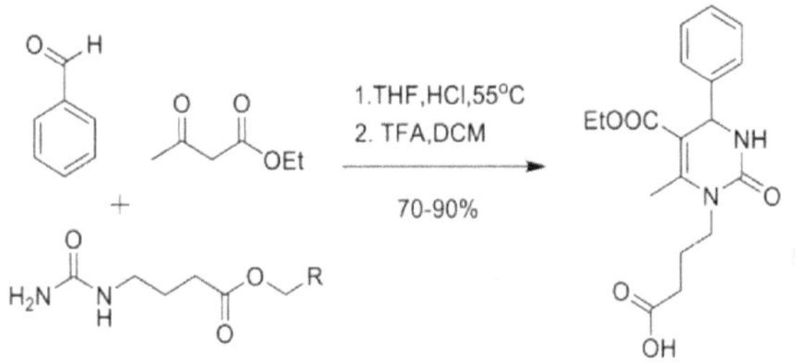

SCHEME 11.11

Kappe further extended the scope of the solid-phase application by using a β keto ester immobilized reagent, which yields 5-carboxylic acid DHPM. Kappe and co-workers therefore developed an alternative approach, in which the acetoacetate building block is linked to the solid support, allowing access to pharmacologically active *N*-1-unsubstituted DHPMs in high overall yield (Scheme 11.12).

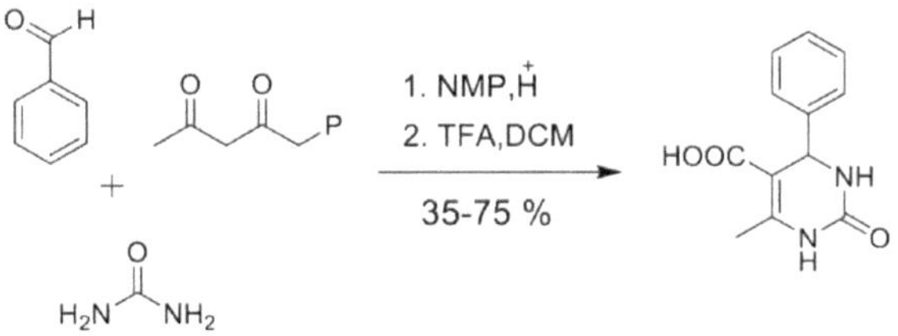

SCHEME 11.12

Enantioselective one-pot synthesis of Biginelli reaction is reported with use of lanthanide triflates ytterbium triflates. These catalyst leads to highly enantioselectivity and up to 99% enantioselective Biginelli reaction can be carried out.[19] Kappe and Stadler reported automated microwave synthesis by utilizing $Yb(OTf)_3$ as a catalyst. They have prepared around 48 compounds within 12 h, which includes a variety of aldehydes N-substituted ureas (Scheme 11.13).

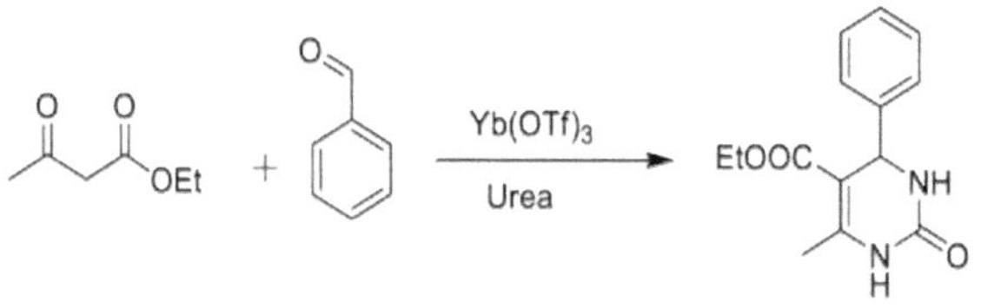

SCHEME 11.13

Yttruim (III) nitrate hexahydrate can also be used as a catalyst under solvent-free condition for one-pot synthesis of large-size dihydropyrimidinones (Scheme 11.14). It shows high catalytic activity and gives a high yield.

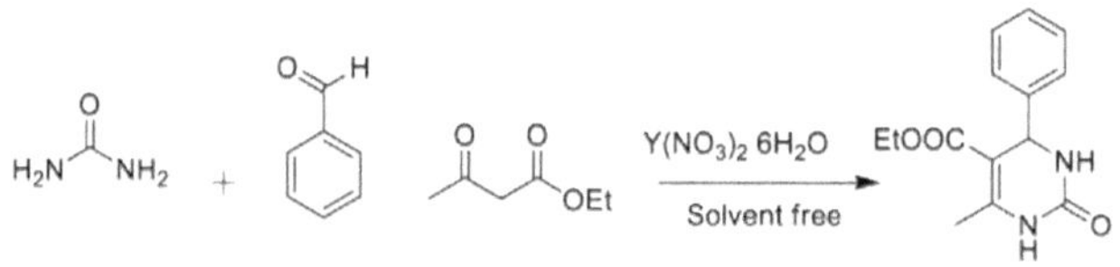

SCHEME 11.14

Capanee reported DHPM synthesis using tetraethyl orthosilicate in presence of ferric chloride improves yield and purity under the mild condition. Russowsky and co-workers described the ability of $SnCl_2\,2H_2O$ as catalyst (Scheme 11.15) to promote three-component condensation of aldehydes, β-keto esters, and urea or thiourea using acetonitrile or ethanol as solvent in neutral media and advantage over $FeCl_3.6H_2O$ and $COCl_2.6H_2O$, which were used as CO catalyst with HCl.

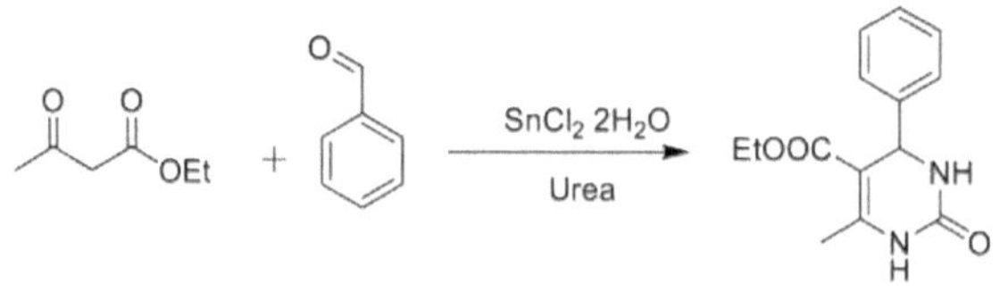

SCHEME 11.15

Debache and co-workers have demonstrated the synthesis of DHPM derivatives using a phenyl bororonic acid $PhB(OH)_2$ as a catalyst (Scheme

11.16) to promote Biginelli three-component condensation. It gives excellent yield of dihydropyrimidinones with diversity of aromatic aldehydes.[20]

Boumoud and co-workers also reported the use of phenylboronic acid as mild and efficient catalyst for Biginelli reaction. Later on, they disclosed use of Ni $(NO_3)_2.6H_2O$ (Scheme 11.17) as a new, mild, and convenient Lewis acid. $Ni(NO_3)_2.6H_2O$ was used as the sole promoter agent in neutral media, while, for the others, previously reported hydrates of metallic halides, such as Fe (III), Ni (II) and CO(II), a catalytic amount of conc. HCl was needed as a Brönsted acid co-catalyst in order to improve the yields by using different quantities of reagents. The best results were obtained with a 0.1:1:1:1.5 ratios of Nickel (II) nitrate hexahydrate, aldehyde, ethylacetoacetate, and urea/thiourea in acetonitrile as solvent at reflux temperature.

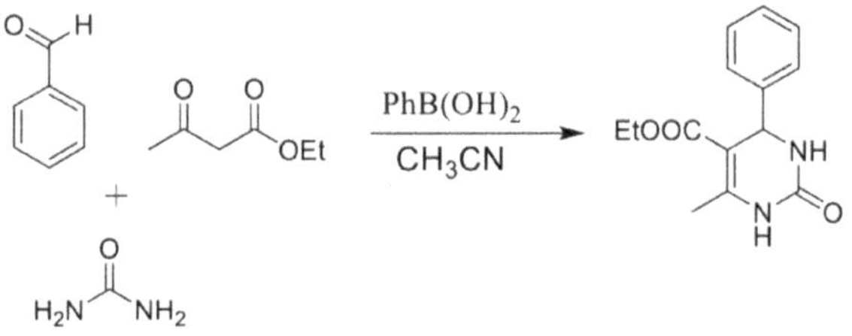

SCHEME 11.16

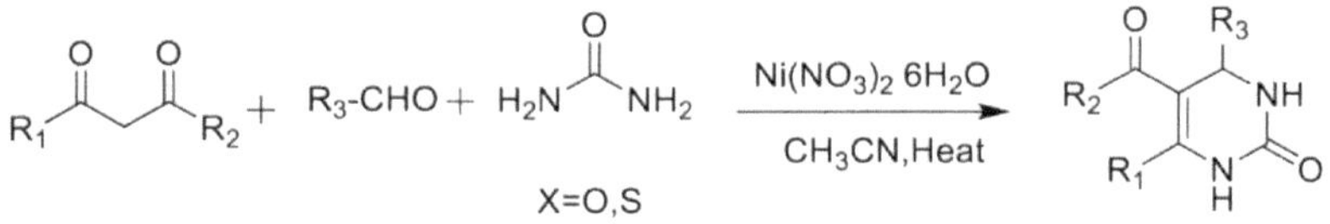

SCHEME 11.17

Rajendra P. Pawar reported in *International Journal of Industrial Chemistry* for in situ generation of HCl from 2, 4, 6 tri chloro[1,3,5] cyanuric chloride (or TCT). It is a stable nonvolatile, inexpensive, and safe reagent used for synthesis of 4-aryl-3,4 dihydropyrimidine. Compared with CuCl, $CuCl_2$, and HCl, TCT acts as best catalyst among tested catalysts as it gives good yield. TCT was used as source of HCl to catalyze this reaction. It reacts incipient moisture and releases three molecules of HCl and cyanuric acid as by product, which is removed by washing with water (Scheme 11.18).

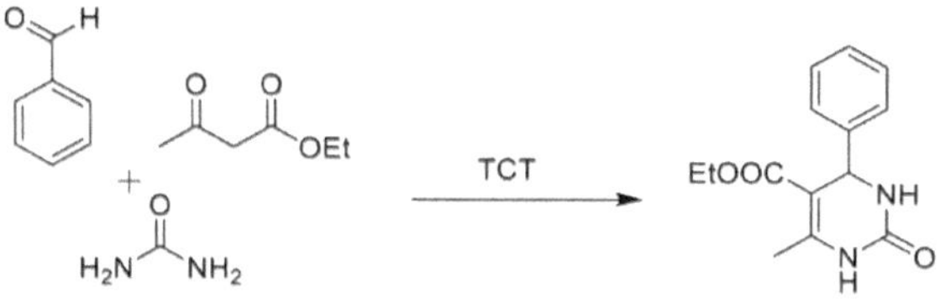

SCHEME 11.18

DHPM synthesis was initiated by exposing a mixture of ethyl acetoacetate, urea, benzaldehyde, and tributyl borate in microwaves in the absence of a solvent for 3 min at a power of 390 W (Scheme 11.19). An efficient reaction occurred and the dihydropyrimidinone was obtained in 95% yield.

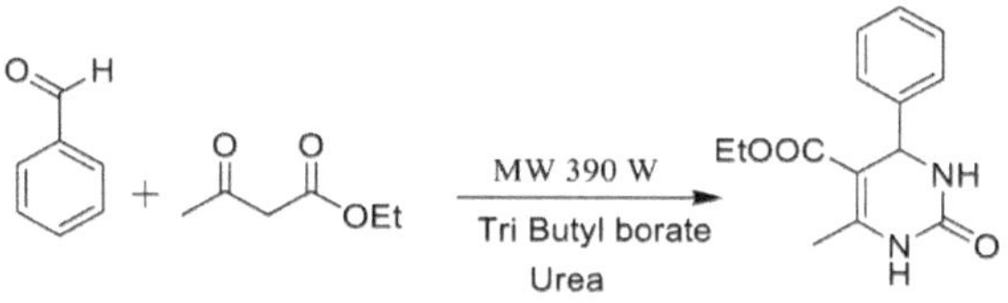

SCHEME 11.19

Wide range of synthesis of 3,4-dihydropyrimidinones (Biginelli compounds) derived from divergent aldehydes, β-keto esters, and urea or thiourea under solvent-free conditions was efficiently performed by Srinivas R. Adapa and coworkers citric acid (0.5 equiv.) at 80°C with good to excellent yields. The proposed mechanism for the citric acid-promoted synthesis of dihydropyrimidinone may tentatively be visualized to occur via a tandem sequence of reactions involving formation of acylimine intermediate, formed by the reaction of the aldehydes and urea activated by citric acid (Scheme 11.20) through intermolecular hydrogen bonding and subsequent addition of β- diketoester enolate to the acylimine followed by cyclo dehydration to give dihydroprimidin-2(1H)-one.

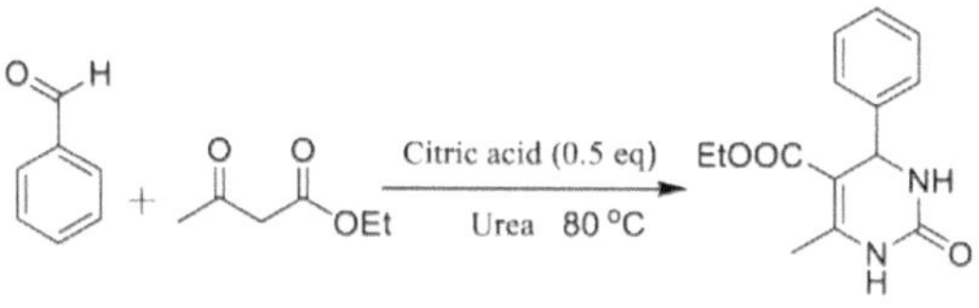

SCHEME 11.20

Study of reaction rate and yield enhancements were also reported for Biginelli reactions under microwave irradiation conditions, in combination with PPE under solvent-free conditions. A large amount of products can be obtained in short reaction times, and with at least 95% purity by a simple aqueous workup procedure (Scheme 11.21).

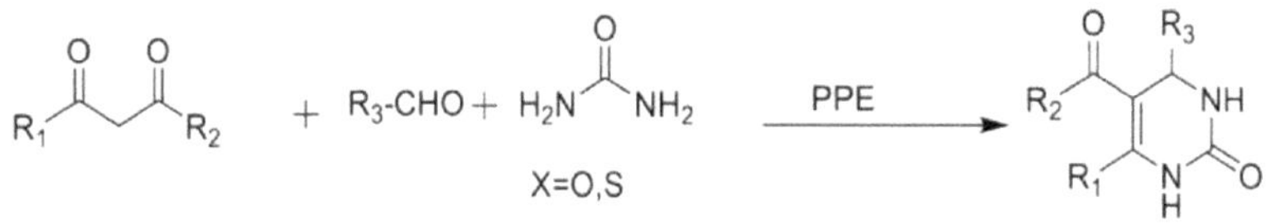

SCHEME 11.21

Sharghi, Mahboubeh Jokar in January 2009 reported that a mixture of Al_2O_3/CH_3SO_3H (AMA) is an inexpensive and effective reagent. It is a reusable catalyst for the one-pot synthesis of 3,4-dihydropyrimidin-2-(1H)-ones and thiones in high yield (Scheme 11.22). The reaction rate was found to be slow at room temperature; therefore, reaction was carried out at high temperature (60°C). At this temperature, reaction rate was found to be maximum, and further increase in temperature did not show any enhancement. Solvents like ethanol, toluene, acetonitrile, and dimethyl formamide afforded the product in high yield. They chose ethanol for its low cost and environmental acceptability.

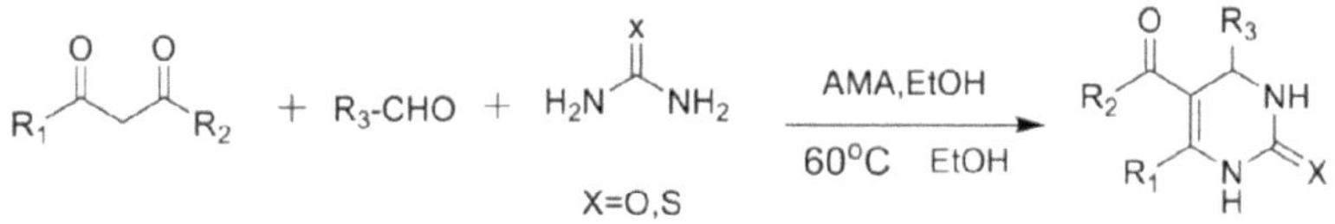

SCHEME 11.22

Dabiri and coworkers use ionic liquid (ILs) as catalyst for synthesis of DHPMs and other organic reactions in that acidic ionic liquids are special interest as it undergoes hydrolysis with water (Scheme 11.23). 1,1,3,3-Tetramethylguanidinium trifluoroacetate at a room temperature ionic liquid promoted preparation of 3,4-dihydropyrimidin-2(1H)-ones has been reported at 100°C.

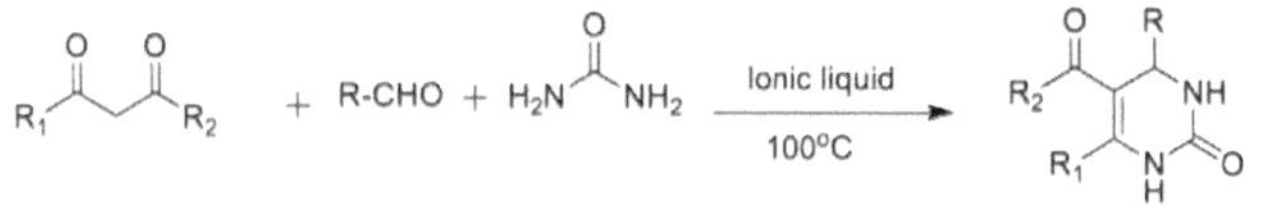

SCHEME 11.23

DHPM can also be synthesized in high yield in the presence of catalytic amounts ionic liquids such as 1-*n*-butyl-3-methylimidazolium tetrafluoroborate ($BMImBF_4$) or hexafluorophosphate ($BMImPF_6$) (Scheme 11.24).

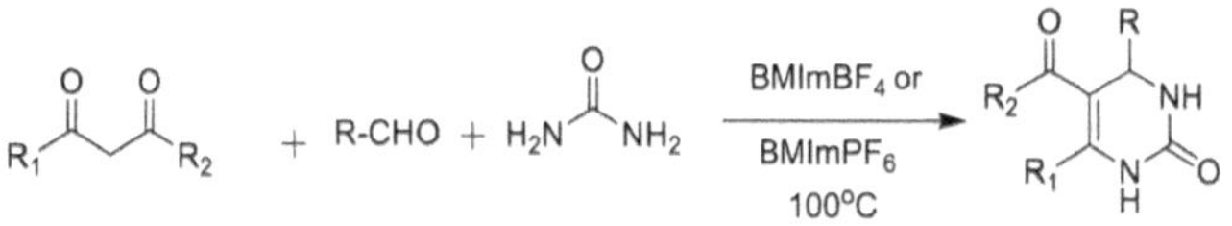

SCHEME 11.24

Interesting reactions are in which the aldehyde is derived from a carbohydrate, affording access to *C*-glycosylated dihydropyrimidinone (Scheme 11.25). Although the Biginelli reaction has mostly been carried out in its chiral version, its noteworthy that in this case, a satisfactory asymmetric induction was observed, affording chiral products with given configurations at the C-4 stereo center of the DHPM ring.

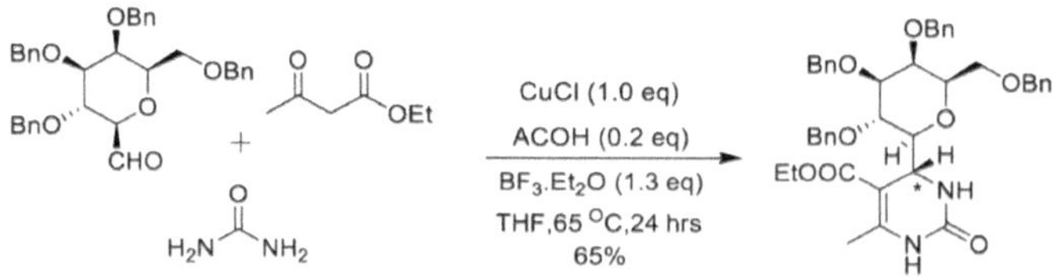

SCHEME 11.25

Hitendra N. Karade and coworkers have used silica chloride as a catalyst in solvent-free condition. As Si-Cl bond is labile, it can give rise to Lewis acid center as silica. Dalip Kumar and coworkers have synthesized DHPMs and parallel synthesis of aryl -14-H dibenzo xanthenes by using sulfated zirconia and phosphotungestic acid as solid acid catalyst in solvent-free conditions.

Farhad Shirini and coworkers used silica triflate as efficient catalyst in solvent-free synthesis. Silica triflate is prepared in situ by the reaction of silica gel with trifluoromethyl sulphonyl chloride. At room temperature, the reaction is easy and clean as there is immediate evolution of HCl gas and does not need any workup procedure (Scheme 11.26). Silica triflate is a reusable catalyst even after 5 runs activity of reagent remains same.

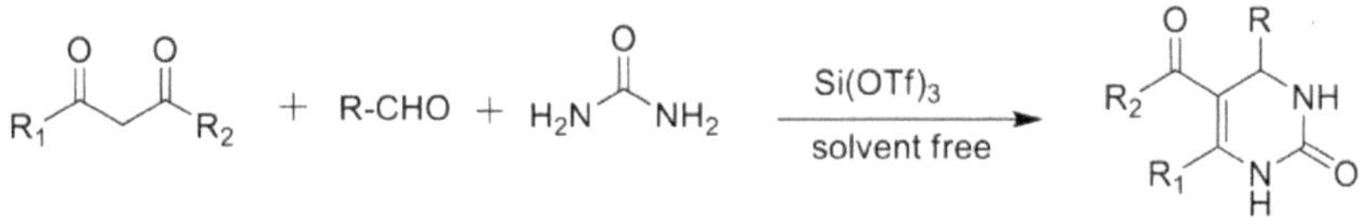

SCHEME 11.26

Bose et al. reported the large-scale synthesis of dihydropyrimidinone derivatives by using water-based biphasic reactions of immiscible organic reagents. The essence of the method was dynamic mixing of the two phases. In all cases, the corresponding N heterocycle crystallized out quickly (<30 min) from the mixture, affording an easy isolation in essentially pure form and in >90% yield.

Prashant V. Anbhule et al. in *Open Catalysis Journal*, in 2010, 3, 83–86 gives simple and efficient approach toward one-step synthesis of 6-amino-5-cyano-4-phenyl-2-mercaptopyrimidine, and its hydroxyl derivatives have been developed by three-component condensations of aromatic aldehydes, malononitrile, and thiourea/urea in presence of phosphorus pentoxide (Scheme 11.27).

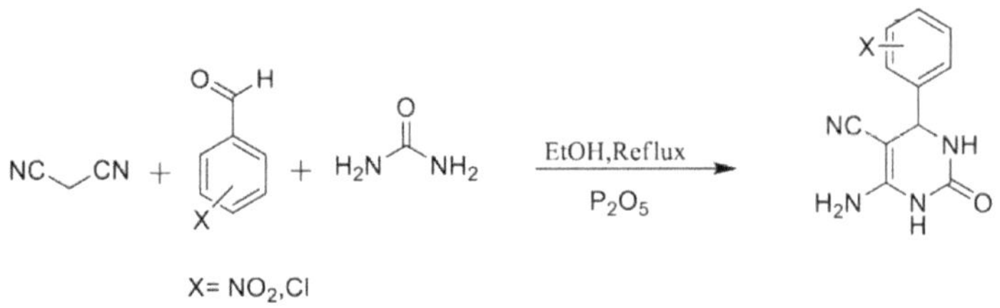

SCHEME 11.27

Aghayan and coworkers, in March 2011, described a simple method for the preparation of 2-amino-5,6 dihydro pyrimidine-4(3H)-ones from the cyclocondensation of Meldrum acid, aldehydes, and guanidinium carbonate using catalytic amount of amino-functionalized MCM-41 (Scheme 11.28). Equimolar solution of meldrum acid aldehydes and guanidinium carbonate in presence of 10 mol % of catalyst MCM 41- NH_2 took place at 120°C in 0.5 mL DMF as solvent under these conditions various aldehydes reacted smoothly to give corresponding 2 amino-5,6 dihydropyrimidin-4 (3H)–one derivatives.

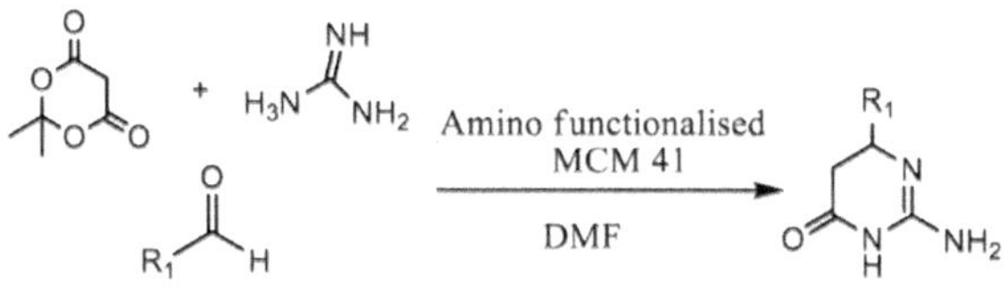

SCHEME 11.28

Dondoni and Massi developed an ecofriendly method of synthesis of DHPM, which involved the use of polymer-bound Yb (III) reagent under solvent-free conditions (Scheme 11.29) with suitable polymer-supported scavengers, which gives moderate yield of DHPM.

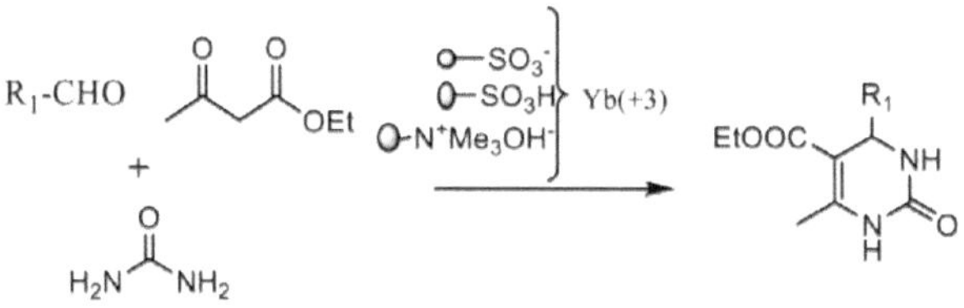

SCHEME 11.29

A large number of investigations reported the use of Lewis acid like catalysts and only very few papers describe basic catalysts. Very recently, Chinese workers have described the use of strong bases and proposed different pathways, as shown in Scheme 11.30.

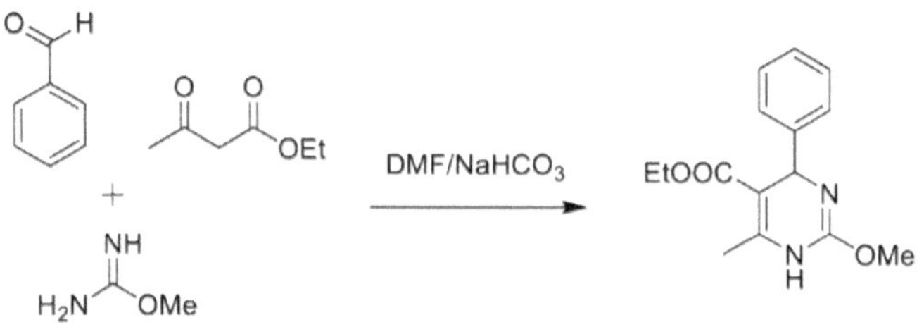

SCHEME 11.30

One-pot condensation uses benzyl halides in place of aldehydes in the presence of bismuth nitrate pentahydrate in molten tetrabutylammonium fluoride (Scheme 11.31). It is an in situ, one-pot oxidation-cyclocondensation and the advantage of this type of reaction is that several commercially available alkyl halides are used rather than aldehydes and the intermediate aldehydes do not require isolation. This reaction is particularly useful for the aldehydes that are unstable (e.g., volatile or prone to polymerization or hydration). Bismuth nitrate pentahydrate catalyzes the three-component condensation reaction of an aromatic aldehyde, urea, and a β-ketoester or a β-diketone under solvent-free conditions to afford the corresponding dihydropyrimidinones (DHPMs) in high yields. The present method is also effective for the selective condensation of aryl aldehydes in the presence of aliphatic aldehydes.[22] .

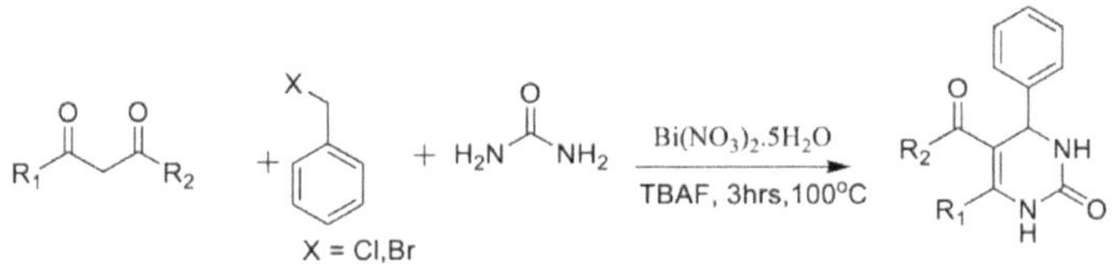

SCHEME 11.31

An efficient synthesis of 5-unsubstituted 3,4-dihydropyrimidin-2(1H)-ones has been published by Wang et al. via iron(III)-catalyzed Biginelli-like cyclocondensation of urea with aldehydes and ketones in acetonitrile. The facile preparation of these compounds by one-pot condensation of aldehydes, urea, and enolizable ketones (Scheme 11.32) has been reported by Sandhu et al. using bimetal system aluminum (III) chloride and potassium iodide in acetonitrile under reflux conditions. Some combinations of catalytic systems have also been studied and out of these, the most effective combination worked out to be the aluminum (III) chloride and potassium iodide system

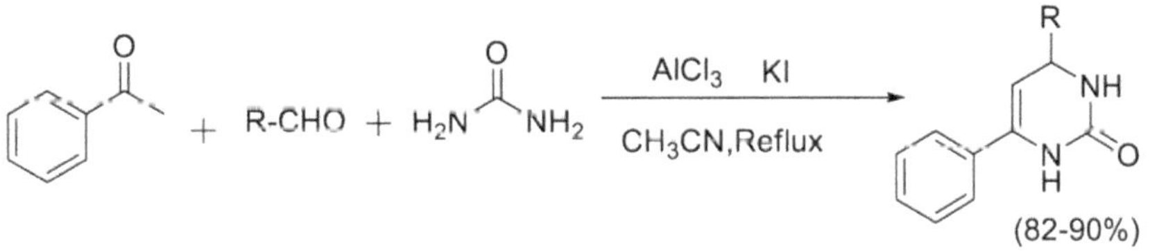

SCHEME 11.32

11.5 DIHYDROPYRIMIDINONES SYNTHESIS USING SOLID-SUPPORTED CATALYST

In view of the continuous demand for combinatorial methods of synthesis, there has been growing interest in the use of solid-supported reagents due to facile work-up by simple filtration, regeneration and recycling of the catalyst, etc. Application of this environmentally conscious protocol to the three-component reactions has been the subject of numerous recent reports. Convenient method for the synthesis of 3,4-dihydropyrimidin-2(1H)-ones has been described by using silica gel-supported sodium bisulfate as a heterogeneous catalyst. Yadav et al. have disclosed a green protocol for the Biginelli reaction (Scheme 11.33) using silver salt of hetero polyacid, that is, $Ag_3PW_{12}O_{40}$, in water. This catalyst was found to be superior in terms of conversion and reaction times when compared to other solid acid catalysts, such as KSF clay, acid resin (Amberlyst-15), and acid-washed silica gel (H_2SO_4-SiO_2). There is reusability of the catalyst in an eco-friendly solvent such as water and gives yield 82–90%.

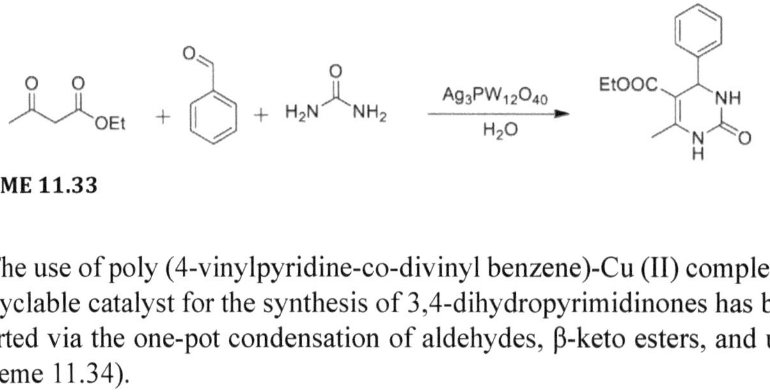

SCHEME 11.33

The use of poly (4-vinylpyridine-co-divinyl benzene)-Cu (II) complex as a recyclable catalyst for the synthesis of 3,4-dihydropyrimidinones has been reported via the one-pot condensation of aldehydes, β-keto esters, and urea (Scheme 11.34).

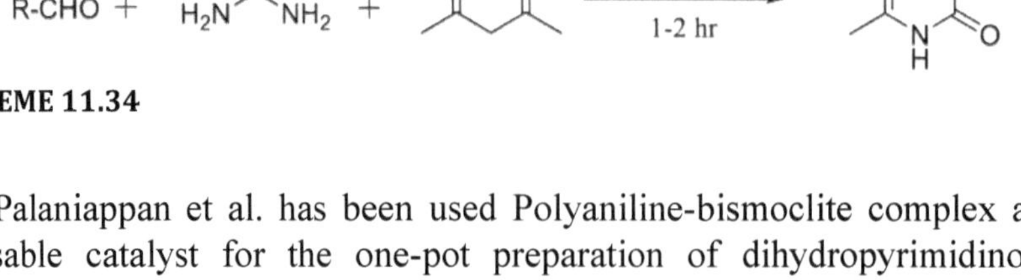

SCHEME 11.34

Palaniappan et al. has been used Polyaniline-bismoclite complex as a reusable catalyst for the one-pot preparation of dihydropyrimidinones

(Scheme 11.35). Polyaniline salts have also been reported as catalysts for Biginelli reactions.

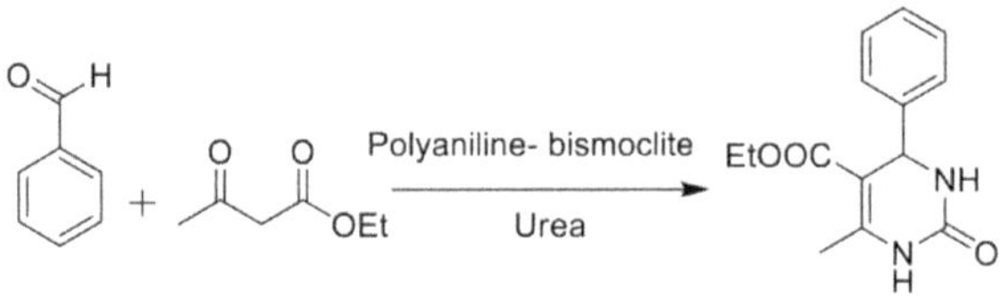

SCHEME 11.35

In recent years, synthesis of small molecular compound libraries using the soluble polymer complex has increasingly become an attractive field due to high reactivity and lack of diffusion phenomena. Soluble polymer like PEG is the most useful catalyst, as reported in 2002 (Scheme 11.36).

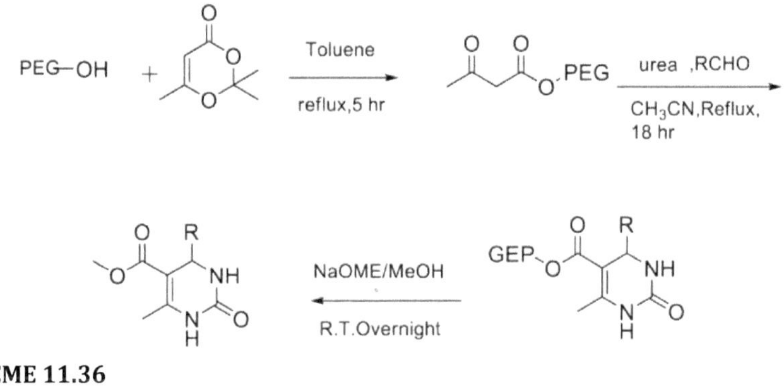

SCHEME 11.36

Bir Sain et al. reported the use of polyethylene glycol (PEG) as catalyst under solvent-free condition synthesis of dihydropyrimidinones (Scheme 11.37). They have also studied effect of solvent in presence of PEG catalyst but DHPM yield decreases due to dilution of PEG.

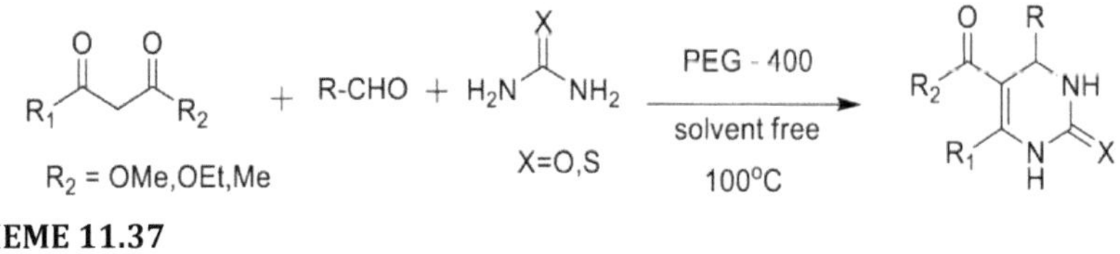

SCHEME 11.37

Gowravaram Sabitha, in 2005, explained the catalytic activity of ceria/vinyl pyridine polymer nanocomposite for the synthesis of 3,4-dihydropyrimidin-2 ones (Scheme 11.38). Catalyst prepared easily and added the advantage of recovery, reusability, and high activity. Reaction was carried out in water as solvent at 80°C.

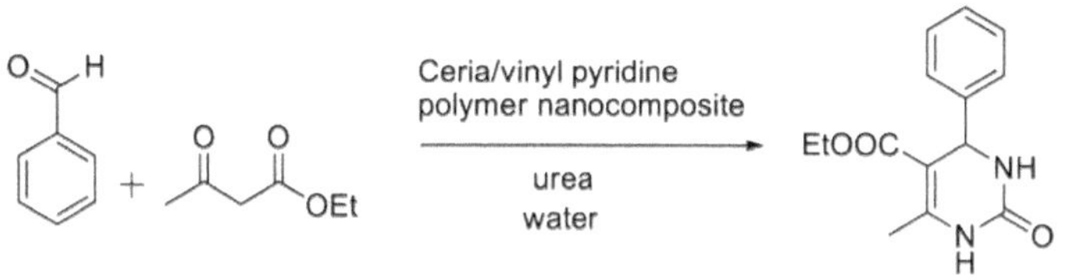

SCHEME 11.38

Heravi and coworkers have reported the use of 12-tungstophosphoric acid as a recyclable catalyst for the one-pot cyclocondensation of a 1,3-dicarbonyl compound, an aldehyde, and urea (or thiourea) to afford 3,4-dihydropyrimidin-2(1H)-ones. The catalyst is noncorrosive and reusable retaining the same activity as that of the freshly used catalyst even after five runs (Scheme 11.39).

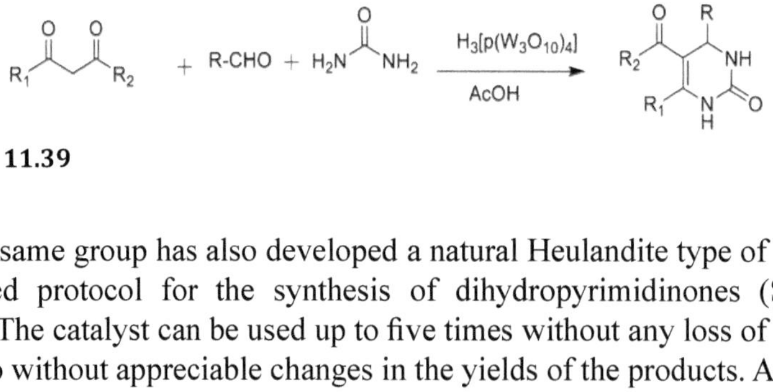

SCHEME 11.39

The same group has also developed a natural Heulandite type of zeolite-catalyzed protocol for the synthesis of dihydropyrimidinones (Scheme 11.40). The catalyst can be used up to five times without any loss of activity and also without appreciable changes in the yields of the products. Aliphatic aldehydes afforded products in lower yields of 44–46% with longer reaction times (12 h) compared to aromatic aldehydes, which gave yields of 60–87% in 4–5 h.

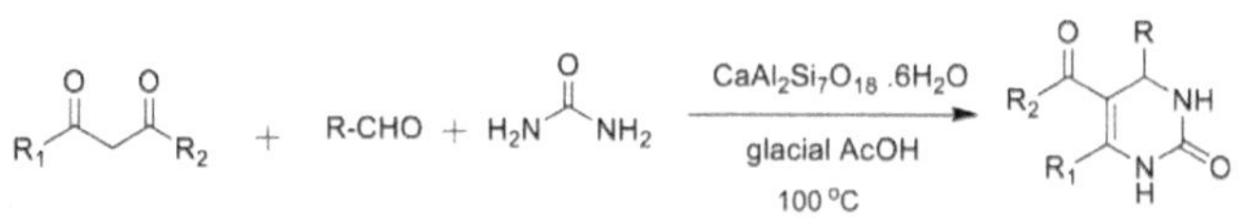

SCHEME 11.40

11.6 THE ASYMMETRIC BIGINELLI REACTION

DHPMs obtained through a Biginelli reaction contain a stereogenic center and therefore inherently asymmetric molecules are usually obtained as a racemic mixtures. The absolute configuration has a considerable influence on the biological activity. Two enantiomers may perform different or even opposite activities. Thus, the development of representative methods to approach enantiomerically enriched DHPMs is a task of primary importance initially and chemical resolutions and enzymatic were the methods of choice. For example, SQ 32926 (R) enentiomer I exhibits 400 timesfold more powerful antihypertensive activity than the other (S) isomer. (S) enantiomer of Monastrol II possess 15-fold more potent anticancer activity than (R) Monastrol.

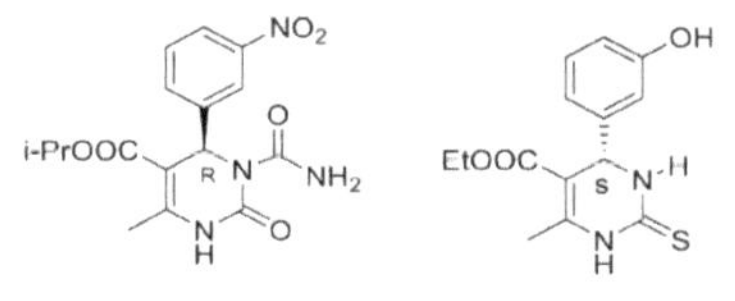

(R) SQ-32926 (S) Monastrol

I II

Biocatalytic strategy is used for the synthesis of enantiomer in pure forme (R) SQ 32926. The key step in the synthesis is the enzymatic resolution of N-3 acetoxymethyl activated dihydropyrimidinone precursor by *Thermomyces lanuginosus lipase.* Racemic DHPM was hydroxyl methylated of N-3 with formaldehyde, followed by standard acetylation with acetyl chloride. The resulting acetoxy methyl-activated DHPM was then cleaved enantioselectively by *Thermomyces lanuginosus lipase* with excellent selectivity and degradation of unreacted 3 hydroxy methylated (R) DHPM, which was converted to desired target (R) 32926 in one step N-3 carbomoylation with trichloro isocyanate (Stadler and Kappe, 2000) (Scheme 11.41).

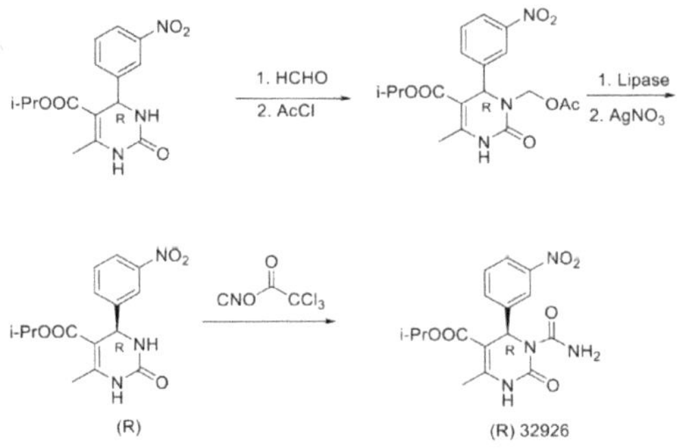

SCHEME 11.41

Synthesis of optically active DHPMs were also conducted through auxiliary-assisted asymmetric synthesis, using chiral starting materials such as C- glycosyl aldehydes (-) menthol-derived acetoacetate. A few years later, significant advances were made in the synthesis of optically active DHPMs with the use of catalytic amounts of chiral metal complexes.

In 2005, Zhu and coworkers described a highly enantioselective multi-component Biginelli reaction using a recyclable Yb triflate coordinated with a novel chiral hexadentate ligand bearing tertiary amine, phenol, and pyridine functional groups (Scheme 11.42). The products were obtained in high yield with enantiomeric excesses up to 99%.

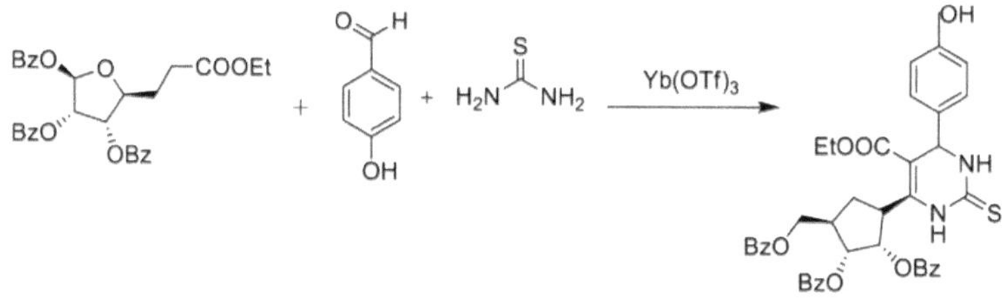

SCHEME 11.42

A highly efficient organic catalytic asymmetric approach was described by Gong et al. in 2006, using chiral phosphoric acids as catalysts. These results opened a window for the development of new optically active DHPMs synthesis. More recently, chiral organic catalysts such as cinchona alkaloids or substituted tetrazoles have also found interesting applications in asymmetric Biginelli transformations.

Recently, M. A. Blasco et al. reported biocatalytic highly enantioselective synthesis of (S)-Monastrol by using enzymatic resolution employing enzyme lipase from Candida antarctica B and lipase from Candida rugosa yielding the (R) enantiomer in 48% yield (66% ee) and (S) in 33% yield (97% ee) (Scheme 11.43).

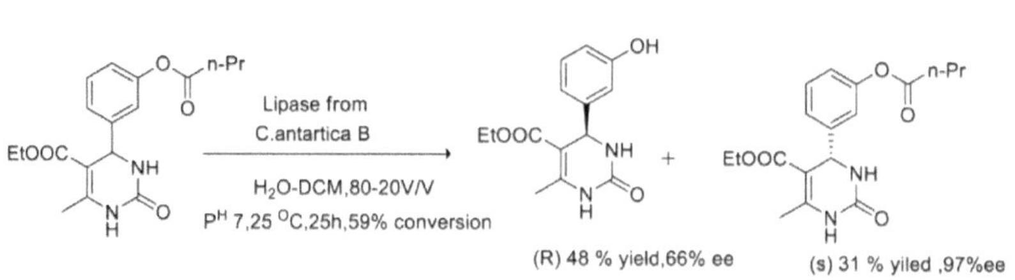

SCHEME 11.43

Dipak Prajapati et al. have given green chemistry approaches to the regioselective synthesis of spiro heterobicyclic rings using iodine as a new catalyst (Scheme 11.44). The reactions are carried out by heating the mixture of barbituaric acid with urea and benzaldehyde.

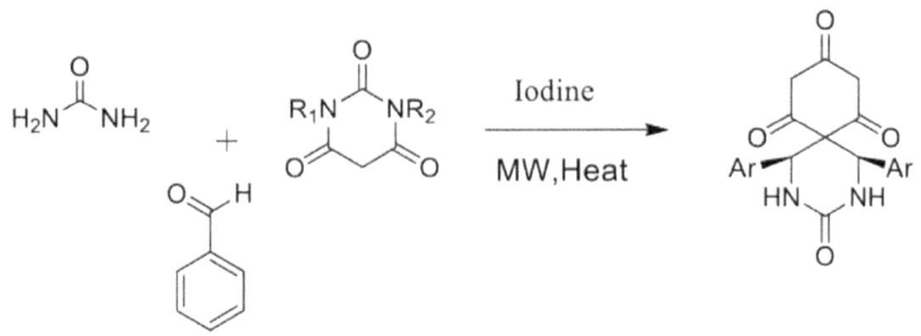

SCHEME 11.44

11.7 CATALYST USED IN THE BIGINELLI REACTIONS

2011:

I_2/MWI[23], sulphonated β-cyclo dextrin,[24] $Cu(OTF)_2$/MWI,[25] Sulphated tungstate,[26] Melamine tri-sulphonic acid,[27] $Yb(PFO)_3$,[28] HCl/ EtOH,[29] N,N' di-chloro-bis(2,4,6)trichloro phenylurea,(CC-2),[30] H_3PO_3/Pdcatalyst, *hv*[31]/ DMSO,[32] TFA/THF/MWI,[33] TMSCl/CAN,[34] HCl,[35] H_2SO_4 or TEBA,[36] $FeCl_3$/ Al-MCM,[37] silica immbolized Ni(II),[38] $Ca(OCl)_2$,[39] BPBPAT-TfOH,[40] L(+)-tartaricacid urea mixture,[41] AcOH,[42] $Mg(NO_3)_2$,[43] Chloroaceticacid,[44] PPA SiO_2, gypsum,[45] Nano $BF_3.SiO_2$,[46] Organo catalytic.[47]

2010:

HCl/MWI,[48] Me_3SiCl,[49] [Hmim]HSO_4NaNO_3,[50] NH_4VO_3/MWI,[51] $CeCl_3.7H_2O$,[52] Me_3SiCl/DMF,[53] PTSA.H_2O,[54] TMSCl/DMF,[55] $Sm(ClO_4)_3$,[56] AcOH,[57] TSIL,[58] VitaminB_1/EtOH,[59] HCl/EtOH,[60–61] BAIL,[62] Cl_3CCOOH,[63] 1-Carboxymethyl-3-methylimadazoliumhydrogensulphate,[64] Sulphamicacid ,[65] ultrasonic radiation,[66] I_2/MWI,neat condition,[67] PEG-400,[68] Copper methane sulphonate(CMS),[69] TEA,[70] diphenyl ammonium triflate,[71] Al-mesoporous silica,[72] aluminum planted mesoporous silica,[73] Cerium(IV) ammonium nitrate,[74] $NbCl_5$/Primary amine,[75] amine derived from quinine QNH_2,[76] Tartar-icacid,[77] $NiCl_2$/KI,[78] Gallium(III)iodied,[79] dioxane/aceticacid,[80] $NaHSO_4$,[81] (diacetoxyiodo)benzene,[82] TMSCl,[83] TiO_2,[84] HBF_4-SiO_2,[85] HCl/EtOH,[86] bronsted acidic ionic liquid, $MgSO_4.7H_2O$,[87] [Bmim]HSO_4,[88] 1,3-Dichloro-5,5-dimethylhydantoin,[89] $Y(OAc)_3$,[90] Nickel Nanoparticles,[91] Tributyl Borate,[92] $V(HSO_4)_3$,[93] Nafion-H,[94] $CuCl_2$,[95] Dodecylphosphonic acid,[96] P-amino benzene sulphonic acid,[97] thiamine hydrochloride,[98] ammonium carbonate,[99] PEG-embeded thiourea dioxide,[100] hexa aaqua aluminum(III) tetrafluroborate,[101] piperdiniumtriflate,[102] Borax,[103] $FeCl_3.6H_2O$,[104] $Cu(NO_3)_2.3H_2O$,[105] $Ce(SO_4)_2$.SiO2,[106] $Fe(NO_3)_2.9H_2O$,[107] $H_3PMO_{12}O_{40}$,[108] silica-supported tin chloride and titanium tetra chloride,[109] and Lewis acid.[110]

2009:

PS-PEG-SO_3H,[111] trifluro acetic acid,[112] Trifluro methane sulphonic acid,[113] PS-AFDPAT,[114] Yttria zirconia based lewis acid,[115] TMSCL,[116] TMSCL and $CO(OAc)_2.4H_2O$,[117] ytterbium chloride,[118] calcium fluoride,[119] mesoporous alumino silicate,[120] $Al_2O_3/MeSO_3$,[121] Acidic ionic liquid,[122] $Al(H_2PO_4)_3$,[123] Alumina sulphuric acid,[124] ZnO,[125] Copper nitrate,[126] Cellulose sulphuric acid,[127] lactic acid,[128] ionic liquids,[129] $NaHSO_4.H_2O$,[130] H_2SO_4 supported on silica gel or alumina,[131] $Al(HSO_4)_3$ and Al_2O_3-SO_3H,[132] NaH,[133] $SnCl_2$,[134] TBAB,[135] Mg/MeOH,[136] PTSA,[137] $Sc(OTf)_3$ or $la(OTf)_3$,[137] L-Proline-TFA,[138] Copper(II) sulphamate,[139] I_2/MWI,[140] UV irriadiation,[138] I_2,[139] amberlyst-70,[140] Etidronic acid,[141] Fe^{+3}, CO^{+2}, Zn^{+2}, Li^+ salt,[142] NBS silica sulphuric acid/ MWI,[143] 1,3di bromo 5,5dimethyl hydantoin(DBH),[145] PdO,[146] Indion130,[147] proline estersalts,[148] CBPA-thiourea,[149] Yb bis(perfluoro octanesulfonyl) imide,[150] I_2,[151] phosphinite ionic liquid,[152] Organocatalyst,[153] alumina-supported trifluoromethane sulfonic acid,[154] etidronic acid,[155] zinc tetrafluoroborate,[156] ammonium dihydrogen orthophosphate,[157] lipase,[158] Na_2SeO_4,[159] Bifunctional Organocatalyst,[160] $NiSO_4.7H_2O$,[161] $NaIO_4$,[162] Phosphoric Acids.[163]

11.8 RATE ENHANCEMENTS

In the past, sonication and microwave irradiations were at the forefront tools employed for time economy and rate enhancement of organic reactions.

11.8.1 SONICATION

Though sonication of reaction mixture proved quite fruitful and there is a large amount of literature available on this topic, a detailed discussion on this topic is beyond the scope of this account as it is on general organic chemistry. As far as Biginelli reaction is concerned, there are several reports[164–167] using this technique and along with suitable catalysts systems.

11.8.2 MICROWAVE IRRADIATIONS

Gedye[168] introduced the use of a domestic microwave oven in organic reaction in 1986. Thereafter, there has been very fast investigation of organic reactions employing this technique and the use of microwave in Biginelli reaction is reported by Gupta and coworkers.[169] As far as our survey is concerned, this seems to be the first-ever application in this reaction. Later on microwave is frequently used in this reaction and several protocols are reported employing various solvents/solvent-free and catalysts/catalysts-free condition.

11.9 CONCLUSION

Initially, there was slow activity in this reaction; later, it picked up to understand the mechanism, etc., in the past two decades or so after the disclosure that this structure is biologically significant. A number of publications give all types of catalysts, acidic, basic Lewis acids, ionic liquids, nano particles, and polymer supported reports on polymer metal composite. Optimization of the synthesis by using microwave and ultrasound under solvent-free conditions are also reported. Report on use of biocatalyst (i.e., yeast, enzymes) for the Biginelli reaction shows that evidently more work is needed in the use of biocatalyst in the reaction.

The real problem in this area was preparation of an optically pure Biginelli scaffold, which was achieved recently and further refinement are being actively pursued by several groups. Biological aspect of these moieties are being examined more intensively and several new activities are being observed; hopefully in near future, some molecules of this class may be in clinical use. Most of the researchers have stopped their work up to dihydropyrimidinone. Very few publications report dihydropyrimidinone reactions, that is, oxidation, coupling, halogenations, and acetylation. There is more scope for researchers to study variety of reactions with DHPM as a starting material.

ACKNOWLEDGMENTS

We sincerely acknowledge financial support from UGC, Nowrosjee Wadia College Pune, BCUD, Pune University and Indian Academy of Science Bangalore. Also, we thank National Chemical Laboratory, Pune, for performing analysis.

KEYWORDS

- **dihydropyrimidinone**
- **anti-inflammatory**
- **solid supported**
- **polyaniline**
- **antihypertensive**

REFERENCES

1. Hantzsch, A. *Ber* **1881,** *14*, 1637.
2. Knoevenagel, E. *Ber* **1894,** *27*, 2345. (b) Knoevenagel, E. *Justus Liebigs Ann. Chem.* **1894,** *281*, 25.
3. Biginelli, P. *Ber.* **1891,** *24*, 317.
4. Kenner, G. W.; Todd, A. Pyrimidine and Its Derivatives. In *Heterocyclic Compounds*; Elderfield, R. C. Ed.; John Wiley and Sons, Inc.: New York, 1957, p 239.
5. Hinkel, L. E.; Hey, D. H. *Recl. Trav. Chim.* **1929,** *48*, 1280.
6. (a) Biginelli, P. *Gazz. Chim. Ital.* **1893,** 23, 360. (b) Kappe, C. O. *Tetrahedron* **1993,** *49*, 6937. (c) Kappe, C. O. *Acc. Chem. Res.* **2000,** *33*, 879.
7. (a) Folkers, K.; Johnson, T. B. *J. Am. Chem. Soc.* **1933,** *55*, 3784. (b) Dallinger, D. Kappe, C. O. *Pure Appl. Chem.* **2005,** *77*, 155.
8. Sweet, F. J.; Fissekis, D. *J. Am. Chem. Soc.* **1973,** *95*, 8741.
9. O'Reilly, B. C.; Atwal, K. S. *Heterocycles* **1987,** *26*, 1185.
10. Atwal, K. S.; O'Reilly, B. C.; Gougoutas, J. Z.; Malley, M. F. *Heterocycles*, **1987,** *26*, 1189.
11. Atwal, K. S.; Rovnyak, G. C.; O'Reilly, B. C.; Schwartz, J. J. *Org. Chem.* **1989,** *54*, 5898.
12. Kappe, C. O. *J. Org. Chem.* **1997,** *62*, 7201.
13. Saloutina, V. I.; Burgarta, V. Ya.; Kuzuevaa, O. G.; Kappe, C. O.; Chupakhin, O. N. *J. Fluor. Chem.* **2000,** *103*, 17.
14. Cepanec, I.; Litvic, M.; Filipan-Litvic, M.; Grüngold, I. *Tetrahedron* **2007,** *63*, 11822.
15. Ma, J. G.; Zhang, J. M.; Jiang, H. H.; Ma, W. Y.; Zhou, J. H. *Chin. Chem. Lett.* **2008,** *19*, 375.

16. De Souza, R. O. M. A.; da Penha, E. T.; Milagre, H. M. S.; Garden, S. J.; Esteves, P. M.; Eberlin, M. N.; Antunes, O. A. C. *Chem. Eur. J.* **2009,** *15*, 9799.
17. Shen, Z-L.; Xu, X-P.; Ji, S-J. *J. Org. Chem.* **2010,** *75*, 1162.
18. Litvic, M.; Vecenaj, I.; Ladisic, Z. M.; Lovric, M.; Vinkovic, V.; Litvic, M. F. *Tetrahedron.* **2010,** *66*, 3463.
19. Yijun, H.; Fengyue, Y.; Chengjian, Z. *J. Am. Chem. Soc.* **2005,** *127*, 16386.
20. Debache, A.; Boumoud, B.; Amimour, M.; Belfaitah, A.; Rhouati, S.; Carboni, B. *Tetrahedron Lett.* **2006,** *47* (32), 5697.
21. Vijay, K.; Ganapaty, S.; Srinivas Rao, A. *Asian J. Chem.* **2010,** *22* (4), 2518.
22. Khodaei, M. M.; Khosropour, A. R.; Beygzadeh, M. *Syn. Comm.* **2004**, *34*, 1551.
23. Prajapati, D.; Bhuyan, D.; Gohain, M.; Hu, W. *Mol. Divers.* **2011,** *15*, 257.
24. Asghari, S.; Tajbakhsh, M.; Kenari, B. J.; Khaksar, S. *Chin. Chem. Lett.* **2011,** *22*, 127.
25. Pasunooti, K. K.; Chai, H.; Jensen, C. N.; Gorityala, B. K.; Wang, S.; Liu, X. W. *Tetrahedron Lett.* **2011,** 52, 80.
26. Salim, S. D.; Akamanchi, K. G. *Catal. Commun.* **2011,** *12*, 1153.
27. Shirini, F.; Zolfigol, M. A.; Albadi, J. *Chin. Chem., Lett.* **2011,** *22*, 318.
28. Wu, M.; Yu, J.; Zhao, W.; Wu, J.; Cao, S. *J. Fluor. Chem.* **2011,** *132*, 155.
29. Svetlik, J.; Kettmann, V. *Tetrahedron Lett.* **2011,** *52*, 1062.
30. Rao, G. B. D.; Acharya, B. N.; Verma, S. K.; Kaushik, M. P. *Tetrahedron Lett.* **2011,** *52*, 809.
31. Kumar, P. M.; Kumar, K. S.; Poreddy, S. R.; Mohakhud, P. K.; Mukkanti, K.; Pal, M. *Tetrahedron Lett.* **2011**, *52,* 1187.
32. Ren, Y.; Wang, X.; Wang, W.; Li, B.; Shi, Z.; Zhang, W. *Tetrahedron Lett.* **2011,** 52, 192.
33. Fang, Z.; Lam, Y. *Tetrahedron.* **2011,** *67*, 1294.
34. Wan, J. P.; Wang, C. Y. *Tetrahedron* **2011,** *67*, 922.
35. Cho, H.; Nishimura, Y.; Yasui, Y.; Kobayashi, S.; Yoshida, S.; Kwon, E.; Yamaguchi, M. *Tetrahedron* **2011,** *67*, 2661.
36. Oskooie, H. A.; Heravi, M. M.; Karimi, N. M.; Monjezy, H. *Synth. Commun.* **2011,** *41*, 826.
37. Zhang, L. Y.; Wang, C. H.; Zhou, Li.; Li, P. H.; Zhang, X. *Li. Sci. Chin. Chem.* **2011,** *54*, 74.
38. Tamaddon, F.; Razmi, Z. *Synth. Commun.* **2011,** *41*, 485.
39. Wang, Y.; Yu, J.; Miao, Z.; Chen, R. *Org. Biomol. Chem.* **2011,** *9*, 3050.
40. Gore, S.; Baskaran, S.; Koenig, B. *Green. Chem.* **2011,** *13*, 1009.
41. Svetlik, J.; Veizerova, L. *Helv. Chim. Acta.* **2011,** *94*, 199.
42. Liu, C.; Zhao, X.; Li, Y. J. *Het. Chem.* **2011,** *48*, 92.
43. Couto, I.; Tellitu, I.; Domínguez, E. *Arkivoc* **2011,** *2*, 115.
44. Dastmalbaf, M. N.; Davoodnia, A.; Heravi, M. M.; Hoseini, N. T.; Khojastehnezhad, A.; Zamani, H. A. *Bull. Korean Chem. Soc.* **2011,** *32*, 656.
45. Boumoud, T.; Boumoud, B.; Mosset, P.; Debache, A. *J. Chem.* **2011,** *8*, 312.
46. Mirjalili, B. F.; Bamoniri, A.; Akbari, A. J. *Iran Chem. Soc.* **2011,** *8*, 135.
47. Saha, S.; Moorthy, J. N. *J. Org. Chem.* **2011,** *76*, 396.
48. Gorobets, N. Y.; Sedash, Y. V.; Ostras, K. S.; Zaremba, O. V.; Shishkina, S. V.; Baumer, V. N.; Shishkin, O. V.; Kovalenko, S. M.; Desenko, S. M.; Eycken, E. V. Van der. *Tetrahedron Lett.* **2010,** *51*, 2095.
49. Ryabukhin, S. V.; Plaskon, A. S.; Bondarenko, S. S.; Ostapchuk, E. N.; Grygorenko, O. O.; Shishkin, O. V.; Tolmachev, A. A. *Tetrahedron Lett.* **2010,** *51*, 4229.

50. Garima, V. P.; Srivastava, L. D.; Yadav, S. *Tetrahedron Lett.* **2010**, *51*, 6436.
51. Niralwad, K. S.; Shingate, B. B.; Shingare, M. S. *Tetrahedron Lett.* **2010**, *51*, 3616.
52. Zych, A. J.; Wang, H. J.; Sakwa, S. A. *Tetrahedron Lett.* **2010**, *51*, 5103.
53. Li, W.; Liu, J. S.; He, P.; Ding, M. W. *Tetrahedron.* **2010**, *66*, 8151.
54. Shaabani, A.; Seyyedhamzeh, M.; Maleki, A.; Hajishaabanha, F. *Tetrahedron* **2010**, *66*, 4040.
55. Ryabukhin, S. V.; Plaskon, A. S.; Boron, S. Y.; Volochnyuk, D. M.; Tolmachev, A. A. *Mol. Divers.* **2011**, *15*, 189.
56. Liu, C. J.; Wang, J. D. *Molecules* **2010,** *15*, 2087.
57. Mousawi, S. M. A.; Apasery, M. A. E.; Elnagdi, M. H. *Molecules* **2010,** *15*, 58.
58. Fang, D.; Zhang, D. Z.; Liu, Z. L. *Monatsh. Chem.* **2010,** *141*, 419.
59. (a) Lei, M.; Ma. L.; Hu, L. *Monatsh. Chem.* **2010,** *141*, 1005. (b) Lebedyeva, I, O.; Povstyanoy, M. V.; Povstyanoy, V. M.; Panasyuk, O. G.; Guban, E. S. *Monatsh. Chem.* **2010,** *141*, 997.
60. Valverde, L. F.; Cedillo, F. D.; Luis, A. C. *Monatsh. Chem.*, **2010,** *141*, 75.
61. Srivastava, R. *Catal. Lett.*, **2010,** *139*, 17.
62. Kurbanova, M. M. *Russ. J. Org. Chem.* **2010,** *46*, 599.
63. Makaev, F.; Styngach, E.; Shargarovskii, V.; Bets, L.; Vlad, L. *Russ. J. Org. Chem.* **2010,** *46*, 610.
64. Luo, H. L.; Yang, W.; Li, Y.; Yin, S. F. *Chem. Natural Comp.*, **2010,** *46*, 412.
65. Prajapati, D.; Bhuyan, D.; Gohain, M.; Hu, W. *Mol. Divers.* **2011,** *15*, 257.
66. Abdolali, A.; Sadegh, R. *Synthesis* **2010,** *23,* 4057.
67. Wang, X. C.; Yang, G. J.; Quan, Z. J.; Ji, P. Y.; Liang, J. L.; Ren, R. G. *Synlett* **2010,** *11,* 1657.
68. Wang, M.; Song, Z.; Jiang, H.; Gong. H. *Prep. Biochem. Biotechnol.* **2010,** *40*, 101.
69. Abdelmadjid, D.; Wassima, G.; Raouf, B.; Ali, B.; Salah, R.; Bertrand, C. *Lett. Org. Chem.* **2010,** *7*, 272.
70. Li, J.; Sun, J.; Su, W. *Lett. Org. Chem.* **2010,** *7*, 314.
71. Murata, H.; Ishitani, H.; Iwamoto, M. *Org. Biomol. Chem.* **2010,** *8*, 1202.
72. Murata, H.; Ishitani, H.; Iwamoto, M. *Phys. Chem. Chem. Phys.* **2010,** *12*, 14452.
73. Cai, Y. F.; Yang, H. M.; Li, L.; Jiang, K. Z.; Lai, G. Q.; Jiang, J. X.; Xu, L. W. *Eur. J. Org. Chem.*, **2010,** *26*, 4986.
74. Ding, D.; Zhao, C. G. *Eur. J. Org. Chem.* **2010,** *20*, 3802.
75. Lu, N.; Chen, D.; Zhang, G.; Liu, Q. *Int. J. Quant. Chem.* **2010,** *111*, 2031.
76. Gupta, P.; Gupta, S.; Sachar, A.; Kour, D.; Singh, J.; Sharma, R. L. *J. Het. Chem.* **2010,** *47*, 324.
77. Ding, D.; Zhao, C. G. *Eur. J. Org. Chem.* **2010,** *20*, 3802.
78. Li, D.; Mao, H.; An, L.; Zhao, Z.; Zou, J. *Chin. J. Chem.* **2010,** *28*, 2025.
79. Rahman, M.; Majee, A.; Hajra, A. *J. Het. Chem.* **2010,** *47*, 1230.
80. Darwish, E. S.; Abdelhamid, I. A.; Nasra, M. A.; Gallil, F. M. A.; Fleita, D. H. *Helv. Chim. Acta.* **2010,** *93*, 1204.
81. Cheng, Q.; Wang, Q.; Xu, X.; Ruan, M.; Yao, H.; Yang, X. J. *Het. Chem.* **2010,** *47*, 624.
82. Karade, N. N.; Gampawar, S. V.; Tale, N. P.; Kedar, S. B. *J. Het. Chem.* **2010,** *47*, 740.
83. (a) Azizian, J.; Mohammadi, M. K.; Firuzi, O.; Mirza, B.; Miri, R. *Chem. Biol. Drug Design.* **2010,** *75*, 375.100. (b) Kassaee, M. Z.; Masrouri, H.; Movahedi, F.; Mohammadi, R. *Helv. Chim. Acta.* **2010,** *93*, 261.
84. Kamble, V. T.; Muley, D. B.; Atkore, S. T.; Dakore, S. D. *Chin. J. Chem.* **2010,** *28*, 388.

85. Nagawade, R. R.; Shinde, D. B. J. *Het. Chem.* **2010,** *47*, 33.
86. Mukhopadhyay, C.; Datta, A. J. *Het. Chem.* **2010,** *47*, 136.
87. Singh, V.; Kaur, S.; Ratti, R.; Kad, G. L.; Singh, J. *Ind. J. Chem.* **2010,** *49B*, 611.
88. Hojati, S. F.; Gholizadeh, M.; Haghdoust, M.; Shafiezadeh, F. *Korean Chem. Soc.* **2010,** *31*, 3238.
89. Aridoss, G.; Jeong, Y. T. Bull. *Korean Chem. Soc.* **2010,** *31*, 863.
90. Sapkal, S. B.; Shelke, K. F.; Shingate, B. B.; Shingare, M. S. *Bull. Korean Chem. Soc.* **2010,** *31*, 351.
91. Ranjith, C.; Srinivasan, G. V.; Vijayan, K. K. *Bull. Chem. Soc.* **2010,** *83*, 288.
92. Shirini, F.; Yahyazadeh, A.; Abedini, M.; Langroodi, D. I. *Bull. Korean Chem. Soc.* **2010,** *31*, 1715.
93. Wang, Q.; Pei, W. J. *Iran. Chem. Soc.* **2010,** *7*, 318.
94. Aghayan, M. M.; Moradi, A.; Bolourtchian, M. J. *Iran. Chem. Soc.* **2010,** *7*, 269.
95. Ghassamipour, S.; Sardarian, A. R. J. *Iran. Chem. Soc.* **2010,** *7*, 237.
96. Wu, M. S.; He, P.; Zhang, X. Z. S. *Afr. J. Chem.* **2010,** *63*, 224.
97. Tamaddon, F.; Razmi, Z.; Jafari, A. A. *Tetrahedron Lett.* **2010,** *51*, 1187.
98. Verma, S.; Jain, S. L.; Sain, B. *Tetrahedron Lett.* **2010,** *51*, 6897.
99. Mandhane, P. G.; Joshi, R. S.; Nagargoje, D. R.; Gill, C. H. *Tetrahedron Lett.* **2010,** *51*, 3138.
100. Ramalingan, C.; Park, S. J.; Lee, I. S.; Kwak, Y. W. *Tetrahedron.* **2010,** *66*, 2987.
101. Murata, H.; Ishitani, H.; Iwamoto, M. *Phys. Chem.* **2010,** *12*, 14452.
102. Nath, J.; Chaudhuri, M. K. *Synth. Commun.* **2010,** *40*, 2976.
103. Majd, M. M.; Saidi, K.; Khabazzadeh, H. *Phosphorus Sulfur Silicon Relat. Elem.* **2010,** *185*, 325.
104. Wang, D. C.; Guo, H. M.; Qu, G. R. *Synth. Commun.* **2010,** *40*, 1115.
105. Pei, W.; Wang, Q. *Synth. Commun.* **2010,** *40*, 1209.
106. Phukan, M.; Kalita, M. K.; Borah, R. *Green. Chem. Lett. Rev.* **2010,** *3*, 329.
107. Murata, H.; Ishitani, H.; Iwamoto, M. *Org. Biomol. Chem.* **2010,** *8*, 1202.
108. Pourghobadi, Z.; Derikvand, F. *Chin. Chem. Lett.* **2010,** *21*, 269.
109. Niknam, K.; Hasaninejad, A.; Arman, M. *Chin. Chem. Lett.* **2010,** *21*, 399.
110. Wang, M.; Song, Z.; Jiang, H.; Gong, H. *Prep. Biochem. Biotechnol.* **2010,** *40*, 101.
111. Quan, Z. J.; Da, Y. X.; Zhang, Z.; Wang, X. C. *Catal. Commun.* **2009,** *10*, 1146.
112. Shobha, D.; Chari, M. A.; Ahn, K. H. *Chin. Chem. Lett.* **2009,** *20*, 1059.
113. Pani, M. S.; Arjun, M.; Sridhar, D.; Srinivas, K. T. *Chin. Chem. Lett.* **2009,** *20*, 909.
114. Lei, M.; Wu, D-D.; Wei, H-G.; Wang, Y-G. *Synth. Commun.* **2009,** *39*, 475.
115. Ramalingam, S.; Kumar, P. *Synth. Commun.* **2009,** *39*, 1299.
116. Guo, S-R.; Yuan, Y-Q. *Synth. Commun.* **2009,** *39*, 2205.
117. Kefayati, H.; Fakhriyannejad, M.; Mohammadi, A. A. *Phosphorus Sulfur Silicon Relat. Elem.* **2009,** *184*, 1796.
118. Zhang, H.; Zhou, Z.; Yao, Z.; Xu, F.; Shen, Q. *Tetrahedron Lett.* **2009,** *50*, 1622.
119. Chitra, S.; Pandiarajan, K *Tetrahedron Lett.* **2009,** *50*, 2222.
120. Shobha, D.; Chari, M. A.; Mano, A.; Selvan, S. T.; Mukkanti, K.; Vinu, A. *Tetrahedron* **2009,** *65*, 10608.
121. Sharghi, H.; Jokar, M. *Synth. Commun.* **2009,** *39*, 958.
122. Gui, J.; Liu, D.; Wang, C.; Lu, F.; Lian, J.; Jiang, H.; Sun, Z. *Synth. Commun.* **2009,** *39*, 3436.

123. Shaterian, H. R.; Hosseinian, A.; Ghashang, M. *Phosphorus Sulfur Silicon Relat. Elem.* **2009,** *184*, 126.
124. Shaterian, H. R.; Hosseinian, A.; Ghashang, M. *Phosphorus Sulfur Silicon Relat. Elem.* **2009,** *184*, 197.
125. Bahrami, K.; Khodaei, M. M.; Farrokhi, A. *Synth. Commun.* **2009,** *39*, 1801.
126. Wang, M.; Jiang, H.; Song, Z.; Gong, H. *Prep. Biochem. Biotechnol.* **2009,** *39*, 372.
127. Saini, S. A.; Kumar, D.; Sandhu, J. S. *Green. Chem. Lett. Rev*. **2009,** *2*, 29.
128. Chavan, S. S.; Sharma, Y. O.; Degani, M. S. *Green. Chem. Lett. Rev.* **2009,** *2*, 175.
129. Zeynizadeh, B.; Dilmaghan, K. A.; Yari, M. *Phosphorus Sulfur Silicon Relat. Elem.* **2009,** *184*, 2465.
130. Dilmaghan, K. A.; Zeynizadeh, B.; Yari, M. *Phosphorus Sulfur Silicon Relat. Elem.* **2009,** *184*, 1722.
131. Shaterian, H. R.; Hosseinian, A.; Ghashang, M.; Khorami, F.; Karimpoor, N. *Phosphorus Sulfur Silicon Relat. Elem.* **2009,** *184*, 2333.
132. Fustero, S.; Catalan, S.; Acena, J. L.; Pozo, C. D. J. *Fluor. Chem.* **2009,** *130*, 1145.
133. Singh, O. M.; Devi, N. S. J. *Org. Chem.* **2009,** *74*, 3141.
134. Ahmed, B.; Khan, R. A.; Habibullah Keshari, M. *Tetrahedron Lett.* **2009**, *50*, 2889.
135. Singh, K.; Singh, K. *Tetrahedron Lett.* **2009,** *50*, 2219.
136. Matache, M.; Dobrota, C.; Bogdan, N. D.; Dumitru, I.; Ruta, L. L.; Paraschivescu, C. C.; Farcasanu, I. C.; Baciu, I.; Funeriu, D. P. *Tetrahedron* **2009,** *65*, 5949.
137. Pandey, J.; Anand, N.; Tripathi, R. P. *Tetrahedron* **2009,** *65*, 9350.
138. Liu, C. J.; Wang, J. D. *Molecules*. **2009,** *14*, 763.
139. Pan, N.; Zhang, W.; Liu, Q. *Molbank* **2009,** *4*, 645.
140. Memarian, H. R.; Farhadi, A. *Monatsh. Chem.* **2009,** *140*, 1217.
141. Ren, Y. M.; Cai, C. *Monatsh. Chem.* **2009,** *140*, 49.
142. Chandak, H. S.; Lad, N. P.; Upare, P. P. *Catal. Lett.* **2009,** *131*, 469.
143. Savant, M. M.; Pansuriya, A. M.; Bhuva, C. V.; Kapuriya, N. P.; Naliapara, Y. T. *Catal. Lett.* **2009,** *132*, 281.
144. Sedova, V. F.; Krivopalov, V. P.; Shkurko, O. P. *Russ. J. Org. Chem.* **2009,** *45*, 1535.
145. Khrustalev, D. P. *Russ. J. Gen. Chem.* **2009,** *79*, 164.
146. Alibeik, M. A.; Zaghaghi, Z. *Chem. Papers.* **2009,** *63*, 97.
147. Horacio, C.; Alexandre, B. D.; Romain, N.; Frank, K.; Bernard, R.; Roman, L. *Synlett* **2009,** *11*, 1737.
148. Sik, K. S.; Seung, C. B.; Hoon, L. J.; Kon, L. K.; Hee, L. T.; Ho, K. Y.; Hyunik, S. *Synlett* **2009,** *168,* 599.
149. Jing, X.; Li, Z.; Pan, X.; Wang, Q.; Yan, C.; Zhu, H. *Synth. Commun.* **2009,** *39*, 3796.
150. Pansuriya, A. M.; Savant, M. M.; Bhuva, C. V.; Kapuriya, N.; Singh, J.; Naliapa, Y. T. *Lett. Org. Chem.* **2009,** *6*, 619.
151. Wan, J-P.; Pan, Y-J. *Chem. Commun.* **2009,** *19,* 2768.
152. Sohn, J. H.; Choi, H. M.; Lee, S.; Joung, S.; Lee, H. Y. *Eur. J. Org. Chem.* **2009,** *23*, 3858.
153. Wang, Y.; Yang, H.; Yu, J.; Miao, Z.; Chen, R. *Adv. Synth. Catal.* **2009,** *351*, 3057.
154. Hong, M.; Cai, C. *J. Het. Chem.* **2009,** *46*, 1430.
155. Valizadeh, H.; Shockravi, A. *Heteroatom. Chem.* **2009,** *20*, 284.
156. Wu, Y. Y.; Chai, Z.; Liu, X. Y.; Zhao, G.; Wang, S. W. *Eur. J. Org. Chem.* **2009,** *6*, 904.
157. Arjun, M.; Sridhar, D.; Chari, M. A.; Sarangapani, M. J. *Het. Chem.* **2009,** *46*, 119.

158. Pansuriya, A. M.; Savant, M. M.; Bhuva, C. V.; Singh, J.; Naliapara, Y. T. Arkivoc **2009,** *vii*, 79.
159. Kundu, S. K.; Majee, A.; Hajra, A. *Indian J. Chem.* **2009**, *48B*, 408.
160. Mase, N.; Nakamura, D.; Kawano, Y.; Suzuki, Y.; Takabe, K. *Heterocycles* **2009,** *78*, 3023.
161. Hekmatshoar, R.; Heidari, M.; Heravi, M. M.; Baghernejad, B. *Bull. Chem. Soc. Ethiop.* **2009,** *23*, 141.
162. Jing, X.; Li, Z.; Pan, X.; Shi, Y.; Yan, C. J. *Iran. Chem. Soc.* **2009,** *6*, 514.
163. Li, N.; Chen, X. H.; Song, J.; Luo, S. W.; Fan, W.; Gong, L. Z. *J. Am. Chem. Soc.* **2009,** *131*, 15301.
164. Kumar, H.; Parmar, A. *Ultrason. Sonochem.* **2008,** *15*, 129.
165. Sakhno, Y. I.; Desenko, S. M.; Shishkina, S. V.; Shishkin, O. V.; Sysoyev, D. O.; Groth, U.; Kappe, C. O.; Chebanov, V. A. *Tetrahedron* **2008,** *64*, 1104.
166. (a) Li, J. T.; Han, J. F.; Yang, J. H.; Li, T. S. *Ultrason. Sonochem.* **2003**, *10* (119), 527. (b) Memarian, H. R.; Farhadi, A. *Ultrason. Sonochem.* **2008,** *15*, 1015.
167. Memarian, H. R.; Soleymani, M. *Ultrason. Sonochem.* **2011**, *18*, 745.
168. Gedye, R.; Smith, F.; Westaway, K.; Ali, H.; Baldisera, L.; Laberge, L.; Rousell, J. *Tetrahedron Lett.* **1986,** *27*, 279.
169. Gupta, R.; Gupta, A. K.; Paul, S.; Kachroo, P. L. *Indian J. Chem.* **1995,** *34B*, 151.

Index

A

Abutilon indicum and *Prosopis juliflora*
material and method
Parthenium hysterophorus, collection, germination and preservation, 18–19
plant species, collection and extract preparation, 13–14
Xanthomonas axonopodis pv. punicae, collection, isolation, and identification, 14–18
observation
morphological characters, *Xanthomonas axonopodis* pv. punicae, 19
Parthenium hysterophorus, effects of plant extracts on seedling, 20–21
Xanthomonas axonopodis pv. punicae, growth rate at 28°C and 37°C, 20
Parthenium hysterophorus L., 13
parthenium weed, 13
pomegranate (*Punica granatum* L.), 12
bacterial blight, management of, 12
stems, 12
Xanthomonas axonopodis, 12
Abutilon indicum and *Prosopis juliflora*
results
acid production, 17
dextrose, 17
germination indices, calculation of, 21
maltose, 17
Parthenium hysterophorus, graphical representation of germination indices of, 24
Parthenium hysterophorus, reduction in germination indices, 23
Parthenium seeds, phytotoxic effects, 21
plant extracts, graphical representation, 23
plant extracts showing zone of inhibition, antibacterial activity, 22
pomegranate fruits and leaves, bacterial blight symptoms, 15
results
minimum inhibitory and bactericidal concentration, 22
Parthenium hysterophorus, germination indices of, 23–24
well diffusion assay, 22
sucrose, 17
tropical medicinal plants, database, 14
Xanthomonas axonopodis pv. punicae
MIC and MBC of plant extracts, 23
Xanthomonas axonopodis pv. punicae
isolate, growth rate of, 20
Xanthomonas axonopodis pv.Punicae
isolate on nutrient agar media, cultural and morphological characteristics of, 19
Xanthomonas on NA medium, pale yellow, glistening colonies of, 16
Activating protein-1 (AP-1), 85

B

B-cell lymphoma (BCL)6 protein, 102–103
Biologically active heterocyclic 1, 2, 3, 4
tetrahydropyrimidine-2, synthesis
asymmetric Biginelli reaction
biocatalytic strategy, 261
DPHM, 261
green chemistry approaches, 263
optically active DHPMs, 262
organic catalytic asymmetric approach, 262
Thermomyces lanuginosus lipase, 261
Biginelli condensation, 248
aldehydes N-substituted, 250
2-amino-5,6 dihydro pyrimidine-4(3H)-ones, 255
β keto ester immobilized reagent, 249
Boumoud and co-workers, 251
Capanee reported DHPM synthesis, 250
C-glycosylated dihydropyrimidinone, 254

Debache and co-workers, demonstrated synthesis of DHPM, 250–251
dihydropyrimidinone derivatives, 255
3,4-dihydropyrimidinones, 252
dihydropyrimidinones (DHPMs), 252, 254, 257
ecofriendly method of, 256
enantioselective one-pot synthesis, 249
equimolar solution, 255
ionic liquid (ILs) as catalyst, 253
Lewis acid, use, 256
one-pot condensation, 257
Open Catalysis Journal, in 2010, 255
Rajendra P. Pawar reported in *International Journal of Industrial Chemistry,* 251
reaction rate and yield enhancements, 253
silica chloride, 254
silica triflate, 254–255
5-unsubstituted 3,4-dihydropyrimidin-2(1H)- ones, 257
Yttruim (III) nitrate hexahydrate, 250
catalyst used in Biginelli reactions
2009, 264
2010, 263–264
2011, 263
discovery
acid catatalyzed dihydropyrimidinones synthesis, 241
Biginelli reaction, 241–242
compounds, 241
DHPM-based calcium channel modulators, structures, 242
mechanistic studies
Atwal and O'Reilly (1987), 245
Cepanec, Lewis acid catalyzed reaction, 246
De Souza (2009), 246
density functional theory (DFT), 246
3,4 dihydropyrimidinone via iminium ion intermediate, formation, 246
Folker and Johnson (1933), 243
intermediates proposed by, 244
Jian-Hua Zhou (DFT STUDY—2008), 246
Litvic 2010 (Brønsted acid), 247
O. Kappe (1997), 245
Shun-Jun Ji 2010 (base catalyzed), 247
Sweet and Fissekis (1973), 244
rate enhancements
microwave irradiations, 265
sonication, 264–265
solid supported catalyst, dihydropyrimidinones synthesis
acid catalysts, 258
convenient method, 258
3,4-dihydropyrimidin-2 ones, synthesis of, 260
dihydropyrimidinones, 260
poly (4-vinylpyridine-co-divinyl benzene)-Cu (II) complex, 258
polyaniline salts, 259
polyaniline-bismoclite complex, 258
polyethylene glycol (PEG), 259
soluble polymer, 259
12-tungstophosphoric acid, 260

C

Cancerous cells, 209
Cerebrospinal fluid (CSF) flow, 157–158
Chalcones
antigout agents
Angelica keiskei, 225
Caesalpinia sappan, 224
natural and semisynthetic, 225
Perilla frutescens leaves, 224
trans-chalcone, 225
Claisen-Schmidt reaction, 224
compounds with notable antigout activity, 227
low molecular weight ligands, 223
Closed-loop systems, 2

D

Density functional theory (DFT), 246
3,4-dihydropyrimidin-2(1H)-ones (DHPMs), 240
Direct electrochemical oxidation
capillary electrolysis in electrolyte flow
use of bipolar polarization for, 236–237
extended wire electrode, polarography with, 232
device description, 233
potentiostat operates, 234

galvanic coatings, device for quantitative measurement of porosity, 234
device description, 235
device design, 235
quantitative measurement, 236
Disease mechanism
advancing biology via understanding metabolism, 71
attrition causes, 53
butin cyto-protective pathway *versus* OS-induced mitochondrial-involved apoptosis, 62
chemical analysis-integrated platform, 50
computational method
automorphism orbit, 52
connectivity–protein function, 52
degree of node, 52
extensive chemical repositories, 53
graphlet degree vector (GDV), 52
network topology–biofunction relation, 52
network types, 52
protein–protein interaction network (PIN), 52
smart/dumb thieves, 53
systems pharmacology, 52
CYP families in humans, 63
Ebola virus disease, 52
frequent radicals and effects, 55
genotype and phenotype
union of, 70
information flow, 70
liquid chromatography, coupled with mass spectrometry (LC/MS), 50
metabolomics, 50
multiple reaction monitoring (MRM), 51
oxidative stress
reactive oxygen and nitrogen species in, 56
results and discussion
autoantibody-PSA, 67–69
diagnostic dilemma in prostate cancer CYP2B6, receptors and genes, 61–66
diseases and mitochondria, 54–61
metabolism platform technology, 70–71
metals, 54–61
microfluidic-based multiplex QRT-PCR, [–2] PRO-PSA, 67–69
O_2 toxicity, 54–61
optical manipulation and sensing, antennas and application to, 70
oxidative stress, 54–61
prostate-specific antigen (PSA) test, 54
prostatic hyperplasia (BPH), 54
proteins and MRNAS in cell, spatial and temporal organization of, 66–67
radicals, 54–61
surface-enhanced laser desorption/ ionization (SELDI)-MS, 54
trichomonas vaginalis, 70
ROS, reactive oxygen species
oxidative stress, epigenetics and ageing, 55
system biology predictive *in silico* approach
modeling/simulation via use of, 53
two independent pathways, 57

E

Ebola virus
disease, 52
Eugenol, 31
source of
antibacterial activity of, 34–36
anti-cancer activity, 40
anti-fungal activity, 37
anti-inflammatory potential, 38–39
antioxidant activity, 36
anti-parasitic activity, 37–38
anti-viral activity, 38
pharmacological activities, 33–34
physical and chemical properties, 32–33
plants, isolation and identification of, 32–33
testified, studies, 32

F

F-box only protein 11 (FBXO11), 102–103

G

Gout, 221
drugs, 222
Green solvents, 2
biosolvents
application, 4–5

fluorous solvents, 5
perfluorinated liquids, 6
stoichiometric and catalytic transformations, 6
ionic liquids
anions, 3
cations, 3
organic carbonates
cyclic organic carbonates, 5
open chain, 5
Pd-catalyzed allylic substitution, 5
propylene carbonate (PC), 5
plants
eco-friendly solvents, 8
polymers, 7–8
vegetable oils, 7
supercritical carbon dioxide ($scCO_2$), 6–7
water
non-polarity, 3

H

Hepatocyte model systems
diverse applications, 210
Human leukocyte antigen (HLA), 123

I

Intracellular proteolysis and neoplasia
activating protein-1 (AP-1), 85
chromosome instability and cancer
activation of DNA damage checkpoint, 92–93
ATM/ATR substrate analyses, 93–94
deficient lung cells, proteomics reveals apoptosis-associated proteins in, 94–95
DNA damage response, 91
DNA damage via yeast arrays, protein phosphorylation in response, 95
etiology of ataxia telangiectasia, neuro-astro-vascular unit in, 98
genomic instability and, 89–90
mitotic chromosomal instability and cancer, 98
rapamycin signalling pathway, mammalian target of, 96
replication regulating endonuclease, WEE1 controls genomic stability in, 97
syndromes, 98
synthetic lethal properties, cell-based screen ATR inhibitor with, 95–96
in vivo formation/repair of DNA double-strand breaks after tomography, 93
computational method, 88
drug design, 89
quantitative structure-activity relationship (QSAR), 89
rat sarcoma (Ras) protein, 86
relationship to cancer
ageing/cancer, 106–107
anaplastic large-cell lymphoma (ALCL), 105
CDK2-dependent FOXO1 phosphorylation, balancing life/death decisions, 99–100
F-box protein FBXO11 targets BCL6, 102–103
genome-wide association study (GWAS), 106
GSK3/SCF^{FBXW7}, 105
jun-B, 103–104
long non-coding RNAS and P53 regulation, 107
manipulating degradation, 108
PI3K-related kinases (PIKKs), 103
PKC-dependent ubiquitylation promotes suppressor FBW7A inactivation, 101–102
prostatic cell cytotoxicity caused, 108
proteasome degrades C-FOS proto-oncoprotein independently, 100–101
proto-oncoprotein c-Fos, 100–101
superoxide dismutase (SOD), 107–108
supress cell proliferation, 106
ubiquitin ligase E3, 104
warsaw breakage syndrome, 100
trichomonacidal drugs, 87

L

Liquid chromatography, coupled with mass spectrometry (LC/MS), 50
Liver tumors, TGF-α death

cancerous cells, 209
hepatocyte model systems
diverse applications, 210
method
label-free detection, 211
real-time (RT)-cell electronic sensing (CES™) system, 211
RTCA, 212–213
normal hepatocite, 214
real-time cell electronic sensing, 209
results
autophagy, 216–217
and carcinogenesis, 216–217
cell response to mitogenic/motogenic stimuli, real-time analysis, 215–216
hepatic fibrosis, 216–217
MEK/ERK pathway over activation in, 214–215
NADPH oxidase NOX4 by TGF-β for pro-apoptotic activity, 213
RTCA, 212–213
studies, 211

M

Monosodium urate (MSU) crystals, 221–222
Multiple reaction monitoring (MRM), 50, 51

N

Neurospheres, 172

P

Parthenium hysterophorus
collection, germination and preservation of
agar, seeding, 18
germination agar, preparation of, 18
seed, treatment with extracts, 19
Pigment epithelium-derived factor (PEDF), 168
Polymerase chain reaction (PCR), 123
Prostate cancer
stem cells and their connection with
acute myeloid leukemia (AML), 154
adult brain, net of tangential pathways for neuronal migration in, 155–156
adult brain, niche interactions in, 173
adult human hippocampus, neurogenesis in, 157
basal radial glia in subventricular zone, 156–157
bioethical considerations, 192
biology and therapy, 171
BMP2 transcriptional repression by P21WAF1/CIP1 links quiescence, 171–172
brain tumors, 194–195
breast cancer cell lines, 181–182
chromosomal instability abrogate neural stem cells proliferation, 167–168
concepts renew cancer research, 179
coordinated dynamical process, 185
and decline during infancy, corridors of, 158
dissecting cancer stem cell theories, 183
DLK1-imprinting postnatal loss in, 170–171
elimination, 179–180
enhancer of zeste homolog 2 (EZH2), role of, 188–189
epigenetic (EPI) phenomenon, 184
Fiona Watt, fun, 183
flow of cerebrospinal fluid in, 157–158
and folding, 156–157
glial precursors clear sensory neuron corpses, 169
human hippocampus, 159
infant human frontal lobe, young neurons extensive migration of, 158–159
kinase-DYRK1A critical regulated segregation, 170
lymph node cancer of prostate (LNCAP) cells, 188–189
mechanisms of cerebral cortex expansion, 156–157
Meselson–Stahl experiment, 154–155
method, 153
migrating neurons in brain, corridors of, 158
mother and daughter cells, stories of, 193
mouse embryonic stem cells, 191–192
nervous system, pattern formation in, 160–167
neural stem cells (NSCs), 157, 174
neuroprotection, 181
neuroprotective, 178

neurorestorative strategies, 178
new hope for therapy, 187
niche signal for neural stem cell renewal, pigment epithelium-derived factor, 168
pattern-recognition receptors (PRRs), 182–183
pluripotent murine embryonic stem cells, membrane proteins expressed in, 179
pluripotent stem cells, 189–191
pluripotentiality/differentiation, gene expression profile of, 185–186
potential reprogramming via MI-2/NuRD modulation, 187
prometheus and regenerative medicine, myth of, 169
quantum jump for regenerative medicine, 192
recent trends in, 186
remote signaling and, 175
(RE)programming director, 184
resistance to conventional cancer treatment, 188
Rett's syndrome, 193–194
SC-niche biology in brain, 175–177
signals of success, 178
stem cell-niche biology in, 175
subependymal zone biopotential progenitors, P27KIP1 regulates population, 172
subgranular zone (SGZ), 159
telomere shortening, 167–168
therapy, implications of, 186
toll-like receptors (TLRs), 182–183
transcriptional activity, discovery of small-molecule inhibitors of, 193
traveling cells, 180–181
variation because of differentiation, 186–187
vascular senescent cells, 177–178
Prostate-specific antigen (PSA) test, 54

R

Rat sarcoma (Ras) protein, 86
Real-time cell electronic sensing, 209

T

Trichomonacidal drugs, 87

U

U.S. Pollution Prevention Act in 1990, 1

V

Vit-D receptors (VDRs)
analogues and prostate cancer preventon/therapy
androgen receptor (AR), 136
androgen-independent prostate cancer, development of, 137
1α,25-dihydroxyvitamin D_3 receptors and actions in protease cancer, 131–132
early recurrent prostate cancer, treatment of, 134–135
expression regulation, inhibition of prostate cancer growth, 138–139
fulvestrant down regulates androgen receptor expression, 139–140
hematopoiesis/immune system and clinical applications, 142
human prostate cancer cell lines, 132–133
$1,25(OH)_2D_3$ inhibits proliferation, 135–136
LNCAP prostate cancer, IGFBP-3 mediates $1,25(OH)_2D_3$ growth inhibition in, 137–138
metabolism and action in, 141–142
and naproxen in recurrent prostate cancer, 140
phase-II trial of $1,25(OH)_2D_3$, 140
prostaglandin synthesis/actions inhibition, 140–141
prostate adenocarcinoma growth and metastasis, inhibition, 134
and prostate cancer, 133–134
prostate cancer cells correlation, reduced 1α-hydroxylase activity in, 137
signaling pathway interaction, nuclear receptor ligands, 139

computational method, 121
algebraic chemistry (AC), 122
self-appointed pseudo-scientists, 122
CYP families in humans, 132
genetic factors, 118
inhibits cell proliferation and promotes apoptosis, 119
molecular basis
cancer incidence and mortality, 123–125
cancers and colonic adenoma, meta-analysis, 126–127
colon cancer, 128, 130–131
death by sunshine, 131
epidemiological studies, 126
epidemiology of vitamin D, 123–125
HLA-A2 gene frequency, 123
latitude, 123
ovarian, 123
potential of vitamin-D to prevent cancer, molecular basis of, 126
prostate cancer mortality rates, 123
protease inhibitors and cancer, 127–128
proteases, 127–128
serum 25(OH)D levels, meta-analysis, 126–127
summary relative risk (SRR), 127
multicellular systems and tumor–immune cells, 121
renal and extrarenal $1,25(OH)_2D_3$ production, 119
Volatile organic compounds (VOC), 2

X

Xanthomonas axonopodis pv. punicae
collection, isolation, and identification of
antibacterial activity, testing, 17–18
bacteria, isolation of, 15–16
bacterial samples, field visit and collection of, 14
biochemical variability, 16–17
storage of sample, 14
isolate growth rate of, 20